Navigation: Science and Technology

Volume 16

This series *Navigation: Science and Technology (NST)* presents new developments and advances in various aspects of navigation - from land navigation, marine navigation, aeronautic navigation to space navigation; and from basic theories, mechanisms, to modern techniques. It publishes monographs, edited volumes, lecture notes and professional books on topics relevant to navigation - quickly, up to date and with a high quality. A special focus of the series is the technologies of the Global Navigation Satellite Systems (GNSSs), as well as the latest progress made in the existing systems (e.g., GPS, BDS, Galileo, GLONASS). To help readers keep abreast of the latest advances in the field, the key topics in NST include but are not limited to:

- Satellite Navigation Signal Systems
- GNSS Navigation Applications
- Position Determination
- Navigational instruments
- Atomic Clock Technique and Time-Frequency System
- X-ray pulsar-based navigation and timing
- Test and Evaluation
- User Terminal Technology
- Navigation in Space
- New theories and technologies of navigation
- Policies and Standards

This book series is indexed in **SCOPUS** and **EI Compendex** databases.

Dongkai Yang · Feng Wang

Fundamentals and Applications of GNSS Reflectometry

Authors
See next page

ISSN 2522-0454　　　　　　　ISSN 2522-0462　(electronic)
Navigation: Science and Technology
ISBN 978-981-96-4553-4　　　ISBN 978-981-96-4554-1　(eBook)
https://doi.org/10.1007/978-981-96-4554-1

Jointly published with Publishing House of Electronics Industry
The print edition is not for sale in China (Mainland). Customers from China (Mainland) please order the print book from: Publishing House of Electronics Industry.

This work was supported by the School of Electronic and Information Engineering, Beihang University, International Innovation Institute, Beihang University, Hangzhou and Sino-African Joint Laboratory for the Beidou Satellite Navigation System and Remote Sensing Applications (2023YFE0126500).

© The Editor(s) (if applicable) and The Author(s) 2026. This book is an open access publication.

Open Access This book is licensed under the terms of the Creative Commons Attribution-NonCommercial-NoDerivatives 4.0 International License (http://creativecommons.org/licenses/by-nc-nd/4.0/), which permits any noncommercial use, sharing, distribution and reproduction in any medium or format, as long as you give appropriate credit to the original author(s) and the source, provide a link to the Creative Commons license and indicate if you modified the licensed material. You do not have permission under this license to share adapted material derived from this book or parts of it.
The images or other third party material in this book are included in the book's Creative Commons license, unless indicated otherwise in a credit line to the material. If material is not included in the book's Creative Commons license and your intended use is not permitted by statutory regulation or exceeds the permitted use, you will need to obtain permission directly from the copyright holder.
This work is subject to copyright. All commercial rights are reserved by the author(s), whether the whole or part of the material is concerned, specifically the rights of reprinting, reuse of illustrations, recitation, broadcasting, reproduction on microfilms or in any other physical way, and transmission or information storage and retrieval, electronic adaptation, computer software, or by similar or dissimilar methodology now known or hereafter developed. Regarding these commercial rights a non-exclusive license has been granted to the publisher.
The use of general descriptive names, registered names, trademarks, service marks, etc. in this publication does not imply, even in the absence of a specific statement, that such names are exempt from the relevant protective laws and regulations and therefore free for general use.
The publishers, the authors, and the editors are safe to assume that the advice and information in this book are believed to be true and accurate at the date of publication. Neither the publishers nor the authors or the editors give a warranty, express or implied, with respect to the material contained herein or for any errors or omissions that may have been made. The publishers remain neutral with regard to jurisdictional claims in published maps and institutional affiliations.

This Springer imprint is published by the registered company Springer Nature Singapore Pte Ltd.
The registered company address is: 152 Beach Road, #21-01/04 Gateway East, Singapore 189721, Singapore

If disposing of this product, please recycle the paper.

Dongkai Yang
School of Electronic Information
Engineering
Hangzhou International Innovation Institute
Beijing, China

Translated by
Shuyan Liu
School of Foreign Languages
Beihang University
Beijing, China

Yi Yang
China Communications Information
and Technology Group Co., Ltd./CCCG
Xingyu Technologies Co., Ltd.
Beijing, China

Feng Wang
School of Electronic and Information
Engineering
Beihang University
Beijing, China

Jie Li
School of Electronic Information
Engineering
Beihang University
Beijing, China

Foreword to the English Edition

The development of multiple global navigation satellite systems (GNSS) has greatly propelled the advancement of the information society, with its spatiotemporal services deeply integrated into all aspects of human life, ushering in a new era of human beings' efficient and precise utilization of time and space information. The BeiDou system, a global navigation satellite system (GNSS), independently developed by China, has undergone nearly 30 years of development, experiencing a journey from inception to strength, and achieving globally recognized accomplishments. It has become one of the country's strategic spatial infrastructure foundations. It plays an extremely important role in the fields of intelligent transportation, unmanned driving, smart city, and disaster monitoring. With the continuous improvement of BeiDou industry chain, fostering new application fields and deepening integration with other disciplines will further enrich the application system of BeiDou and flourish the application ecology of BeiDou.

Navigation and remote sensing are two means of acquiring spatial information; the former addresses the issue of continuous time and space positioning, while the latter solves measurement problems of physical properties in space. Remote sensing technology inverts physical parameters of targets using electromagnetic signals radiated, refracted, or scattered by them, making the naturally emitted electromagnetic signals from navigation satellites inherent remote sensing resources. GNSS Reflectometry (GNSS-R) is a typical representative of the application of Reflected GNSS Signals, becoming a global research hotspot in the last 30 years. Based on the constellation characteristics of the Global Navigation Satellite System and the bistatic or multistatic configuration features of separate transceivers, GNSS-R technology, with its advantages of multiple radiation sources and low power consumption of equipment, has been applied in oceanography, meteorology, and agricultural remote sensing, and has achieved good results, such as sea wind and wave and soil moisture monitoring.

Professor Yang Dongkai of Beihang University has been researching the theoretical methods of Reflected GNSS Signal application since 2003. Leading his team in the research on physical mechanisms, mathematical model construction, receiver software and hardware development, field experiment verification, and application promotion, they have accomplished significant work. With the support of the National

Science and Technology Academic Works Publication Fund, he published a book titled *GNSS Reflection Signal Processing: The Fundamentals and Applications* in 2012, the first monograph systematically discussing the GNSS-R signal processing and application in China, and played an important role in promoting the development of this technology. After more than a decade of development, the team continues to delve deeply in the field of GNSS-R technology, achieving a series of new results in signal processing methods, receiver design, and inversion applications. This book is a summary and compendium of their work over the past decades. The second edition of Reflected GNSS Signal Processing: Fundamentals and Applications of GNSS-Reflectometry not only reorganizes the content of the first edition but also includes the team's recent research achievements, expanding into new content to provide readers with a more comprehensive and systematic understanding of GNSS-R technology. The technologies and results covered in the book are of great significance to promote the remote sensing application of GNSS and to prosper the application ecology of China's BeiDou system.

I have known Prof. Yang Dongkai for many years, and I am fully aware of the hardships of persistent research in one field day after day for a decade. I am writing this Preface for his book and sincerely congratulate the publication of this book. I believe the publication will provide a systematic reference for teachers, students, and researchers engaged in GNSS and remote sensing technology, as well as their interdisciplinary integration. It is also hoped that more readers will come to understand, love, and engage in GNSS-R technology research through reading this book, collectively promoting the flourishing development of China's BeiDou Navigation Satellite System (BDS).

December 2022

Yuanxi Yang
Academician
Chinese Academy of Sciences
Beijing, China

Foreword to the Chinese Edition

The Global Navigation Satellite System (GNSS) has evolved over more than 40 years, increasingly playing a vital role in the economic and social aspects of nations, and in people's daily lives, gradually becoming an important component of the national spatial information infrastructure. With the accelerated construction of China's Beidou system, the implementation of the EU Galileo program, and the advancement of the US GPS and Russian GLONASS, the applications of GNSS are progressively permeating the lives of people. It can be said that the applications of GNSS have far exceeded people's initial imagination.

Among the various GNSS application fields, GNSS-R is a branch that has been developed since the early 1990s. It involves using reflected navigation satellite signals to invert the physical properties and parameters of reflective surfaces, a typical inverse problem. The United States and the European Union are taking the lead in the development of this field, having installed GNSS-R receivers on low Earth orbit satellites, and acquired a vast amount of observational data on sea surface wind field, soil moisture, sea ice, through airborne and land experiments to develop corresponding inversion models. The reflection mechanism of navigation satellite signals, mathematical models of signal energy, phase, and frequency variations with changes in physical characteristics of the reflective surface, the algorithm of reflective signal reception processing, and various parameters retrieval models, have become significant scientific issues in the field of GNSS-R. These issues, as research hotspots, have also attracted wide attention from domestic and foreign scholars.

Professor Yang Dongkai, under the guidance of the nationally prominent Prof. Zhang Qishan, a national outstanding expert, has conducted a large amount of in-depth and detailed research work in GNSS-R-related fields in recent years. They have undertaken the National 863 Program "Public Platform for Reflected GNSS Signal Reception Processing," and championed and participated in the application demonstration of BeiDou system for sea breeze and sea wave detection. Through continuous research, a series of innovative research achievement have been made in areas of refined tracking of reflected signals, power waveform generation, and related reception processing.

The book *Processing: Fundamentals and Applications* authored by the two professors, based on a summary of their research achievements over the years and referring to the latest research progress in related fields at home and abroad. It comprehensively, systematically, and thoroughly elaborates on the theories and methods of GNSS-R, covering the electromagnetic wave theory of Reflected GNSS Signals, the reflected signal reception processing methods, ocean wind field detection, effective wave height measurement, and soil moisture measurement. It has preliminarily formed a theoretical and methodological system for Reflected GNSS Signal reception processing, representing an important contribution made by Chinese scholars in the field of GNSS-R research. The technical methods and research results introduced in the book also hold significant engineering reference value for promoting the application of BeiDou in China.

I sincerely congratulate the publication of this book Reflected GNSS Signal and the achievements of its authors, believing that the publication of this book will play an important role in the development of China's satellite navigation endeavor.

October 2011

Jingnan Liu
Academician
Chinese Academy of Engineering
Beijing, China

Preface to the English Edition

A great era creates great endeavors.

It is the year 2020 when China's BeiDou 3 was fully operational, providing satellite navigation services to the whole world. Our team, Supported by the National Science and Technology Academic Works Publication Fund, published "GNSS-Reflectometry Ocean Remote Sensing: Methods and Applications," which gained wide attention from the society and provided a reference for expanding the innovative applications of BeiDou. This book was written on the basis of extending the achievements of "GNSS Reflective Signal Processing: Fundamentals and Applications," the drafting of which can be traced back to around 2008, and which was published in 2012 when BeiDou officially began serving the Asia-Pacific region. A decade has passed since then, and the hardware and software designs, the signal processing methods, and the applications have all evolved to varying degrees. Most of the signal processing methods have matured and are gradually starting to be commercialized. Some innovative applications, such as imaging and target detection, have moved from conceptual demonstrations to preliminary recognition, with significant progress in algorithms, models, and even data-processing results.

Considering the above, the authors contacted Publishing House of Electronic Industry to plan a second edition of "Fundamentals and Applications" to update the current understanding of GNSS reflection signal processing and applications. The new version reflects the latest domestic and international research frontiers, and expanded the breadth and depth of the application of BeiDou in agriculture, meteorology, oceanography, water conservancy, disaster and environmental monitoring. This idea was soon supported by the Editor Zhang Laisheng, leading to today's second edition.

With the continuous advancement of the BeiDou satellite navigation system, the advantages of its constellation and signal characteristics are also becoming increasingly apparent, showing a notable performance in its use for remote sensing detection. Following the research direction of many years ago, the authors' team has taken

BeiDou as a key area of expansion, and compared with the 2012 edition, the following aspects have been reorganized and supplemented.

First, the description of navigation satellite signals has shifted from the commonly used system-based arrangement to one starting from signal systems, focusing on the essential characteristics of spread-spectrum ranging codes, with new signal systems like Binary Offset Carrier (BOC) appearing as extensions. Signals from major navigation systems including GPS, GLONASS, GALILEO, and BDS are all implementations of the universal signal system.

Second, with the rapid development of integrated circuits, the software and hardware design of GNSS reflection signal receivers has also changed dramatically. The book elucidates in detail the new generation of receiver designs and results from the research team, providing a foundation for industry applications and promotion.

Third, soil moisture application, which was only introduced as a test in the first edition, is discussed in depth in an entire chapter in this version, analyzing and exploring soil moisture retrieval applications based on single- and dual-antenna modes and different observations.

Fourth, the section on imaging in the first edition has been expanded into a chapter, and the research results of GNSS reflection signal imaging conducted by the group are summarized, including geometric configuration, signal model, processing algorithms, and experimental demonstration.

Finally, some new applications, such as aerial target detection, inland water body detection, and river boundary parameter measurement, are given from different perspectives, such as mathematical model, system configuration, algorithm implementation, simulation demonstration, and experimental validation.

It can be said that the content of this edition, in comparison with the previous one, is more comprehensive in coverage, more thorough in analysis and more novel in design, fully reflecting the latest research results of the current author's team, and hopefully giving a refreshing experience.

This book was published with the help of several graduate students in the research team. Xing Jin, Han Mutian, Wu Shiyu, Zhang Guodong Miao Duo, Yang Pengyu, Kuang Keyuan, Zhen Jiahuan, and Tan Chuanrui all contributed much valuable time and energy in literature review, text organization, and figure preparation. The first edition of this book was strongly supported by Prof. Liu Jingnan, an academician of the Chinese Academy of Engineering, who wrote the Preface, and this edition is fortunate to have been supported by Yang Yuanxi, an academician of the Chinese Academy of Sciences, who was invited to write the preface of this book, for which we would like to express our heartfelt thanks. Teacher Zhu Yunlong of the BUAA PNaRL laboratory focused on organizing the content of Chap. 2, and Postdoctoral Fellow Hong Xuebao organized and edited the content of Chap. 7, for which thanks are also extended.

Despite our best efforts, there is still great room for improvement from a scientific perspective, and we sincerely welcome criticism and corrections from our peers. We will work together to promote the innovative application of the BeiDou system, and expand the application range and depth of Reflected GNSS Signals.

Beijing, China
December 2022

Dongkai Yang
Feng Wang

Preface to the Chinese Edition

The use of reflected signals from Global Navigation Satellite Systems (GNSS) for inversion is one of the current research hotspots in the GNSS domain. Since the discovery of this phenomenon by overseas scholars in 1993, researchers from NASA and the National Weather Service, the European Space Agency, and the Satellite Meteorological Center, as well as scholars from Australia, Japan, and China have been carrying out research on the Reflected GNSS Signals at various levels. Its application fields involve the inversion of marine meteorological parameters (e.g., sea waves, sea winds, sea surface salinity, and sea ice), soil moisture, forest coverage, mobile target detection, and Earth surface imaging. The research in this field has mostly focused on the development of processing equipment, signal processing algorithms, and physical inversion models. The United States and Europe are at the forefront of this field, having equipped terrestrial, aerial, and satellite platforms with GPS reflective signal receivers, conducted a large number of data collection and processing experiments, and preliminarily obtained effective sea surface wind field inversion models.

Research on GNSS-R remote sensing and detection technology in China is still in its initial stage. The first paper was published at the "High Technology Forum on Ocean Monitoring" in 2002, followed by a publication in the March 2003 special issue of "High Technology Letters." Subsequently, under the support of the National 863 Program (2002AA639190), the author led the research team to independently develop a 12-channel airborne serial delay mapping receiver system for the first time in China, and successfully completed an airborne test in August 2004. In 2006 and 2007, under the continuous support of the National 863 Program (2006AA09Z137 and 2007AA127340), respectively, the research team conducted coastal ground experiments and air flight tests in the Yellow Sea, Bohai Sea, South China Sea, etc., and obtained a large amount of original data obtaining a large amount. In 2008, supported by the National Natural Science Foundation of China (60742002), the research team carried out the theoretical analysis of the moving target detection based on the GNSS reflection signals and the preliminary data acquisition experiments. Targeting application areas such as ocean wind measurement, sea surface height measurement, target detection, and soil moisture, the research team published a series of academic papers and applied for and were granted several national invention patents. Currently, the GPS reflection signal receivers are conducting real-time observation at ocean meteorological observation stations under the China Meteorological Administration's, with data processing underway and preliminary operational data analysis results having been obtained.

The Chinese Academy of Sciences, the General Staff Hydrology and Meteorology Bureau as well as several domestic universities in China have also carried out in-depth follow-up research in this field, completing the actual data acquisition and analysis processing experiments, and achieving some academic results that are instructive for further research. Supported by the National Science and Technology Academic Woks Publication Fund, this book summarizes the research achievements of the National 863 Program and the National Natural Science Foundation projects undertaken by the authors. Starting from the current status of GNSS, it analyzes the characteristics of reflected GNSS signal and the basic methods of reception processing, and discusses the design of reflected signal receiver from the perspectives of hardware and software in details. The application model of GNSS reflective signal and the actual test results obtained by the research group are comprehensively summarized and discussed in two application fields: sea surface wind measurement and sea surface altimetry, and the preliminary exploration of soil moisture measurement, moving target detection and surface imaging is also carried out.

This book is based on the achievements in Reflected GNSS Signal reception processing technology and application research obtained by the authors in recent years, and aims to reflect the latest achievements domestically and internationally. It is hoped that through this book, readers will gain a comprehensive and systematic understanding of the current research status in this field. The book was primarily written by Dr. Yang Dongkai, a professor at Beihang University, with Prof. Zhang Qishan reviewing the entire book. Dr. Bo Zhang, doctoral students Li Weiqiang, Li Mingli, Lu Yong, Zhang Yijiang, Yao Yanxin, and Guo Jia, as well as master's degree students Wu Hongjia and Tang Yangyang participated in some simulation analyses and text editing work. The publication of this book was strongly supported by the Electronic Industry Publishing House, and Editor Zhang Laisheng contributed significant hard work, for which I express my sincere thanks and respect. The publication of this book has also been supported by Prof. Zhang Yanzhong, an academician at Beihang University, Li Shujian, a senior engineer, Cao Chong, researcher of China Electronics Technology Group, Zhang Mengyang, researcher of Aerospace Science and Technology Group

Li Ziwei, researcher of the Institute of Remote Sensing Technology and Application of the Chinese Academy of Sciences, Prof. Wang Hongxing of Naval Aviation Engineering Institute, Prof. Zhang Shufang of Dalian Maritime University, Prof. Li Xiaomin of Shijiazhuang College of Ordnance Engineering, and Li Huang, deputy director of the China Meteorological Administration, and Prof. Xia Qing and Cao Yunchang, researcher, expressing my gratitude to them as well. Professor Liu Jingnan, an academician of the Chinese Academy of Engineering and former president of Wuhan University, has written a preface for this book in his busy schedule, for which I extend my heartfelt thanks.

Due to the limited level of the author, there are inevitably errors or inadequacies in this book. I welcome criticism and corrections from my colleagues.

Beijing, China
October 2011

Dongkai Yang
Feng Wang

Introduction

This book presents the fundamental methods and applications of Reflected GNSS Signal reception processing. On the basis of systematic analysis of GNSS signals, it delves the typical characteristics of signals after reflection, discussing the software and hardware design of Reflected GNSS Signal receivers from different perspectives, including RF reception, IF processing, and extraction of observations. It offers a comprehensive introduction and exploration of ocean remote sensing, land remote sensing, and several new applications (such as airborne moving target detection, river boundary monitoring, and surface water body identification). The content covers a variety of different configuration modes such as shore-based, ground-based, airborne and satellite-based, and also includes two signal processing methods: single-antenna interferometry and dual-antenna cooperative method. The data used in the theoretical modeling and algorithm performance simulation validation are both from experiments conducted by the authors' research team and from experiments carried out with partners, as well as results processed from publicly available datasets domestically and internationally.

This book is suitable for faculty and students in the fields related to satellite navigation (such as electronic communication, radar remote sensing, aerospace, and computer), Earth observation, and can also serve as a reference book for engineers and scientific management personnel engaged in application research in the fields of oceanography, meteorology, agriculture, water conservancy, and environment studies.

Contents

1 **Introduction** 1
  1.1 Overview of Satellite Navigation Systems 1
      1.1.1 Beidou-3 Satellite Navigation System 1
      1.1.2 GPS 2
      1.1.3 GLONASS 3
      1.1.4 Galileo System 4
      1.1.5 QZSS System 4
      1.1.6 Indian Regional Navigation Satellite System 4
  1.2 Global Navigation Satellite System-Reflectometry
      or GNSS-Reflectometry Technology Overview 5
      1.2.1 The Forward and Inverse Problems 5
      1.2.2 GNSS-R Technical Definition 5
      1.2.3 GNSS-R Technology Advantages 7
  1.3 Development of GNSS-R Technology 8
      1.3.1 Ocean Remote Sensing 8
      1.3.2 Soil Moisture Detection 12
      1.3.3 Bi-SAR Imaging 15
      1.3.4 Target Detection 17
  1.4 Structure of the Book 18
  References 19

2 **Overview of GNSS Satellite Navigation Signals** 25
  2.1 Spread Spectrum Communication Principle 25
      2.1.1 Basic Concepts of Spread Spectrum 25
      2.1.2 Gold Sequence 27
      2.1.3 Spread Spectrum Ranging Principle and Performance 32
  2.2 Modulation Method 35
      2.2.1 Binary Phase Shift Keying (BPSK) 35
      2.2.2 Quadrature Phase Shift Keying (QPSK) 36
      2.2.3 Binary-Offset-Carrier 39

	2.3	Satellite Navigation Signal	45
		2.3.1 Beidou Signal	45
		2.3.2 GPS Signal	55
		2.3.3 Other Satellite Navigation Signals	62
	2.4	Summary	64
	References		64
3	**GNSS Signal Reception and Processing**		**67**
	3.1	Signal Model	67
		3.1.1 Mathematical Description	67
		3.1.2 Reception and Digitization	68
		3.1.3 Correlation Operations	72
	3.2	Acquisition and Tracking	75
		3.2.1 Acquisition	75
		3.2.2 Code Tracking	82
		3.2.3 Carrier Tracking	82
	3.3	Positioning Solving	86
		3.3.1 Data Synchronization	86
		3.3.2 Positioning Principles and Methods	88
	3.4	Summary	94
	References		94
4	**Reflected GNSS Signal Fundamentals**		**97**
	4.1	GNSS-R Geometric Relations	97
		4.1.1 Macroscopic Geometric Relations	97
		4.1.2 Antenna Coverage Area	99
		4.1.3 Isochronous Zone	101
		4.1.4 Iso-Doppler Lines	104
	4.2	Characterization of the Reflected Signal	107
		4.2.1 Polarization Characteristics	107
		4.2.2 Reflection Coefficient	111
		4.2.3 Description of Reflected Signals	113
	4.3	Correlation Function of the Reflected Signal	116
		4.3.1 One-Dimensional Time Delay Correlation Function	117
		4.3.2 One-Dimensional Doppler Correlation Function	118
		4.3.3 Time-Delay-Doppler Two-Dimensional Correlation Function	119
	4.4	Summary	121
	References		121
5	**Reception and Processing of Reflected GNSS Signals**		**123**
	5.1	Generalized Model of a Reflected Signal Receiver	123
		5.1.1 General Model	123
		5.1.2 Implementation	124
	5.2	Reflected Signal Processing Methods	127
		5.2.1 Discrete Forms of Reflected Signals	127

		5.2.2	Multi-channel Correlation Processing Algorithm	129
		5.2.3	Calculation of Specular Reflection Point	132
		5.2.4	Methods for Improving the Signal-To-Noise Ratio	135
	5.3	Hardware Receiver		138
		5.3.1	Overall Architecture of the Receiver	138
		5.3.2	Main Components of the Receiver	141
		5.3.3	Multi-Channel Correlation Unit	147
		5.3.4	Multi-channel Control Unit	153
	5.4	Software Receivers		156
		5.4.1	Basic Structure	156
		5.4.2	Processing Flow	158
		5.4.3	Software Functionality	158
		5.4.4	Efficiency Analysis of Implementation	160
		5.4.5	Real-Time Solution Research	161
	5.5	Summary		162
	References			163
6	**Marine Remote Sensing Applications**			165
	6.1	Wind Field Inversion		167
		6.1.1	Sea Surface Wind and Waves	167
		6.1.2	Ocean Wave Spectrum Model	170
		6.1.3	Electromagnetic Scattering Models	173
		6.1.4	Wind Speed Inversion Based on Waveform Matching	174
		6.1.5	Wind Speed Inversion Based on Model Functions	187
	6.2	Sea Surface Height Measurement		192
		6.2.1	Principle of Direct and Reflected Synergy for Measuring Sea Surface Height	193
		6.2.2	Code Delay Height Measurement Method	197
		6.2.3	Carrier Phase Altimetry Method	199
		6.2.4	Carrier Frequency Altimetry Method	202
		6.2.5	Interferometric Sea Surface Height Measurement	204
	6.3	Shore-Based GNSS-R Typhoon Wind Field Reconstruction and Storm Surge Simulation		209
		6.3.1	Typhoon Wind Field Reconstruction Method	209
		6.3.2	Coastal GNSS-R Typhoon Wind Field Reconstruction	210
		6.3.3	Establishment and Verification of the Storm Surge Model	211
		6.3.4	Storm Surge Simulation with Reconstructed Wind Fields	214
	6.4	Summary		216
	References			220

7	**Land Soil Moisture Retrieval**		223
	7.1 Basic Principles of Soil Moisture retrieval		225
	7.2 Dual Antenna Soil Moisture retrieval		227
		7.2.1 Extraction of Observables	227
		7.2.2 Estimation of Soil Reflection Coefficients	229
		7.2.3 Soil Moisture retrieval Tests	231
	7.3 Single Antenna Soil Moisture retrieval		237
		7.3.1 Extraction of Observables	238
		7.3.2 Estimation of Soil Reflection Coefficients	240
		7.3.3 Soil Moisture retrieval Experiment	241
	7.4 Example of a Soil Moisture Monitoring System		245
		7.4.1 Overall Design	246
		7.4.2 Terminal Hardware Design	247
		7.4.3 Terminal Software Design	248
		7.4.4 Experimental Results	250
	7.5 Summary		253
	References		253
8	**GNSS-R Imaging**		255
	8.1 System Configuration and Signal Modeling		255
		8.1.1 System Configuration	255
		8.1.2 Two-Dimensional Signal Model	257
	8.2 Spatial Resolution		260
		8.2.1 Resolution Definition	260
		8.2.2 Relationship Between Resolution and Geometric Configuration	263
	8.3 Backward-Projection Imaging Algorithm		265
		8.3.1 Range Compression and Synchronization	266
		8.3.2 Back-Projection	268
	8.4 Imaging Algorithm Implementation		270
		8.4.1 Heterogeneous Architecture Parallel Platforms	270
		8.4.2 Range Compression	270
		8.4.3 Back Projection	272
		8.4.4 Stream Processing Structure Optimization	273
	8.5 Simulation Analysis and Experimental Verification		274
		8.5.1 Simulation Analysis	274
		8.5.2 Test Validation	277
	8.6 Summary		279
	References		281
9	**Exploring New Applications of GNSS Reflectometry**		283
	9.1 Mobile Target Detection		284
		9.1.1 Introduction to Detection Mechanisms	284
		9.1.2 Field Test Analysis and Validation	289

9.2	River Boundary Detection	296
	9.2.1 Detection Principle Analysis	296
	9.2.2 Simulation Analysis and Verification	303
9.3	Surface Water Body Identification	306
	9.3.1 Water Body Identification Process	307
	9.3.2 CYGNSS Data Validation	309
9.4	Summary	316
References		316

Chapter 1
Introduction

1.1 Overview of Satellite Navigation Systems

Since the development of the Navy Navigation Satellite System (NNSS) in 1958 by the United States Navy and Johns Hopkins University cooperated to provide global positioning navigation for Polaris nuclear-powered submarines, research and development of satellite navigation systems have undergone more than half a century of evolution. Currently, satellite navigation is still the most significant and valuable field of research among various navigation methods. As human society progresses and science and technology advance, its value will be increasingly evident. Corresponding to the Global Positioning System (GPS) of the United States, the Soviet Union established the GLONASS (Global Orbiting Navigational Satellite System) system and Galileo, a civil satellite navigation system, was constructed by the European Union. In October and December 2000 and May 2003, China launched three geosynchronous orbit satellites for the self-developed "Beidou-1" positioning navigation system, marking a new chapter in China's research in the field of satellite navigation. After nearly 20 years of development, China completed the BeiDou Navigation Satellite System (BDS) in July 2020, offering global users positioning, velocity, timing and short message services to global users.

1.1.1 Beidou-3 Satellite Navigation System

The BeiDou-3 Satellite Navigation System is China's independently developed global satellite navigation system, utilizing the CGCS2000 coordinate system and BeiDou system time. It has a total of 30 satellites in Geosynchronous Earth Orbit (GEO), Medium Earth Orbit (MEO) and Inclined Earth Orbit (IGSO). Among these, the MEO orbit includes 24 satellites with an orbital altitude of 21,528 km, distributed across three orbital planes with an inclination of 55°, evenly spaced 120° apart. There

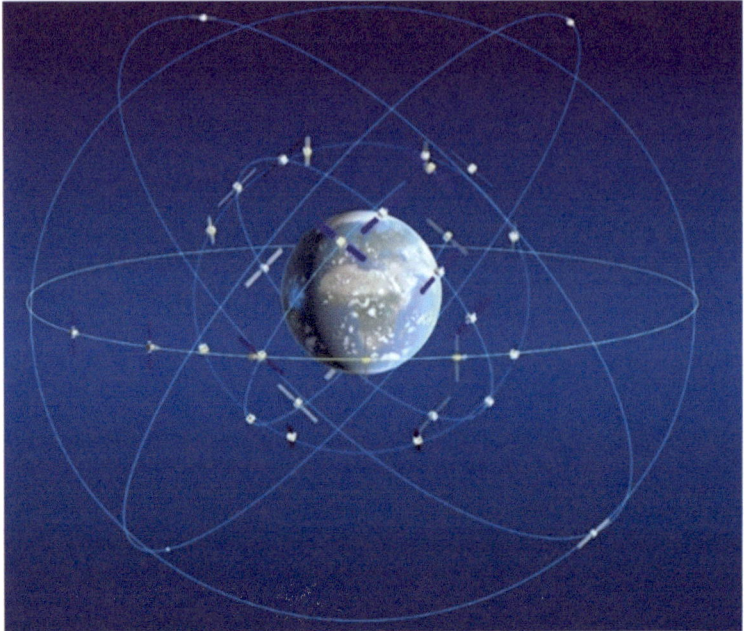

Fig. 1.1 BeiDou navigation system constellation

are three satellites in GEO orbit with an orbital altitude of 35,786 km, positioned at 80°E, 110.5°E, and 140°E, respectively. The three IGSO satellites are placed in orbital planes with an inclinations of 55° and an altitude of 35,786 km. The specific constellation structure of BeiDou system is shown in Fig. 1.1.

1.1.2 GPS

From an academic perspective, GPS is defined as a system capable of providing global positioning services. Since the United States was the first to build and utilize this term, "GPS" has been recognized as a designation for the specific global satellite navigation system constructed and maintained by the United States.

In the 1950s and 1960s, the United States began to develop satellite-based positioning and navigation system. Due to limitations of aerospace technology development at the time, its service capability and performance could not meet the needs of global users. To coordinate the navigation requirements of various departments and to research and develop a new generation of satellite navigation systems, the U.S. Department of Defense set up a special Joint Program Office (JPO) in 1973. After a comprehensive analysis of satellite navigation technologies and concepts available at the time, a new satellite navigation system was proposed: Navigation System Timing and Ranging/Global Positioning System (NAVSTAR/GPS), commonly referred to as GPS (Global Positioning System). Construction of GPS started in the 1970's. It took 20 years, costing 20 billion dollars, and was fully operational and completed in 1995 to provide services.

The GPS space segment consists of 24 satellites distributed in 6 orbital planes, with four satellites per orbital plane. The inclination of the satellite orbital plane relative to the Earth's equatorial plane is approximately 55°, with ascending node longitudes of adjacent orbital planes separated by 60°. For satellite in adjacent orbits, the ascending node right ascension angles are spaced by 30°. The orbital planes are approximately 20,200 km above the Earth's surface, and the satellites have an orbital period of about 11 h and 58 min. The Current GPS constellation is composed of 30 operational satellites, including 7 Block IIR, 7 Block IIR-M, 12 Block IIF, and 4 Block III and IIIF satellites serving as on-orbit spares.

1.1.3 GLONASS

The Soviet Navy started a satellite navigation system in 1965, called CICADA, which, similar to the American Transit Navigation System, based on the principle of Doppler shift measurements, marking the first-generation satellite navigation system. This system consisted of 12 satellites known as cosmic apparatuses forming its satellite constellation, with an orbital altitude of 1000 km and an orbital period of 105 min, each satellite transmitting navigation signals at frequencies of 150 MHz and 400 MHz. In the 1980s the USSR initiated the second generation of the satellite navigation system-the GLONASS, which provided three-dimensional positioning, velocity and time broadcasting services around the world. GLONASS was very similar to GPS in many respects, and its Position, Velocity, and Time(PVT) determines also performed using PRN (pseudo-random number) ranging signals. Its constellation consists of 24 satellites distributed over three equally spaced circular orbits with orbital planes inclined at 120° to each other and an orbital inclination of 64.8°, each orbit hosting eight satellites, averaging an orbital altitude of 19,100 km, and a satellite operational period of 11 h and 15 min. Unlike GPS, GLONASS adopts the FDMA (Frequency Division Multiple Access) system to distinguish between different satellites, with each satellite using the same ranging code but transmitting on a different frequency. Each satellite broadcast two carrier waves, L1 and L2, with their frequencies being $f_{L1} = 1602$ MHz + k-0.5625 MHz and $f_{L2} = 1246$ MHz + k-0.4375 MHz respectively, where k is the number of each satellite. Following multiple directives from the Government of Russia, GLONASS is in the process of comprehensive recovery and enhancement of its service capabilities and system performance. Specific measures include the development of third-generation (GLONASS-K) long-life satellites, successive launches of new navigational satellites, and the development of CDMA signals compatible with GPS.

1.1.4 Galileo System

The Galileo system, led by the European Union, is a global navigation satellite system, and it offers more comprehensive functions than those of the U.S. GPS and the Russian GLONASS. In addition to the general positioning, velocity measurement and timing public service (Open Service, OS), it also specifically provides high-performance services for life safety (Safety-of-Life Service, SOL), search and rescue (Search-and-Rescue Service, SAR), commercial users (Commercial Service, CS), and certain government users (Public Regulated Service, PRS). The Galileo satellite constellation consists of 30 satellites uniformly distributed across three medium earth orbits with each orbit plane hosting ten satellites, nine operational and one spare, with an orbital plane inclination of 56°, covering latitudes up to 75° north and south. The Galileo system was developed to address the shortcomings of GPS in global full-time availability by: (1) enhancing the availability of satellite navigation in high latitudes areas of the northern hemisphere; (2) improving the accuracy of satellite navigation services; (3) increasing the automation of logistics support; and (4) improving availability of satellite navigation in urban areas.

1.1.5 QZSS System

Quasi-Zenith Satellite System (QZSS) is characterized by high-elevation angle services and large elliptical asymmetric geosynchronous orbit in figure-of-eight pattern, which serves the communication and positioning in the central urban area of Japan and mid-latitude mountainous regions. It is a regional enhancement system for GPS, including of L1, L2, and L5 frequencies, broadcasting signals fully compatible or interoperable with GPS L1 C/A, L1C, L2C, and L5. Additionally, it features the L1-SAIF signal, specially compatible with GPS-SBAS, enabling sub-meter level enhancement with integrity function. Moreover, the Lex (1278.75 MHz) signal is used for high rate information transmission experimental verification. The system is jointly constructed by four Japanese government departments and dozens of private enterprises, and consists of three satellites positioned in orbit spaced 120° apart, with an orbital period of 23 h 56 min, an inclination of 45°, an eccentricity of 0.1, and an orbital altitude of 31,500–40,000 km.

1.1.6 Indian Regional Navigation Satellite System

On May 9, 2006, the Government of India officially approved the implementation of the major project (IRNSS, India Regional Navigational Satellite System), implemented by the Indian Space Research Organization (ISRO). The IRNSS constellation consists of seven satellites, including three GEO satellites and four IGSO satellites, with the GEO satellites being an extension of GPS-augmented geostationary orbit navigation satellites. The IRNSS system provides Standard Positioning Service and Restricted Service, utilizing frequencies in the L5 (1176.45 MHz) and S (2492.08 MHz) bands with a signal bandwidth of 1 MHz and employing BPSK modulation. The restricted service uses the same frequency bands, but with a BOC (5,2) structure.

1.2 Global Navigation Satellite System-Reflectometry or GNSS-Reflectometry Technology Overview

1.2.1 The Forward and Inverse Problems

The forward problem involves determining the spatiotemporal distribution characteristics and frequency domain relationships of fields when the electromagnetic and geometric properties of media and objects are known by studying their response to electromagnetic waves. Conversely, the inverse problem (or reverse problem) involves inferring the geometric and electrical parameter structure characteristics of the tested media and objects based on the known (detected) spatiotemporal and frequency domain distribution characteristics of the "scattered" external wave field. The term "scattering" here is encompassed a broad concept of electromagnetic responses, including transmission, reflection, refraction, and diffraction. Information extraction and processing in the inverse problem is accordingly referred to as remote sensing and inversion (or reconstruction). For actual problems, the choice of remote sensing parameters and inversion methods requires specific analysis on a case-by-case basis, offering a variety of different options. The measurement of parameters include not only the amplitude and phase of the wave field but also polarization, time delay, ray arrival angles, bending angles, and Doppler shifts. Moreover, the results of the forward problem are often utilized in inversions, setting a certain model for the inversion object for iterative repetition.

1.2.2 GNSS-R Technical Definition

Satellite positioning and navigation have been profound and impactful events in the field of space and navigation technology in the second half of the twentieth century, and representing the integration of modern space technology, radio communication technology and computer technology. The satellite positioning and navigation systems, using artificial satellites as navigation beacons, are space-based radio navigation system, capable of providing high-precision three-dimensional position, speed and precision time information for all kinds of military and civilian carriers on land, at sea, air and space all round the clock or space 24 h a day. The Global Positioning

System not only provides users of spatial information with navigation and positioning and precise timing information, but also provides highly stable, long-term use of L-band microwave free signal resources. Satellite navigation systems have a large number of wide-ranging applications in surveying and mapping, seismic monitoring, geological surveys, offshore and desert oil development, fisheries, civil engineering, archaeological excavations, iceberg tracking, search and rescue, resource surveys, forest and mountain tourism, intelligent transportation, as well as in missile guidance in the military field, and in many other areas. Many of these applications go beyond the original function and design objectives of the navigation system and, in terms of scientific applications, enabling mankind to enhance its knowledge of the Earth and its environment.

GNSS-R (GNSS-Reflections; GNSS-Reflectometry; GNSS-Remote Sensing) technology is a new branch that has been gradually developed since the 1990s, and has become one of the research hotspots of research in the field of remote sensing detection and navigation technology in China and other parts of the world. Fig. 1.2 shows, while receiving direct signals from navigation satellites, a GNSS receiver also captures reflected signals from the surface. These signals, often considered harmful for positioning solutions due to multipath interference, are usually estimated and suppressed or eliminated using various methods in the receiver. Alternatively, the signals can be suppressed or eliminated directly using signal processing methods designed to reduce multipath effects without the need for accurate estimation of the multipath signals. However, from the perspective of the basic theory of electromagnetic wave propagation, the reflected signal carries the characteristic information of the reflective surface. Changes of the of the reflected signal waveform, the polarization characteristics, amplitude, phase and frequency parameters all directly reflect the physical properties of the reflective surface, or in other words, are directly related to the reflective surface. Accurate estimation and reception processing of the reflected signal can achieve estimation and inversion of the physical properties of the reflective surface. In this sense, GNSS-R represents a typical inverse problem. It uses the L-band signal from navigation satellites as the transmission source and installs reflective signal reception devices on land, aircraft, satellites, or other platforms. By receiving and processing the reflected signals from oceans, land, or moving targets, this technology can extract the characteristic elements of the measured medium or detects moving targets. Unlike radars and altimeters that use back scattering signals to detect the physical information of target objects, GNSS-R typically utilizes forward scattering to obtain the physical information of target objects, representing a typical passive bistatic (or multistatic) remote sensing technology.

As shown in Table 1.1, GNSS-R technology is capable of being applied to sea surface altimetry, sea surface wind fields, soil moisture, sea water salinity and sea ice.

1.2 Global Navigation Satellite System-Reflectometry ...

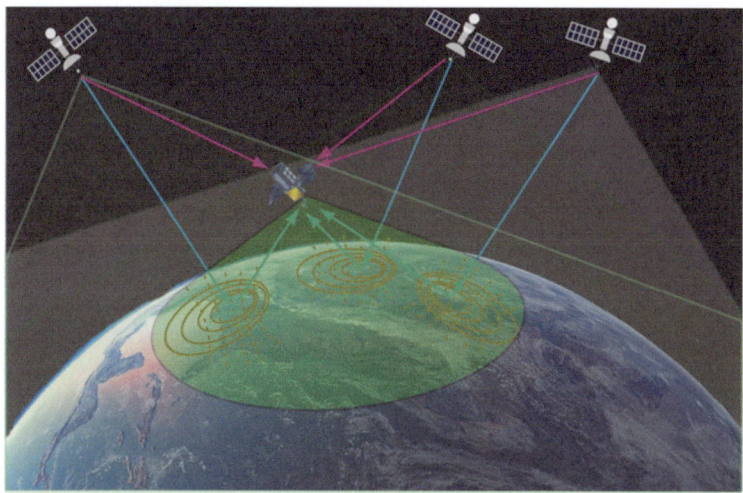

Fig. 1.2 Schematic diagram of GNSS-R principle

Table 1.1 GNSS-R technology application areas

Application areas	Ground-base	Airborne	Satellite-base
Sea height	✓	✓	✓
Sea surface wind	✓	✓	✓
Soil moisture	✓	✓	✓
Sea ice	✓	✓	✓
Snow	✓	N/A	N/A

1.2.3 GNSS-R Technology Advantages

GNSS-R technology, compared to traditional single-placement altimeters and scatterometers, has the following advantages:

1. Abundant signal sources. Various navigation systems have a large number of in-orbit or planned signal sources. With the continuous improvement and increasing maturity of GNSS systems, the number of navigation satellites in orbit will reach more than 150 in future applications. The multitude of navigation satellites not also offers different coverage, systems, facilitating large-scale, high-spatial-resolution detection.
2. The relatively simple detection equipment is more conducive to the application of airborne and satellite platforms; GNSS-R adopts the heterogeneous observation mode, requiring no transmitter, and only a relatively low-power consumption and low-cost reception equipment. Therefore, the weight, power consumption, cost and software and hardware complexity of the observation equipment are lower than those of the scatterometers and the altimeters.

3. Minimal weather effects such as clouds and rain. The L-band signals used by GNSS-R have longer wavelength compared to the C-band or Ku-band signals used by other remote sensing means (e.g., satellite altimeters, microwave scatterometers, etc.), and are essentially affected by weather conditions like clouds and rain.
4. Wide range of applications. GNSS-R signals contain a large amount of physical information about reflective surfaces, and the physical characteristics of reflective surfaces can be obtained by processing the direct and reflected signals from GNSS-R receivers. This has a wide range of applications in the fields of ocean remote sensing, land remote sensing, as well as in the fields of agriculture, meteorology, environmental protection and disaster reduction.

1.3 Development of GNSS-R Technology

1.3.1 Ocean Remote Sensing

The concept of utilizing Reflected GNSS Signals for oceanic remote sensing was among the earliest applications proposed for GNSS-R technology and remains one of its most developed areas. Specifically, applications in sea surface wind field and sea level measurement are progressively moving towards operational use.

1.3.1.1 Sea Surface Wind Speed Inversion

The idea of remote sensing of sea surface wind fields using GNSS reflection signals originated from Katzberg's report [1]. Subsequent airborne experiments in 1998 and 2000 validated the feasibility of the technology [2, 3]. Many research institutions have since conducted extensive experiments to explore the viability and methods of GNSS-R technology in ocean remote sensing applications [4]. The National Aeronautics and Space Administration (NASA), the University of Colorado, the European Space Agency (ESA), and Starlab of Spain, have also carried out numerous experiments, including airborne, satellite-borne, hot air balloon to match theoretical models with actual waveforms and obtain wind field information for comparison and analysis. In 2004, Soulat utilized a ground-based GNSS-R device for related experiments to remotely sense ocean state parameters [5]. In 2008, Wang and others conducted the first ground-based GNSS-R experiment in China's coastal areas, effectively inverting wave height parameters and achieving consistency with onsite observations [6]. In 2009, Lu and others designed a ground-based receiver and theoretically demonstrated the feasibility of a GNSS-R wind field observation system [7]. In 2014, Li and others used data from the 2013 Shenzhen typhoon experiment (code: TIGRIS) to invert the sea surface wind speed during Typhoons "Haiyan" and "Utor," with a Root Mean Square Error (RMSE) of less than 2.4 m/s [8]. In 2015, Martin proposed the concept of effective incoherent accumulation times, applied to TIGRIS typhoon data

1.3 Development of GNSS-R Technology

processing, showing that the proposed parameter has a linear relationship with the corresponding period [9]. In 2018, Kasantikul and others integrated neural network and particle filtering technology, reprocessing TIGRIS typhoon data with an inversion accuracy of 1.9 m/s under high wind speed conditions [10].

Recent years have seen the development of wind field inversion methods under airborne conditions [11], including waveform matching and the introduction of Delay-Doppler Maps (DDM) convolution and the extraction of DDM geometric parameters [12]. Additionally, Unmanned Aerial Vehicle (UAV) platforms, with their low cost and capability to carry a variety of payloads, are particularly suitable for GNSS-R applications, with successful trial cases already reported [13]. In 2019, Juang and others conducted airborne tests to assess the performance of a new generation of receivers [14]. Gao and others used airborne data combined with neural network technology to comprehensively invert sea surface wind speed, showing good consistency with NCEP reanalysis data [15].

Regarding satellite-based conditions, in 2002, Lowe detected GPS reflected signals from a satellite platform, marking the beginning of satellite-based GNSS-R research. Subsequent initiatives by various countries have continuously proposed GNSS-R satellite observation plans [16]. In September 2003, UK-DMC, one of the DMC constellations of the Disaster Monitoring Constellation (DMC), was launched by the United Kingdom, with a GPS receiver (SGR, space GPS receiver) developed by Surrey Satellite Science and Technology Ltd (SSTL) [17]. UK-DMC initially conducted experiments to confirm the usability of GPS reflected signals received from LEO satellites, and the results of the experiments have fully demonstrated their feasibility. In July 2014, a new generation of SGR-ReSI receivers were carried on board the TechDemoSat-1 (TDS-1) satellite launched by the UK to conduct GNSS-R LEO measurements, providing the fundamental observation DDMs for technical validation. The TDS-1 data showed that high quality DDM data could still be obtained during the processing of sea surface wind speeds up to 27.9 m/s [18]. ESA's 3Cat-2 project plans to carry the GNSS-R payload PYCARO (P(Y) & C/A ReflectOmeter) on a CubeSat to measure sea level and other surface parameters with dual-frequency, dual-polarization Reflected GNSS Signals [19]. ESA also launched the GEROS-ISS (GNSS rEflectometry, Radio Occultation and Scatterometry onboard International Space Station) program in 2011 to conduct GNSS signal reflection, occultation and scatterometry measurements on the International Space Station, supporting global climate change research. NASA launched the Cyclone Global Navigation Satellite System (CYGNSS) constellation of eight low-orbiting small satellites in October 2016, which also carried SGR-ReSI receivers for global tropical storm and hurricane detection at $\pm 35°$ latitude, improving the forecasting capability of extreme weather events [20–22]. Currently, the Wind Catcher I A and B satellites, developed by the Fifth Academy of China Aerospace Science and Technology Group (CASTG) and the East Red Satellite Limited Company (Aerospace ORSAT), were successfully launched in June 2019 over China's Yellow Sea area, to achieve broader range and higher precision monitoring in the sea surface wind field [23].

The satellite-based GNSS-R differs from the traditional active detection methods of scatterometer remote sensing, with its bistatic radar scattering model nature its observational performance is determined not only by the design parameters of the satellite observation equipment but also by its orbital design and visible GNSS satellites. GNSS satellites, as a signal transmitting source, and the visible satellites for in-orbit observation are not fixed, leading to a significant randomness in the observation area. The detection area cannot be calculated based on continuous swath observation ranges and times, but can only be statistically analyzed globally. The CYGNSS mission has a calculated average revisit time of 7.2 hours at observation points.

GNSS-R technology is developing rapidly in China, and the research work mainly focuses on sea surface wind field, effective wave height and tide level. Beihang University took the lead in developing a delay-mapping receiver for GPS reflection signals, successfully applied for the related patents, and, jointly with the Institute of Remote Sensing and Digital Earth Research of the Chinese Academy of Sciences (IRDRDR), carried out airborne experiments in Bohai Sea in 2004, using the experimental data to conduct research on sea surface wind speed and altimetry measurement. Subsequently, a validation test for the inversion of sea surface wind field using airborne GNSS-R was carried out in the South China Sea, and the corresponding results were summarized in the literature. Zhou Zhaoming, Fu Yang, et al. summarized the characteristics of reflected signals, scattering areas, and related power waveforms, analyzed the impact of wind speed, wind direction, and navigation satellite elevation angle on waveforms, and studied wind field inversion methods using NOAA hurricane experiment data, verifying the feasibility of GPS-R technology in the application of sea surface wind field detection.Liu Jingnan and Shao Lianjun et al.analyzed the key technology of GNSS-R from the aspects of satellite reflection signal software receiver and the establishment of sea surface parameter inversion models,pointing out further development directions of this technology. Wang Yingqiang and Yan Wei et al. proposed a two-dimensional interpolation algorithm waveform matching method for wind speed inversion, with actual data validation showing desirable inversion results.

In recent years, especially after the invention of Resnet neural network in 2013, deep learning technology has been developed rapidly. The previous research has provided new ideas for the research work of sea surface wind field, and many international scholars have started to conduct extensive research related to wind speed inversion using deep neural network in the field of GNSS-R. References [24–27] have all presented satisfactory inversion results.

1.3.1.2 Sea Surface Height Measurements

GNSS-R measures sea surface height by measuring the propagation delay of the reflected signal from the sea surface relative to the direct signal. Depending on the antenna device, GNSS-R altimetry is categorized into single-antenna and dual-antenna modes. The single antenna measures the sea surface height by utilizing the oscillation phenomena of signal-to-noise ratio, pseudorange, and carrier phase sequences output by the navigation receiver. Since this method requires that the time delay of reflected and direct signals cannot exceed one code piece length, and

1.3 Development of GNSS-R Technology

the direct and reflected signals need to form an obvious interference pattern, this method is only applicable to low-altitude scenarios, and has low temporal resolution. Compared with the single-antenna mode, the dual-antenna mode has a wide range of applications and is suitable for mounting on platforms at different altitudes.

In 1993, Marti-Neira, an ESA scientist, proposed the concept of "Passive Reflectometry and Interferometry System" (PARIS) to measure sea surface heights using GPS scattering signals. It was pointed out that the inversion of the physical properties of the reflecting surface could be realized by using a bistatic radar to receive the direct reflection signals separately [28]. In the following decade, the European Space Research and Technology Center (ESTEC) and the Jet Propulsion Laboratory (JPL) in the United States conducted multiple space-based experiments to study the accuracy performance of GPS reflection signals for altitude measurement at high altitudes, achieving an accuracy of 5 cm from a 20 km altitude. [29, 30]. In 2002, Lowe et al. processed data from two airborne GPS-R sea surface height measurement experiments, with the best height measurement accuracy reaching 5cm, a spatial resolution of about 5km, which can meet the high-accuracy sea surface and spatial resolution requirement for mesoscale eddy detection [31]. In 2006, Wilmhoff et al. conducted a GPS-R altimetry test on a NOAA WP-3D aircraft, collecting GPS scattering signal data in a calm sea conditions in the northeast Gulf of Mexico to investigate the performance for sea surface height measurement using GPS encrypted P(Y) code signals, with a root mean square error (RMSE) of about 20 cm. [32]. In 2016, Clarizia et al. demonstrated the feasibility of inverting sea surface heights using UK-TDS-1 data, integrating six months of TDS-1 satellite data to generate sea surface height mappings for the South Atlantic and North Pacific seas, with root mean square errors of 8.1 m and 7.4 m, respectively, larger than theoretical error values [33]. In 2018, Li et al. proposed two delay estimation methods based on leading-edge derivative and waveform fitting, verified using TDS-1 satellite data [34]. In 2019, Li further processed CYGNSS intermediate frequency data to obtain more complete time delay waveforms for GPS L1 as well as Galileo E1, achieved meter-level height measurement using three waveform retracking algorithms , corrected errors such as ionosphere, troposphere, and antenna baseline, and after data quality control, the height measurement error was reduced to 2.5–3.9 m [35]. In 2020, Mashburn et al. further analyzed the performance of satellite-borne sea surface height measurement using CYGNSS L1 DDM data, focusing on precise delay re-tracking, ionospheric delay correction, orbit error correction, proposing a delay re-tracking method based on reflection signal delay model, analyzing height measurement accuracy in coherent/non-coherent scenario [36]. Although literature [37–39] have improved GNSS-R height measurement accuracy using the cross-correlation of direct and reflected signals across the full GNSS spectrum , it is susceptible to inter-satellite interference in low-altitude scenarios [40]. The method of GNSS-R measurement accuracy has been improved by correlating direct and reflected signals from the full spectrum of GNSS. With the modernization of GNSS and the continuous improvement of the BeiDou and Galileo constellations, the signals of new high-bandwidth signals, such as GPS L5, BeiDou B3I, and Galileo E5, are gradually being used for surface parameter measurements [41, 42]. The errors in sea surface height measurement are closely related to satellite elevation and azimuth angles due to the random and time-varying GNSS-R geometric configurations brought about by satellite motion. Generally, the larger the elevation angle, the smaller the measurement error [43]. The GEO satellites of the BeiDou system are fixed relative to the Earth's surface, providing stable observation geometry for long-term stable terrestrial observations. The detection technology using electromagnetic signals emitted by geostationary satellites as an external radiation source has become one of the research hotspots [44–46]. Similar to the principle of GNSS positioning, in addition to using code phase, sea surface height measurement could also use carrier phase to obtain accuracy improvement. Literature provides corresponding system structures and signal processing methods. In terrestrial scenarios, Wu et al. [46] and Yun et al. [47] used the carrier phase of BeiDou GEO reflected signals for sea surface height measurement, with the experimental results showing height measurement accuracy reaching centimeter-level precision.

An alternative method for sea surface height measurement is to utilize the interference of direct and reflected signals, known as GNSS-IR (GNSS-Interferometric Reflectometry). This method can make full utilization of existing GNSS Continuously Operating Reference Stations (CORS) with survey-grade receivers to provide centimeter-level accuracy [48–50]. GNSS-IR can be further utilized in monitoring storm surge anomalous water gain . Reference [51] obtained storm surge information during Hurricane "Harvey" using multi-constellation GNSS data. To improve the temporal resolution, reference [52] proposed a method to artificially generate rapidly oscillating interference patterns, with simulation results showing that centimeter-level height measurement accuracy could be achieved within 5-minute intervals. Survey-grade receivers suppress ground-reflected multipath signals, especially for the signals with high altitude angle. It becomes challenging for survey-grade receivers to form an effective interference pattern when the altitude angle is greater than 30°. Standard navigation antennas and receivers have poor multipath suppression performance but can still form interference patterns in high altitude angle scenarios, and can play an important role in improving the time sampling rate of sea surface height measurements, and exhibit performance comparable to survey-grade receivers in sea surface height measurements in environments without strong interference [53, 54]. With the popularization of smartphones, applications based on smartphones are expanding rapidly. Reference [55] has preliminarily demonstrated the concept of using GNSS data from smartphones to measure the height of reflective surfaces.

1.3.2 Soil Moisture Detection

In 2000, V. Zavorotny et al. simulated the GPS signals scattered from the rough surface by using the double-base scattering model. They analyzed the feasibility of using GPS signals scattered by the ground for soil moisture observation from a theoretical perspective. In the same year, they conducted airborne observation experiments to validate their findings [56]. In 2002, Zavorotny and others conducted

1.3 Development of GNSS-R Technology

ground-based GPS-R soil moisture observation experiments at the NOAA Boulder Atmospheric Observatory. The results showed a significant enhancement in the reflected signal after rainfall, and the correlation between the reflected signal power and the measured soil moisture values depended on the soil's dryness/wetness level [57]. In the same year, NASA carried out the SMEX02 space-based experiment, one of which involved using the airborne GPS reflective signal receiver to collect ground-reflected GPS signals for soil moisture remote sensing research [58]. From 2003 to 2006, Masters et al. continuously processed and analyzed the GPS reflectance signals acquired by SMEX02 and demonstrated the feasibility of remote sensing of soil surface moisture changes using GPS bistatic radar [59–61]. In 2006, Ticconi et al. focused on land surface remote sensing applications using opportunistic sources (e.g., Reflected GNSS signals), and analyzed the sensitivity of bistatic scattering coefficients of linearly polarized waves to soil moisture by using the AIEM model. Initial results indicated that the forward-scattered signals of linearly polarized waves could be used for soil moisture inversion [62]. In 2009, ESA carried out a six-month ground-based GNSS-R observation experiment, which showed that the reflectivity calculated from GPS reflection signals was sensitive to soil moisture content. It also pointed out that in some cases, although there was no significant increase in the measured soil moisture, the signal reflectivity exhibited strong fluctuations [63, 64]. In 2013, A. Camps et al. conducted ground-based observation experiments using an a set of orthogonally linearly polarized antennas to receive reflection signals, which yielded a better correlation between the cross-polarized reflectivity and the measured soil moisture values, somewhat attenuating the effects of vegetation [65]. In the same year, Egido et al. carried out an airborne observation experiment using an orthogonal a set of orthogonally circularly polarized antennas to receive the reflection signals. The results showed that the reflectivity coefficients of the two polarization states were highly sensitive to both soil moisture and surface roughness, whereas the polariza tion ratio was less affected by the surface roughness [66]. In 2016, Camps et al. investigated large-scale surface soil moisture sensitivity using TDS-1 satellite data. The results indicated that GNSS-R observations with low NDVI values remained sensitive to soil moisture, with a good Pearson correlation coefficient between the two. With vegetation height increased, the reflectivity and sensitivity to soil moisture and the Pearson correlation coefficient decreased, but remained significant [67, 68]. In 2017, Carreno-Luengo et al. processed and analyzed the SMAP observations data and calculated the polarization ratio of the GNSS-R signals from the satellite platform for the first time. The analysis results showed that the polarization ratio was significantly sensitive to soil moisture [69]. In addition, the sensitivity of the leading and trailing edge widths of the GNSS-R waveform to aboveground biomass and rough terrain was investigated, showing promising results. The validation of these features is instrumental in advancing land parameter inversion algorithms. In 2018, the team analyzed the sensitivity of CYGNSS GNSS-R reflectance and SMAP radiometer brightness temperature to soil moisture, and the results showed that both GNSS-R reflectivity and radiometer brightness temperature are sensitive to soil moisture, while compared to the radiometer, GNSS-R sensitivity to soil moisture was less influenced by wet biomass than the radiometer [70]. They further investigated the influence of surface roughness parameters on CYGNSS DDM features, achieving preliminary results [71]. In 2019, Eroglu et al. inverted surface soil moisture in the CYGNSS coverage using an artificial neural network approach [72], achieving a root mean square error of 0.05cm^3/cm^3 with a 9 km spatial resolution. In 2020, Gleason and others performed geolocation and data calibration processing on CYGNSS land observation Level 1 data products. They also conducted a spatial resolution analysis of GNSS-R coherent reflections combined with river data [73]. In the same year, Volkan Senyurek et al. constructed a model between CYGNSS observations and SMAP soil moisture using the random forest [74], which could realize soil moisture inversion at 9 km × 9 km spatial resolution at a quasi-global scale, showing a correlation coefficient of 0.66 with reference values and a root mean square error of 0.044m^3/m^3.

Research in this field is gradually increasing in China, and most of the research work focuses on ground-based observation, occasionally touching on space-based and satellite-based aspects. From 2006 to 2007, Guan Zhi, Song Dongsheng, and others discussed the use of GPS reflected signals for soil moisture inversion, and verified it using SMEX02 data. In 2008–2009, Mao Kebiao et al. used the AIEM electromagnetic scattering model to simulate and analyze the impact of the incident angle and surface roughness on the forward-scattering coefficient. They processed the GPS reflection signals collected during the SMEX02 experiment, achieving a high linear correlation between the signal-to-noise ratio (SNR) of the reflected signal and the actual soil moisture values, with an average correlation coefficient exceeding 0.85. In 2009, Wang Yingqiang et al. discussed the use of GPS reflection signals for soil moisture inversion, and verified it with the SMEX02 data [75]. In 2009, Wang et al. analyzed and verified the feasibility of GNSS-R soil moisture inversion. From 2009 to 2011, Yan Songhua et al. analyzed the process of GNSS-R soil moisture inversion and carried out a ground test for verification. In 2012, Wan Wei and others discussed errors in the processing of SMEX02 data based on different NDVI regions, with errors of 7.04% for bare soil, 12% for moderate vegetation cover, and 32% for high vegetation cover [76]. In 2014, Liu et al. used ICF to estimate surface reflectance for GNSS-R soil moisture inversion and verified it using actual data [77]. The inversion was validated using actual data. In the same year, the National Space Science Center of the Chinese Academy of Sciences (NSSC), the Institute of Remote Sensing of the Chinese Academy of Sciences (IRS), Tsinghua University, and the Atmospheric Detection Center of the China Meteorological Administration jointly carried out China's first airborne GNSS-R soil moisture remote sensing detection experiment [24]. The results show good consistency between the remote sensing results and ground-based measurement, with an absolute deviation of 4% for bare soil volumetric soil moisture content and increased errors under vegetation cover. In 2015, Peng Xuefeng and others processed ground-based GNSS-R soil moisture retrieval experimental data by fitting the histogram of the results to a Gaussian function to separate soil moisture estimates from bare soil and vegetation reflection signals, enabling more accurate estimation of soil moisture on uniform surfaces [78]. In 2016, Zou Wenbo et al. proposed long-term continuous soil moisture retrieval method based on reflected signals from BeiDou GEO satellites, with experimental results showing good temporal and numerical continuity in the soil moisture retrieval results, matching the soil moisture reference values, and achieving a root mean square error of 5%. In the same year, Yang Lei, Wu Qiulan, and others proposed a soil moisture retrieval method based on SVRM-assisted reflected BeiDou GEO satellite signals. The data processing results showed that the soil moisture results obtained from the retrieval were within 3% of the soil moisture reference values obtained through drying-weighing methods, with a linear regression equation determination coefficient close to 0.9 and a root mean square error of about 1.5%. Yin Cong and others conducted ground-based GNSS-R soil moisture observation experiments, proposing to use L-band microwave radiometer observations to calibrate GNSS-R reflectivity. The results showed a strong negative correlation between the calibrated reflectivity and the brightness temperature data obtained from the radiometer, with the retrieved soil moisture matching the measured values well [79]. In 2017, Li Wei et al. constructed a GNSS-R soil moisture estimation system and realized a GNSS-R soil moisture estimation software based on combining existing GNSS-R soil moisture estimation methods. In 2019, Tu Jinsheng et al. analyzed the potential of TDS-1 GNSS-R DDM for soil moisture retrieval. In the same year, Jing Cheng and others used CYGNSS data to remotely sense soil moisture in the Chinese region in 2018, analyzing the reasons for abnormal soil moisture changes in Guangdong Province in conjunction with meteorological data, validating the effectiveness of using GNSS-R technology for soil moisture assessment. In 2020, Yan et al. used a linear regression method to establish a model between SMAP soil moisture and CYGNSS observation data, achieving nearly global soil moisture retrieval at a spatial resolution of 36 km x 36 km. The correlation coefficient between the retrieved results and SMAP data was 0.80, with a root mean square error of 0.07 m^3/m^3 [80].

1.3.3 Bi-SAR Imaging

The Microwave Integrated Systems Laboratory (MISL) at the University of Birmingham, UK, proposed the concept of SS-BSAR (Space-Surface Bistatic Synthetic Aperture Radar) system in 2002: using a ground receiver to receive surface reflection signals from navigation satellites for SAR imaging [81]. In 2005, the team completed preliminary experiments using GLONASS navigation satellites and ground-based fixed stations, obtaining bistatic SAR images [82]. The data acquisition system used in the experiment consisted of three channels: the GNSS direct wave channel for receiving direct signal from GNSS for system positioning, the Radar Channel (RC) for receiving surface Reflected GNSS Signal, and the Heterodyne Channel (HC) for receiving direct waves from designated navigation satellites. In 2007, the laboratory carried out SSBSAR validation experiments on the a ground-based sliding track[83, 84], demonstrating the feasibility of ground-based mobile platform navigation satellite radiation source bistatic SAR. From 2006 to 2008, M. Usman and others at the

University of Manchester in the UK reported research results on SAR imaging technology based on GPS reflection signals [85–87], proposing a method to increase the azimuth scanning angle to expand the azimuth dimension, increase the cross-coupling of range and azimuth resolution, reduce the impact of range resolution on spatial resolution, and improve the impact of azimuth resolution on spatial resolution, and focusing the image using the back-projection (BP) algorithm. In 2009, Usman and others proposed a point target reconstruction method based on deconvolution to address the "dragging shadow" phenomenon caused by long synthetic apertures, improving image signal-to-noise ratio, and conducting related experimental verification. The experimental target scene used corner reflectors to simulate point targets and employed the BP (Back Projection) imaging algorithm. The results showed that the positions of the corner reflectors in the image had strong reflection signals and could be focused, but due to the influence of other building echoes in the scene, the image signal-to-noise ratio was low. After deconvolution reconstruction, the image quality significantly improved. In 2009, M. Usman and others proposed a point target reconstruction method based on deconvolution to address the dragging shadow phenomenon caused by long synthetic apertures, improving image signal-to-noise ratio, and conducting related experimental verification. The experimental target scene used corner reflectors to simulate point targets and employed the BP (Back Projection) imaging algorithm. The results showed that the positions of the corner reflectors in the image had strong reflection signals and could be focused, but due to the influence of other building echoes in the scene, the image signal-to-noise ratio was low. After deconvolution reconstruction, the image quality significantly improved. In 2009, MISL conducted vehicle-mounted tests in the Kilsheel area of Loddon, UK, using the Galileo navigation system as the radiation source. In the target area of the experiment, four independent building clusters were clearly distinguishable in the obtained SAR images, consistent with the positions given by optical images of the target area. Due to the significantly increased speed of the vehicle-mounted receiver relative to the ground, a higher azimuth resolution was obtained in a shorter time, with a measured range resolution of approximately 25.2 m and a azimuth resolution of about 1 m. In 2012, MISL demonstrated the effects of bistatic angle variations and long synthetic apertures on the system using GLONASS system L1 signal P code (code rate of 5.11 MHz). The results showed that in quasi-monostatic mode (bistatic angle close to 0), the system had a range resolution of about 30 m, which degraded with increasing bistatic angle; under a 5-min synthetic aperture condition, the system could provide azimuth resolutions on the order of ~4 m [84]. The same year, MISL conducted airborne bistatic SAR imaging experiments using Galileo signals as external radiation sources at East Fortune Airport in Scotland [88, 89]. Data acquisition equipment was mounted on a helicopter targeting the East Fortune Airport area. During the experiment, the helicopter's trajectory was provided by an external high-rate GPS system with a 1 Hz output rate. However, due to the low refresh rate of the airborne receiver's position information, the image exhibited significant defocusing. This experiment verified that in airborne mode, the SSBSAR system could achieve higher azimuthal resolution than fixed and vehicle-mounted receivers. In 2015, The MISL team carried out multistatic SAR studies using reflected signals from the GLONASS system. [90]. In ground fixed mode, non-coherent fusion imaging analysis was performed using multiple navigation satellites with different geometric configurations, analyzing the Multistatic Point Spread Function (MPSF). Simulation results of point targets showed that the non-coherent combination of multi-baseline SAR images obtained from multiple spatially separated satellites could produce composite images, significantly improving resolution compared to single bistatic angle SAR images. The essence of this method is to constrain the range resolution of GNSS signals by extending or adding azimuth dimensions to improve system spatial resolution.

Domestic research on GNSS bistatic SAR started relatively late. In 2014, Beijing Institute of Technology conducted theoretical and experimental research on bistatic angle SAR imaging of BeiDou navigation satellites reflection signals. The team developed a data acquisition device for reflected signals from the BeiDou-2 navigation system, and repeated tests were conducted, producing 26 bistatic SAR images using multiple BeiDou navigation satellites [91]. Based on the above experiments results, a multi-angle observation and data processing method for bistatic angle SAR was proposed, achieving multi-angle observations of the target area and obtaining images in 26 different geometric configurations, [92] proving that multi-angle fusion is an effective way to expand GNSS-BSAR remote sensing applications.

1.3.4 Target Detection

In 1995, Koch and Westphal of the German GmbH company first proposed using global navigation satellite signals for passive multistatic target detection [93]. Kabakchiev and others conducted corresponding experiments on airborne, land, and maritime targets using the diffraction effect of forward scattering, achieving certain research results [94–96]. In 2012, Suberviola I realized the detection of airborne targets by using GPS L1 signals using the forward-scattering method and carried out validation tests [97]. The experiments used a right-hand circularly polarized (RHCP) omnidirectional antenna to receive signals, and observed the power change of the signals received when the aircraft crossed the baseline to determine the presence of the target. The analysis results show that depending on the aircraft's crossing position, satellite signals, and receiver position, the received signals may experience varying degrees of attenuation, enhancement, or oscillation. Chow et al. investigated an external radiation source radar using a phased-array antenna to receive GPS reflected signals, improving the performance of target detection through the calibration of antenna array phase error, array gain optimization, and interference signal cancellation techniques [98]. In 2019, Santi et al. used GNSS signals as an external radiation source to detect ships at sea and proposed a single-step joint detection and localization algorithm with long-time integration, which was analyzed in theoretical simulation and verified in experiments [99]. In 2021, the team proposed a new target detection algorithm to obtain well-focused ship images while maximizing the signal-to-clutter ratio (SCR) for ships. In addition, they also analytically derived the scaling factor required to map the backscattered energy in the distance and cross-distance domains, from which the length of the target was estimated, and verified the reliability of the algorithm by conducting experiments using the Galileo satellite [100].

Domestic research on GNSS external radiation source radar has also been gradually developed. Yang Jinpei et al. analyzed the feasibility of passive radar detection using GPS signals and BeiDou satellite signals, respectively and proved that satellite navigation signals could be used as a third-party non-cooperative radiation source of passive radar for target detection. Liu Changjiang et al. investigated forward scattering target detection based on GNSS, and successfully detected the target by selecting an aluminum plate as the detection target with an omnidirectional GPS antenna and a high-gain horn antenna, respectively [101]. In 2020, Chen Wu et al. used GNSS external radiation sources to form a space-surface bistatic radar system, and with the help of the target's motion characteristics and the combination of SAR imaging technology, they focused the target's energy echoes to detect targets on the sea surface ,and then inverted to get the distance from the target to the receiver, and judged the direction of the target's motion. Additionally, the team proposed a GNSS-SAR dynamic target imaging algorithm, achieving good consistency between the detected positions of two cargo ships and the actual situation through experiments [102]. Zeng et al. proposed an improved unmanned aerial vehicle (UAV) detection algorithm based on GNSS external radiation source radar, which achieved focusing of UAV targets in the range-Doppler plane, and verified the algorithm through simulation [103].

1.4 Structure of the Book

This chapter starts with an introduction to satellite navigation systems, covering the basic concepts of direct and inverse problems. It provides a comprehensive overview of GNSS-R technology and its applications, as well as a summary of the current research status of GNSS-R both domestically and internationally. Chapters 2 and 3 delve into the details of GNSS signal structure and processing fundamentals, focusing on mathematical models related to direct signals in GNSS and discussing receiver processing techniques. This lays the theoretical groundwork for the processing of navigation satellite reflection signals and the design of receivers. Chapter 4 describes the characteristics of GNSS reflection signals, including signal waveforms and polarization states. Chapter 5 analyzes the basic principles of receiving and processing reflection signals in light of their characteristics, with in-depth discussions on antenna design, direct signal-reflection signal cooperative processing, and reflection signal correlators. Chapters 6 and 7 introduce the practical applications of the author's research group in the field of GNSS-R, focusing on measurements of sea surface wind speed, sea surface height, and soil moisture content. Chapters 8 and 9 showcase two current research areas in GNSS-R: imaging and target detection technologies, along with the actual data results from the author's research group. The conclusion offers a glimpse into the future of the field.

References

1. Katzberg SJ, Garrrison JL. Utilizing GPS to determine ionospheric delay over the ocean. NASA Tech Rep. 1996.
2. Garrison JL, Katzberg SJ, Hill MI. Effect of sea roughness on bistatically scattered range coded signals from the global positioning system. Geophys Res Lett. 1998;25(13):2257–60.
3. Zavorotny VU, Voronovich AG. Scattering of GPS signals from the ocean with wind remote sensing application. IEEE Trans Geoence Remote Sens. 2000;38(2):951–64.
4. Picardi G, Seu R, et al. Bistatic model of ocean scattering. IEEE Trans Antennas Propag.1998.
5. Soulat F. Sea state monitoring using coastal GNSS-R. Geophys Res Lett. 2004;31(21):133–47.
6. Xin W, Qiang S, Xunxie Z, et al. First China ocean reflection experiment using coastal GNSS-R. Chin Sci Bull. 2008.
7. Lu Y, Yang D, Xiong H, et al. Study of ocean wind-field monitoring system based on GNSS-R. Geomat Inf Ence Wuhan Univ. 2009.
8. Li W, Yang D, Fabra F, et al. Typhoon Wind speed observation utilizing reflected signals from BeiDou GEO satellites. In: Lecture notes in electrical engineering;2014. p. 191–200.
9. Martin F, Camps A, Martin-Neira M, et al. Significant wave height retrieval based on the effective number of incoherent averages. IEEE Int Geosci Remote Sens Symp. 2015;3634–3637.
10. Kasantikul K, Yang D, Wang Q, et al. A novel wind speed estimation based on the integration of an artificial neural network and a particle filter using BeiDou GEO reflectometry. Sensors. 2018;18(10).
11. Park H, Valencia E, Rodriguez-Alvarez N, et al. New approach to sea surface wind retrieval from GNSS-R measurements. Geosci Remote Sens Symp. 2011.
12. Valencia E, Zavorotny VU, Akos DM, et al. Using DDM asymmetry metrics for wind direction retrieval from GPS ocean-scattered signals in airborne experiments. IEEE Trans Geoence Remote Sens. 2014;52(7):3924–36.
13. Micaela TG, Gianluca M, Marco P, et al. Prototyping a GNSS-based passive radar for UAVs: An instrument to classify the water content feature of lands. Sensors. 2015;15(11):28287–313.
14. Juang J C, Ma S H, Tsai Y F, et al. Function and performance assessment of a GNSS-R receiver in airborne tests. In: ION 2019 Pacific PNT Meeting;2019.
15. Gao H, Yang D, Wang F, et al. Retrieval of ocean wind speed using airborne reflected GNSS signals. IEEE Access. 2019;(99):1–1.
16. Lowe ST, Labrecque JL, Zuffada C, et al. First spaceborne observation of an earth-reflected GPS signal. Radio Sci. 2002;37(1):1–28.
17. Unwin M, Gleason S, Brennan M. The space GPS reflectometry experiment on the UK disaster monitoring constellation satellite BIOGRAPHY. In: Proceedings of ION-GPS/GNSS;2003.
18. Foti G, Gommenginger C, Jales P, et al. Spaceborne GNSS reflectometry for ocean winds: First results from the UK TechDemoSat-1 mission. Geophys Res Lett. 2015;42.
19. Carreno-Luengo H, Camps A, Ramos-Perez I, et al. 3Cat-2: A P(Y) and C/A GNSS-R experimental nano-satellite mission. In: Geoscience and remote sensing symposium (IGARSS), 2013 IEEE international;2014.
20. Ruf C S. Storm Surge Modeling with CYGNSS Winds. AGU Fall Meet. 2016.
21. Ruf CS, Balasubramaniam R. Development of the CYGNSS geophysical model function for wind speed. IEEE J Sel Top Appl Earth Obs Remote Sens. 2018;12(1):66–77.
22. Ruf CS, Gleason S, Mckague DS. Assessment of CYGNSS wind speed retrieval uncertainty. IEEE J Sel Top Appl Earth Obs Remote Sens. 2018;1–11.

23. Jing C, Niu X, Duan C, et al. Sea surface wind speed retrieval from the first Chinese GNSS-R mission: Technique and preliminary results. Remote Sens. 2019;11(24):3013.
24. Bai W, Xia J, Wan W, et al. A first comprehensive evaluation of China's GNSS-R airborne campaign: Part II—river remote sensing. 科学通报(英文版). 2015;060(017):1527–1534.
25. Reynolds J, Clarizia M P, Santi E. Wind speed estimation from CYGNSS using artificial neural networks. IEEE J Sel Top Appl Earth Obs Remote Sens. 2020;(99):1–1.
26. Asgarimehr M, Zhelavskaya I, Foti G, et al. A GNSS-R geophysical model function: machine learning for wind speed retrievals. IEEE Geosci Remote Sens Lett. 2019;(99):1–5.
27. Liu Y, Collett I, Morton YJ. Application of neural network to GNSS-R wind speed retrieval. IEEE Trans Geoence Remote Sens. 2019;(99):1–11.
28. Han G, Zhaoguang B, Dongdong F, et al. GNSS-R sea surface wind speed inversion based on BP (Back Projection) neural network. Acta aeronautica et astronautica sinica. 2019.
29. Martin-Neira M. A passive reflectometry and interferometry system (PARIS): Application to ocean altimetry. ESA J. 1993;17(4):331–55.
30. Roussel N, Ramillien G, Frappart F, et al. Sea level monitoring and sea state estimate using a single geodetic receiver. Remote Sens Environ. 2015;171:261–77.
31. Lowe S T, Zuffada C, Chao Y, et al. 5-cm-precision aircraft ocean altimetry using GPS reflections. Geophys Res Lett. 2002;29(10):13-1–4.
32. Wilmhoff B, Lalezari F, Zavorotny V, et al. GPS ocean altimetry from aircraft using the P (Y) code signal. IEEE Int Geos Remote Sens Symp. 2007;2007:5093–6.
33. Clarizia MP, Ruf C, Cipollini P, et al. First spaceborne observation of sea surface height using GPS-reflectometry. Geophys Res Lett. 2016;43(2):767–74.
34. Li W, Rius A, Fabra F, et al. Revisiting the GNSS-R waveform statistics and its impact on altimetric retrievals. IEEE Trans Geosci Remote Sens. 2018;56(5):2854–71.
35. Li W, Cardellach E, Fabra F, et al. Assessment of spaceborne GNSS-R ocean altimetry performance using CYGNSS mission raw data. IEEE Trans Geosci Remote Sens. 2019;58(1):238–50.
36. Mashburn J, Axelrad P, Zuffada C, et al. Improved GNSS-R ocean surface altimetry with CYGNSS in the seas of Indonesia. IEEE Trans Geosci Remote Sens. 2020.
37. Rius A, Nogues-Correig O, Serni R, et al. Altimetry with GNSS-R interferometry: first proof of concept experiment. GPS Solut. 2012;16(2):231–41.
38. Lowe ST, Meehan T, Young L. Direct signal enhanced semi-codeless processing of GNSS surface-reflected signals. IEEE J Sel Top Appl Earth Obs Remote Sens. 2017;7(5):1469–72.
39. Li W, Yang D, Addio S, et al. Partial interferometric processing of reflected GNSS signals for ocean altimetry. IEEE Geosci Remote Sens Lett. 2014;11(9):1509–13.
40. Onrubia R, Pascual D, Park H, et al. Satellite Cross-talk impact analysis in airborne interferometric global navigation satellite system-reflectometry with the microwave interferometric reflectometer. Remote Sens. 2019;11:1120.
41. Powell S J, Akos D M, Backen S. Altimetry using GNSS reflectometry for L5. In: Workshop on satellite navigation technologies and European workshop on GNSS signals and signal processing. IEEE;2015.
42. Munoz-Martin JF, Onrubia R, Pascual D, et al. Experimental evidence of swell signatures in airborne L5/E5a GNSS-reflectometry. Remote Sens. 2020;12:1759.
43. Carreno-Luengo H, Camps A, Ramos-Persz I, Rius A. Experimental evaluation of GNSS-reflectometry altimetric precision using the P (Y) and C/A signals. IEEE J Sel Top Appl Earth Obs Remote Sens. 2014;7:1493–500.
44. Ribo S, Arco JC, Oliveras S, Cardellach E, Rius A. Experimental results of an X-band PARIS receiver using digital satellite TV opportunity signals scattered on the sea surface. IEEE Trans Geosci Remote Sens. 2014;52(9):5704–11.
45. Shah R, Garrison JL, Egido A, et al. Bistatic radar measurements of significant wave height using signals of opportunity in L-, S-, and Ku-bands. IEEE Trans Geosci Remote Sens. 2016;54(2):826–41.
46. Wu J, Chen Y, Gao F, et. al. Sea surface height estimation by ground-based BDS GEO satellite reflectometry. IEEE J Sel Top Appl Earth Obs Remote Sens. 2020;13:5550–9.

47. Yun Z, Binbin L, Luman T, et al. Phase altimetry using reflected signals from BeiDou GEO satellites. IEEE Geosci Remote Sens Lett. 2016;13(10):1410–4.
48. Löfgren JS, Haas R. Sea level measurements using multi-frequency GPS and GLONASS observations. Eurasip J Adv Signal Process. 2014;2014(1):1–13.
49. Larson KM, Ray RD, Williams SDP. A 10-year comparison of water levels measured with a geodetic GPS receiver versus a conventional tide gauge. J Atmos Ocean Technol. 2017;34(2).
50. Stranderg J, Hobiger T, Hass R. Real-time sea-level monitoring using Kalman filtering of GNSS data. GPS Solut. 2019;23:61.
51. Kim SK, Park J. Monitoring a storm surge during hurricane Harvey using multi-constellation GNSS-Reflectometry. GPS Solut. 2021;25(2):63.
52. Yamawaki MK, Geremia-Nievinski F, Galera MJF. High-rate altimetry in SNR-based GNSS-R proof-of-concept of a synthetic vertical array. IEEE Geosci Remote Sens Lett. 2021.
53. Fagundes MRR, Tmendoca-Tinti I, Iescheck AK, Akos DM, Geremia-Nievinski F. An open-source low-cost sensor for SNR-based GNS reflectometry: Design and long-term validation towards sea-level altimetry. GPS Solut. 2021;25:73.
54. Purnell DJ, Gomes N, Minarik W, Porter D, Langston G. Precise water level measurements using low-cost GNSS antenna arrays. Earth Surf Dyn. 2021;9:673–85.
55. Altunta C, Tunalioglu N. Feasibility of retrieving effective reflector height using GNSS-IR from a single-frequency android smartphone SNR data. Digit Signal Process. 2021;112(1):103011.
56. Zavorotny VU, Voronovich AG. Bistatic GPS signal reflections at various polarizations from rough land surface with moisture content. In: IEEE international geoscience & remote sensing symposium "Taking the pulse of the planet: The role of remote sensing in managing the environment". IEEE;2000. p. 2852–4.
57. Zavorotny V, Masters D, Gasiewski A, et al. Seasonal polarimetric measurements of soil moisture using tower-based GPS bistatic radar. In: Geoscience and remote sensing symposium, 2003. IGARSS'03. Proceedings. 2003 IEEE International, vol. 2. IEEE;2004. p. 781–3.
58. SMEX02 Experiment Plan, http://hydrolab.arsusda.gov/smex02/smex60302.pdf.
59. Masters D, Katzberg S, Axelrad P. Airborne GPS bistatic radar soil moisture measurements during SMEX02 geoscience and remote sensing symposium, 2003. In: Proceedings of 2003 IEEE International, vol. 2. IEEE;2003. p. 896–8.
60. Masters D, Axelrad P, Katzberg S. Initial results of land-reflected GPS bistatic radar measurements in SMEX02. Remote Sens Environ. 2004;92(4):507–20.
61. Katzberg SJ, Torres O, Grant MS, et al. Utilizing calibrated GPS reflected signals to estimate soil reflectivity and dielectric constant: Results from SMEX02. Remote Sens Environ. 2006;100(1):17–28.
62. Gleason S, Hodgart S, Sun Y, et al. Detection and processing of bistatically reflected GPS signals from low Earth orbit for the purpose of ocean remote sensing. IEEE Trans Geosci Remote Sens. 2005;43(6):1229–41.
63. Alejandro Egido Egido. GNSS reflectometry for land remote sensing applications. Barcelona: Starlab; 2013. p. 112.
64. Guerriero L, Pierdicca N, Egido A, et al. Modeling of the GNSS-R signal as a function of soil moisture and vegetation biomass. In: Geoscience and remote sensing symposium. IEEE;2014. p. 4050–3.
65. Alonso-Arroyo A, Forte G, Camps A, et al. Soil moisture mapping using forward scattered GPS L1 signals. In: IEEE international geoscience & remote sensing symposium. 2014.
66. Egido A, Paloscia S, Motte E, et al. Airborne GNSS-R polarimetric measurements for soil moisture and above-ground biomass estimation. IEEE J Sel Top Appl Earth Obs Remote Sens. 2017;7(5):1522–32.
67. Camps A, Park H, Pablos M, et al. Sensitivity of GNSS-R spaceborne observations to soil moisture and vegetation. IEEE J Sel Top Appl Earth Obs Remote Sens. 2016;9(10):4730–42.
68. Camps A, Park H, Pablos M, et al. Soil moisture and vegetation impact in GNSS-R TechDemosat-1 observations. In: Geoscience and remote sensing symposium. IEEE;2016. p. 1982–4.

69. Hugo CL, Stephen L, Cinzia Z, et al. Spaceborne GNSS-R from the SMAP mission: First assessment of polarimetric scatterometry over land and cryosphere. Remote Sens. 2017;9(4):362.
70. Carreno-Luengo H, Luzi G, Crosetto M. Geophysical relationship between CYGNSS GNSS-R bistatic reflectivity and smap microwave radiometry brightness temperature over land surfaces. IEEE International geoscience and remote sensing symposium: IEEE; 2018.
71. Eroglu O, Kurum M, Boyd D, et al. High spatio-temporal resolution CYGNSS soil moisture estimates using artificial neural networks. Remote Sens. 2019;11(19):2272.
72. Carreno-Luengo H, Luzi G, Crosetto M. Impact of the elevation angle on CYGNSS GNSS-R bistatic reflectivity as a function of effective surface roughness over land surfaces. Remote Sens. 2018;10(11).
73. Gleason S, O'Brien A, Russel A, et al. Geolocation, calibration and surface resolution of CYGNSS GNSS-R land observations. Remote Sens. 2020;12(8):1317.
74. Senyurek V, Lei F, D Boyd, et al. Evaluations of machine learning-based CYGNSS soil moisture estimates against SMAP observations. Remote Sens. 2020;12 (Applications of GNSS Reflectometry for Earth Observation).
75. Mao K, Zhang M, Wang J, et al. The study of soil moisture retrieval algorithm from GNSS-R. In: International workshop on education technology and training, 2008 and 2008 International workshop on geoscience and remote sensing. ETT and GRS. IEEE;2008. p. 438–42.
76. Wan W, Chen X, Zhao L, et al. Near-surface soil moisture content measurement by GNSS reflectometry: An estimation model using calibrated GNSS signals;2012. p. 7523–6.
77. Liu W, Yang D, Gao C, et al. Soil moisture observation utilizing reflected global navigaiton satellite system signals. In: International conference on electrical engineering and information & communication technology. IEEE;2014. p. 1–5.
78. Peng X, Chen X, Xiao H, et al. Estimating soil moisture content using GNSS-R technique based on statistics. In: Geoscience and remote sensing symposium. IEEE;2015. p. 2004–7.
79. Yin C, Lopez-Baeza E, Martin-Neira M, et al. Intercomparison of soil moisture retrieved from GNSS-R and passive L-band radiometry at the Valencia anchor station. In: Geoscience and remote sensing symposium. IEEE;2016. p. 3137–9.
80. Yan Q, Huang W, Jin S, et al. Pan-tropical soil moisture mapping based on a three-layer model from CYGNSS GNSS-R data. Remote Sens Environ. 2020;247: 111944.
81. Cherniakov M. Space-surface bistatic synthetic aperture radar - prospective and problems. RADAR 2002. Edinburgh:IEEE;2002. p. 22–5.
82. Cherniakov M, Saini R, Zuo R, et al. Space surface bistatic SAR with space-borne non-cooperative transmitters. In: European radar conference, 2005 EURAD 2005. La Defense:IEEE;2005. p. 9–12.
83. Antoniou M, Cherniakov M, Saini R, et al. Modified range-doppler algorithm for space-surface BSAR imaging. In: International conference on radar. Edinburgh:IET;2007. p. 1–4.
84. Cherniakov M, Saini R, Zuo R, et al. Space-surface bistatic synthetic aperture radar with global navigation satellite system transmitter of opportunity-experimental results. IET Radar Sonar Navig. 2007;1(6):447–58.
85. Usman M, Armitage D. A remote imaging system based on reflected GPS signals. Int Conf Adv Space Technol. 2006;2006:173–8.
86. Usman M, Armitage DW. Acquisition of reflected GPS signals for remote sensing applications. In: 2008 2nd international conference on advances in space technologies. Islamabad: IEEE;2008. p. 131–6.
87. Usman M, Armitage D. Details of an imaging system based on reflected GPS signals and utilizing SAR techniques. J Glob Position Syst. 2009;8:87–99.
88. Antoniou M, Cherniakov M. GNSS-based bistatic SAR: A signal processing view. EURASIP J Adv Signal Process. 2013;2013(1):1–16.
89. Zhang Q, Cherniakov M, Antoniou M, et al. Passive bistatic synthetic aperture radar imaging with Galileo transmitters and a moving receiver: Experimental demonstration. IET Radar Sonar Navig. 2013;7(9):985–93.

90. Santi F, Antoniou M, Pastina D. Point spread function analysis for GNSS-based multistatic SAR. IEEE Geosci Remote Sens Lett. 2015;12(2):304–8.
91. Zeng T, Zhang T, Tian W, et al. Permanent scatterers in space-surface bistatic SAR using Beidou-2/Compass-2 as illuminators: Preliminary experiment results and analysis. In: 2016 European radar conference (EuRAD). London:IEEE;2016. p. 169–72.
92. Zeng T, Ao D, Hu C, et al. Multi-angle BiSAR images enhancement and scatting characteristics analysis. In: 2014 International radar conference. Lille:IEEE;2014. p. 1–5.
93. Koch V, Westphal R. New approach to a multistatic passive radar sensor for air/space defense. IEEE Aerosp Electron Syst Mag. 1995;10(11):24–32.
94. Behar V, Kabakchiev C, Rohling H. Air target detection using navigation receivers based on GPS L5 signals. In: 24th International technical meeting of the satellite division of the institute of navigation 2011, ION GNSS 2011. Porland: ION;2011. p. 333–7.
95. Behar V, Kabakchiev C. Detectability of air targets using bistatic radar based on GPS L5 signals. In: International radar symposium, IRS 2011 - proceedings. Leipzig: IRS;2011. p. 212–7.
96. Kabakchiev C, Behar V, Garvanov I, et al. Detection, parametric imaging and classification of very small marine targets emerged in heavy sea clutter utilizing GPS-based Forward Scattering Radar. In: ICASSP, IEEE international conference on acoustics, speech and signal processing-proceedings. Florence: Institute of Electrical and Electronics Engineers Inc.;2014. p. 793–7.
97. Suberviola I, Mayordomo I, Mendizabal J. Experimental results of air target detection with a GPS forward-scattering radar. IEEE Geosci Remote Sens Lett. 2012;9(1):47–51.
98. Chow Y P, Trinkle M. GPS bistatic radar using phased-array technique for aircraft detection. In: 2013 International conference on radar - beyond orthodoxy: New paradigms in radar, RADAR 2013. Adelaide: IEEE;2013. p. 274–9.
99. Santi F, Pieralice F, Pastina D. Joint detection and localization of vessels at sea with a GNSS-based multistatic radar. IEEE Trans Geosci Remote Sens. 2019;57(8):5894–913.
100. Pastina D, Santi F, Pieralice F, et al. Passive radar imaging of ship targets with GNSS signals of opportunity. IEEE Trans Geosci Remote Sens. 2021;59(3):2627–42.
101. Liu C, Hu C, Wang R, et al. GNSS forward scatter radar detection: Signal processing and experiment. In: Proceedings international radar symposium. Prague: IEEE Computer Society;2017. p. 1–9.
102. He ZY, Yang Y, Chen W, et al. Moving target imaging using GNSS-based passive bistatic synthetic aperture radar. Remote Sens 2020;12(20):1–21.
103. Zeng HC, Zhang HJ, Chen J, et al. UAV target detection algorithm using GNSS-based bistatic radar. In: International geoscience and remote sensing symposium (IGARSS). Yokohama: Institute of Electrical and Electronics Engineers Inc.;2019. p. 2167–70.

Open Access This chapter is licensed under the terms of the Creative Commons Attribution-NonCommercial-NoDerivatives 4.0 International License (http://creativecommons.org/licenses/by-nc-nd/4.0/), which permits any noncommercial use, sharing, distribution and reproduction in any medium or format, as long as you give appropriate credit to the original author(s) and the source, provide a link to the Creative Commons license and indicate if you modified the licensed material. You do not have permission under this license to share adapted material derived from this chapter or parts of it.

The images or other third party material in this chapter are included in the chapter's Creative Commons license, unless indicated otherwise in a credit line to the material. If material is not included in the chapter's Creative Commons license and your intended use is not permitted by statutory regulation or exceeds the permitted use, you will need to obtain permission directly from the copyright holder.

Chapter 2
Overview of GNSS Satellite Navigation Signals

2.1 Spread Spectrum Communication Principle

2.1.1 Basic Concepts of Spread Spectrum

Spread spectrum communication system, based on Shannon's information theory, broaden the spectrum of baseband signals over a wide frequency band before transmission, is one of the most effective ways to address issues in wireless communication like multiple access, interference resistance, anti-interception, multipath resistance, confidentiality, positioning, ranging, and identification capabilities.

A typical spread spectrum system is shown in Fig. 2.1. A typical spread spectrum system consists of six main parts: original information, source encoding, channel encoding (error control), carrier modulation and demodulation, spread spectrum modulation and demodulation, and channel. The purpose of source coding is to remove the redundancy from information, compress the code rate of the source and increase channel transmission efficiency. Error control increases redundancy to information during the channel transmission to enable error detection or correction and improve the quality of the channel transmission. Modulation is done to transmit symbols after channel encoding in appropriate frequency bands like microwave and shortwave bands. Spread spectrum modulation and demodulation are techniques used to widen and restore signal spectra for specific purposes. Unlike conventional communication systems, a broadband, low spectral density signal is transmitted in the channel. Shannon, the founder of information theory, pointed out that "the best signal for effective communication is a signal transmitted in the form of white noise" and proposed the famous Shannon theorem, with Formula (2.1) as the basis for spread spectrum theory.

$$C = W \log_2(1 + S/N) \quad (2.1)$$

© The Author(s) 2026
D. Yang and F. Wang, *Fundamentals and Applications of GNSS Reflectometry*,
Navigation: Science and Technology 16,
https://doi.org/10.1007/978-981-96-4554-1_2

where C is the channel capacity, equivalent to the information transmission rate (bit/s), W is the channel bandwidth, S is the signal power, and N is the noise power. This formula points out that, with the channel capacity remaining constant, increasing the channel bandwidth can reduce the signal-to-noise ratio requirement, allowing bandwidth and signal-to-noise ratio to be interchangeable under certain conditions. Even at low signal-to-noise ratios or when the signal is completely submerged in noise, reliable communication can be ensured with a sufficiently wide signal bandwidth at the same information transmission rate. Therefore, transmitting information using a much wider bandwidth signal than the information bandwidth can enhance the communication system's interference resistance, which is the theoretical basis and prominent advantage of spread spectrum communication. The interference resistance of spread spectrum communication systems is typically represented by the system's spread spectrum gain G, which is the ratio of the signal bandwidth B_s after spectrum expansion to the signal bandwidth B_d before expansion, numerically equal to the ratio of signal-to-noise ratios before and after spread spectrum demodulation, and also equal to the number of pseudo-random codes contained within the information bits, as shown in Formula (2.2).

$$G = \frac{B_s}{B_d} = \frac{(S/N)_o}{(S/N)_I} = N \qquad (2.2)$$

Formula (2.2) reflects the extent to which spread spectrum demodulation improves the signal-to-noise ratio in spread spectrum communication, being the main factor in enhancing interference resistance.

As shown in Fig. 2.2, the signal power density decreases after spreading and can be completely submerged in noise when transmitted in the channel, improving the confidentiality of signal transmission. After demodulation, the signal power density matching the spreading code of the transmitted signal increases, surpassing the noise. Simultaneously, narrowband interference in the signal band can be spread during signal demodulation, weakening its intensity, thereby improving the signal transmission's interference resistance.

Spread spectrum technology not only improves the signal-to-noise ratio and interference resistance but also allows different users' information to be carried using different code types in the same frequency band, achieving code division multiple access and increasing the bandwidth reuse rate. GNSS commonly employs direct sequence spread spectrum (DSSS) technology, where each satellite's navigation message is spread using pseudo-random noise (PRN).(PRN) codes with excellent

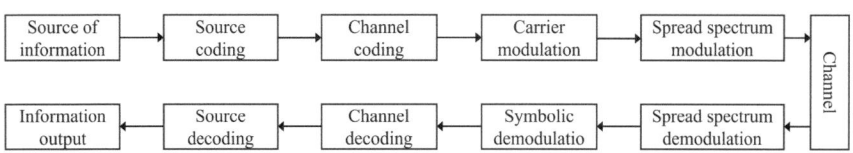

Fig. 2.1 Typical extended spectrum system block diagram

2.1 Spread Spectrum Communication Principle

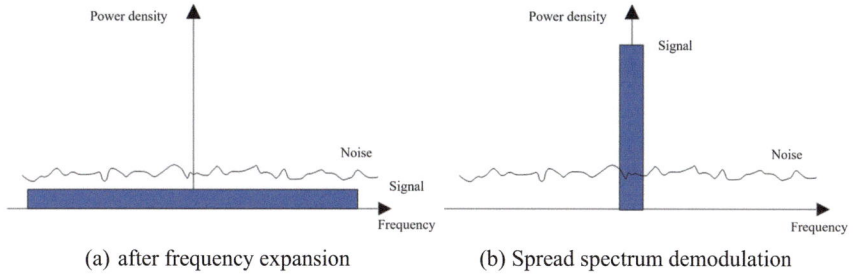

Fig. 2.2 Signal spreading and demodulation

autocorrelation and cross-correlation properties, then modulated on the satellite carrier frequency through binary phase shift keying or other methods. The PRN codes and carriers used for modulation also carry ranging, speed measurement, and time information required for navigation positioning and timing.

2.1.2 Gold Sequence

A GNSS satellite navigation signal mainly consists of three parts: carrier, spreading code, and navigation data. The pseudo-random code is the core component, used for signal spreading, tracking, locking, and distance measurement. This section takes the example of the L1 C/A code (Coarse/Acquisition, known as coarse capture code in Chinese, or coarse code for short) transmitted by GPS to introduce its generation method and related characteristics.

2.1.2.1 C/A Code Generation

A PRN code is a predetermined, periodic binary sequence with good autocorrelation properties, close to a binary random sequence, which is generated by a multi-stage feedback shift register. A sequence generated by an n-stage feedback shift register with a period equal to the maximum possible value (i.e., $2^n - 1$ code chips) is called an m-sequence. A Gold code is formed by the modulo-2 addition (linear combination) of a pair of m-sequences of the same length. The C/A code is a Gold code formed by the optimal combination m-sequences, where two m-sequences of equal length with the smallest maximum value of the correlation are added bitwise modulo 2. Different codes can be obtained by changing the relative phases of the two m-sequences generating it. For an m-sequence of length $N = 2^m - 1$, every two codes can be used in this way to generate N Gold codes, where the maximum cross-correlation value between any two codes is equal to the maximum cross-correlation value of the two m-sequences that make them up. The sidelobes of the cross-correlation function fluctuate, but its peak does not exceed the maximum value of autocorrelation. This is

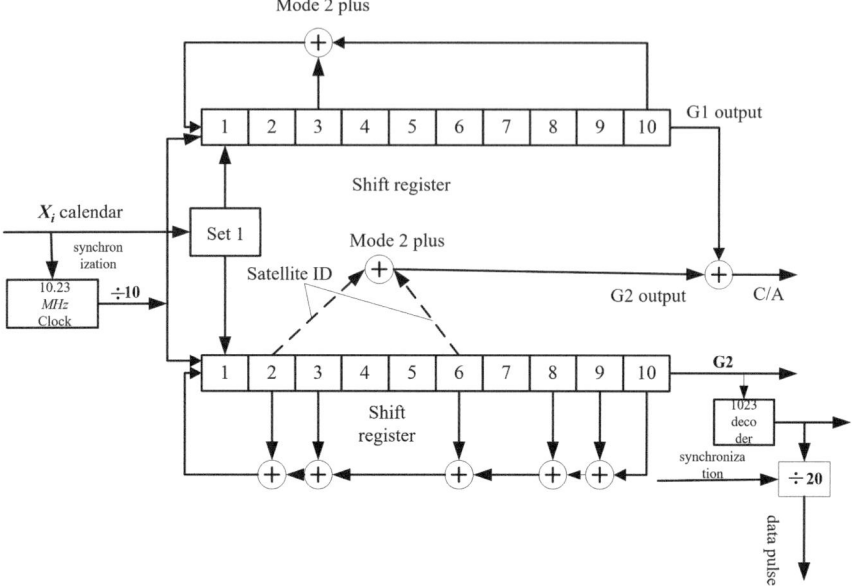

Fig. 2.3 C/A code generator

why Gold codes are widely used in multiple access communication and a primary consideration for GPS to adopt Gold codes as C/A codes.

The C/A code is generated by two 10-stage feedback shift registers. The two shift registers are all in the state of 1 under the action of a +1 pulse every Sunday midnight, and driven by a code rate of 1.023MHz, the two shift registers respectively generate two m-sequences G1(t) and G2(t) with a code length of $N = 2^{10} - 1 = 1023$ and a period of 1 ms. The G2(t) sequence is passed through a phase selector, outputting an m-sequence equivalent to a shift of G2(t), then summed with G1(t) modulo 2 to obtain the C/A code, as shown in Fig. 2.3.

In the actual generation of the C/A code, the output of the G2 is not directly from the final stage of the shift register, but based on the additivity, two stages are selected for modulo and operation before output. The effect of doing this is to generate a sequence to the original G2 sequence shift, where the shift depending on which two stages are selected for the modulo-2 and operation. This C/A code generator yields $C_{10}^2 + 10 = 55$ different C/A codes. From these G(t) codes, 32 codes are selected to be used for each GPS satellite under the names PRN1,..., PRN32, and PRN33,..., PRN37 are reserved for the ground signal transmitter. Since the C/A code is relatively short and can be searched 1000 times in 1 s, it is used not only to capture satellite signals and provide pseudorange observations but also to assist in capturing the P code (another pseudo-code in GPS, Precision code, abbreviated as P code).

2.1.2.2 C/A Code Related Characteristics

One of the most important properties of C/A codes is their correlation property. High autocorrelation peaks and low cross-correlation peaks provide a wide dynamic range for signal acquisition. In order to detect weak signals against a strong noise background, the autocorrelation peak of the weak signal must be greater than the cross-correlation peak of the strong signal. If the code is orthogonal, the theoretical value of the cross-correlation result should be 0. However, Gold codes are not fully orthogonal but only quasi-orthogonal, so their cross-correlation values are not zero but are relatively small. The cross-correlation function for Gold codes is shown in Table 2.1. For C/A codes, $n = 10$, and therefore, $P = 1023$. Using the relationship in Table 2.1 the cross-correlation values can be calculated as: $-65/1023$ (probability of 12.5%), $-1/1023$ (probability of 75%), and $63/1023$ (probability of 12.5%).

Let $PRN_j(t)$ be the C/A code sequence of satellite j consisting of $\{+1, -1\}$ with the autocorrelation function shown in Eq. (2.3).

$$R_j(\delta\tau) = \frac{1}{T}\int_0^T PRN_j(t)PRN_j(t+\delta\tau)dt \qquad (2.3)$$

where T is the code period and $\delta\tau$ is the code delay. Figure 2.4 shows the C/A code sequence (PRN sequence) of GPS satellite 1, with code period T of 1 ms.

According to Eq. (2.3), the autocorrelation function of the satellite PRN sequence can be calculated. Figure 2.5 shows a schematic diagram of the autocorrelation function of the C/A code of satellite 1, in which only the results of the code delay in the range of ± 1000 code chips are plotted.

For two satellites i and j, the cross-correlation function of their C/A codes is shown in Eq. (2.4).

$$R_{i,j}(\delta\tau) = \frac{1}{T}\int_0^T PRN_i(t)PRN_j(t+\delta\tau)dt \qquad (2.4)$$

Table 2.1 Cross-correlation values of Gold codes

Code period	Shift register order	Standardized correlation values	Probability of occurrence
$P = 2^n - 1$	$n =$ odd-number	$-\frac{2^{(n+1)/2}+1}{P}$	0.25
		$-\frac{1}{P}$	0.5
		$\frac{2^{(n+1)/2}+1}{P}$	0.25
$P = 2^n - 1$	$n =$ even number but not a multiple of 4	$-\frac{2^{(n+2)/2}+1}{P}$	0.125
		$-\frac{1}{P}$	0.75
		$\frac{2^{(n+2)/2}+1}{P}$	0.125

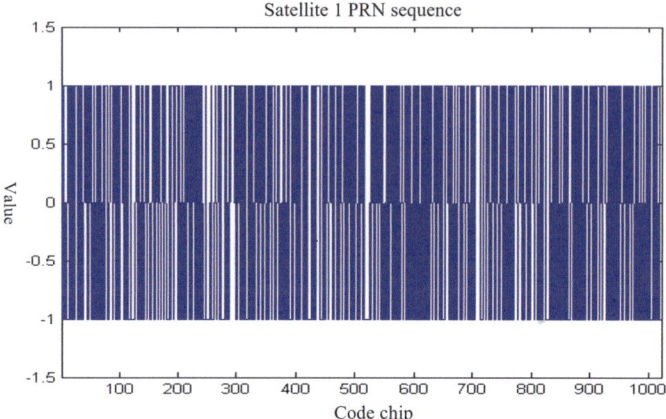

Fig. 2.4 PRN sequence of Satellite 1

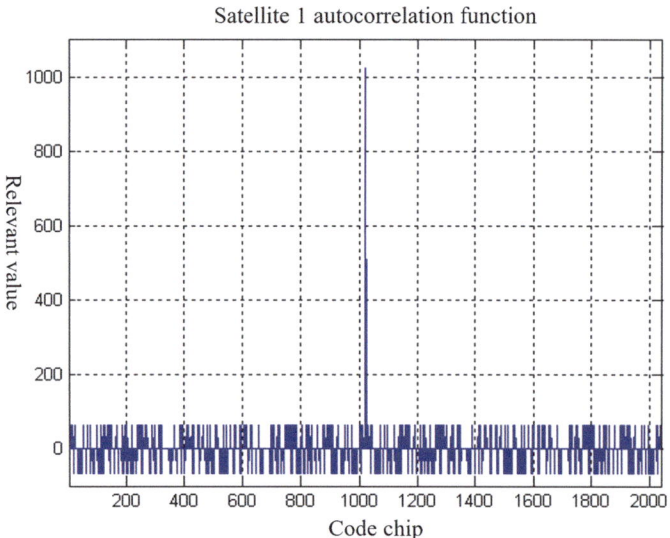

Fig. 2.5 Autocorrelation function schematic diagram of satellite1 C/A code

Figure 2.6 illustrates the C/A code cross-correlation function for satellites 1 and 4, with the same range of code delays ±1000 code chips) on the x-axis as in Fig. 2.5. By comparing Figs. 2.5 and 2.6, it can be seen that the Gold code autocorrelation function used by GPS satellites are similar to the cross-correlation function, both of which are not entirely zero.

To clearly express the sidelobes of the autocorrelation function, Fig. 2.7 takes satellites 7 and 4 as examples, showing only the autocorrelation function values within ±10 code chips of code delay.

2.1 Spread Spectrum Communication Principle 31

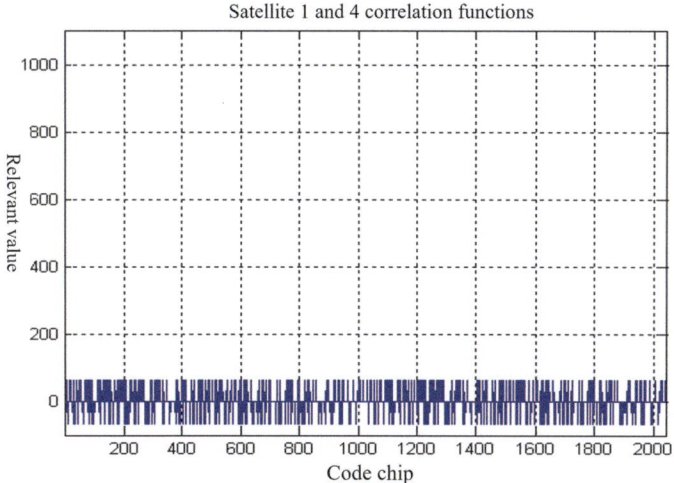

Fig. 2.6 Intercorrelation function of C/A codes for satellites 1 and 4

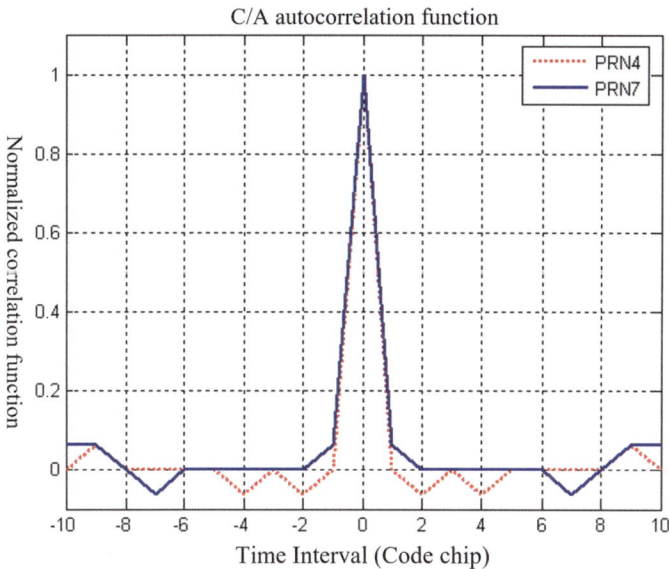

Fig. 2.7 GPS C/A code autocorrelation function and its sidelobes

Note: Sidelobe positions for PRN4 are ±2/4/9 (dashed lines); Sidelobe positions for PRN7 are ±1/7/9/10 (solid lines).

Let the value of the correlation function be expressed as $R_j(i)$ when $\delta\tau/\tau_c$ is the integer $i = [0, 1, ..., 1022]$, which takes the value $\{1, -\tau_c/T, 63\tau_c/T, -65\tau_c/T\}$ ($\tau_c/T = 1/1023$) for GPS C/A code. Then the autocorrelation function value at any moment is calculated as shown in Eq. (2.5).

$$\Lambda_j(\delta\tau) = [R_j(i) - R_j(i+1)](1 - \Delta t) + R_j(i+1) \tag{2.5}$$

where $i = \text{INT}(|\delta\tau/\tau_c|)$, $\text{INT}(\cdot)$ denotes the rounding function and $\Delta t = |\delta\tau/\tau_c| - i$.

Typically, the simplified autocorrelation function is assumed to be

$$R_j(i) = \begin{cases} 1 & i = 0 \\ -\tau_c/T & i \neq 0 \end{cases} \tag{2.6}$$

At this point, Eq. (2.5) is expressed as

$$\Lambda_j(\delta\tau) = \begin{cases} 1 - |\delta\tau|/\tau_c - |\delta\tau|/T & |\delta\tau|/\tau_c \leq 1 \\ -\tau_c/T & |\delta\tau|/\tau_c > 1 \end{cases} \tag{2.7}$$

Since $\tau_c \ll T$, it is generally assumed that $-\tau_c/T \approx 0$, further simplified as

$$\Lambda_j(\delta\tau) = \begin{cases} 1 - |\delta\tau|/\tau_c & |\delta\tau|/\tau_c \leq 1 \\ 0 & |\delta\tau|/\tau_c > 1 \end{cases} \tag{2.8}$$

Equation (2.8) represents an ideal triangular function with a base width of two code chips. This simplified expression of the autocorrelation function is commonly used for signal capture and tracking in ordinary GPS receivers. In the processing and application of reflected signals, the more accurate autocorrelation function given by Eq. (2.5) should be used, taking into account the effects of the autocorrelation function sidelobes.

2.1.3 Spread Spectrum Ranging Principle and Performance

The C/A code can solve the phase of the received C/A code signal based on the position of the main peak of the autocorrelation function during the despreading process. This code phase can reflect the propagation time of the signal from the satellite to the receiver, which is then converted into a distance measurement value from the satellite to the receiver, achieving ranging function.

2.1.3.1 Basic Concepts of Pseudorange

The distance between a satellite and a receiver should be the product of the signal propagation time and the speed of electromagnetic wave propagation (i.e., the speed of light). As shown in Fig. 2.8, due to the asynchrony between the satellite clock and

2.1 Spread Spectrum Communication Principle

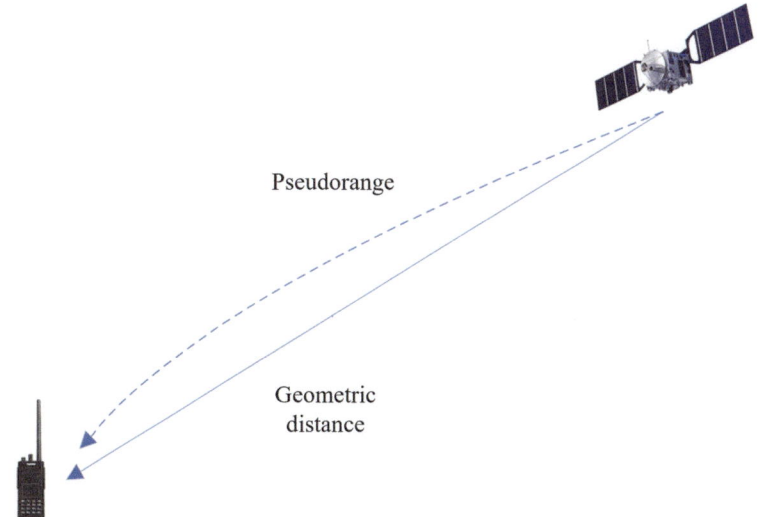

Fig. 2.8 Pseudorange

the receiver clock, as well as various measurement errors, the calculated distance value is often referred to as "pseudorange" rather than the true geometric distance.

The basic pseudorange observation equation is shown in Eq. (2.9) [1]

$$\rho = r + c(\delta_{t接} - \delta_{t卫}) + cT + cI + \varepsilon_\rho \tag{2.9}$$

where ρ denotes the pseudorange, r denotes the geometric distance, c denotes the speed of light,

$$\delta_{t接}$$

denotes the receiver clock bias,

$$\delta_{t卫}$$

denotes the satellite clock bias, T denotes the tropospheric delay, I denotes the ionospheric delay, and ε_ρ denotes pseudorange measurement noise. Clock bias refers to the difference between the clock and GPS time (GPST).

2.1.3.2 Relationship Between Pseudorange and Spread Spectrum Code Phase

Pseudorange is obtained by multiplying the difference between the signal reception time and the transmission time by the speed of light in vacuum. The reception time

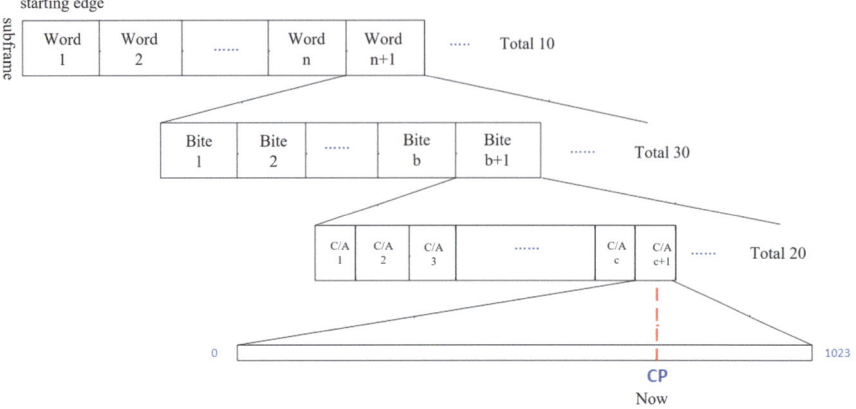

Fig. 2.9 Code phase and transmission time

can be directly read from the receiver clock, while obtaining the satellite transmission signal time is achieved through the receiver's analysis of the received signal.

As shown in Fig. 2.9, the GPS signal contains a lot of time-related information. The code phase, as the smallest time-measurement component, is obtained through C/A code correlation processing in the receiver, indicating the position of the latest reception moment within a complete cycle of C/A code, with values ranging from 0 to 1023. It is known that the C/A code code slice length corresponds to a distance of about 293m ($\frac{1}{1.023 \times 10^6 \text{ Hz}} \times 3 \times 10^8$ m/s $\approx$ 293 m). Assuming a code phase measurement error of \(1/10\), \(1/20\),..., \(1/100\) of a chip, the corresponding range error is 29.3 m to 2.93 m

After the signal undergoes correlation processing and carrier removal, the remaining data bits undergo bit synchronization and frame synchronization processing. Decoding the navigation message in this subframe allows obtaining the Time of Week (TOW) corresponding to the previous subframe, which, when multiplied by 6 after subtracting 1 from it, gives the starting edge time of the current subframe. Then, the word number and bit number in the current subframe at the current moment are determined (a GPS navigation message word contains 30 bits, and each bit is 20 ms long). Finally, the signal transmission time is obtained from the whole number of C/A code cycles and the phase value, thereby obtaining the pseudorange.

The formula for calculating the transmission time is shown in Eq. (2.10) [2]

$$t_{\text{send}} = (\text{TOW} - 1) \times 6 + (30n + b) \times 0.020 + (c + \frac{\text{CP}}{1023}) \times 0.001 \ (s) \quad (2.10)$$

2.1.3.3 Spread Spectrum Ranging Performance Analysis

Spread codes are crucial for ranging, as they use the navigation message format, information, and code phase to ultimately determine the corresponding signal transmission time and pseudorange, without any ambiguity issues. Additionally, spread code ranging, based on the C/A code chip length of 293m, can achieve meter-level accuracy in ranging.

Apart from spread code ranging, there is also the carrier phase ranging method, which is based on the carrier wavelength (the GPS L1 carrier frequency is 1575.42 MHz, and the wavelength is about $\frac{3\times 10^8 \text{ m/s}}{1575.42\times 10^6 \text{ MHz}} \approx 0.19 \text{ m} = 19 \text{ cm}$), and can achieve millimeter-level accuracy, and is less affected by multipath. However, due to the presence of unknown integer cycle ambiguities, spread code ranging is usually needed as a supplementary method.

2.2 Modulation Method

2.2.1 Binary Phase Shift Keying (BPSK)

Initially, GNSS signals used Binary Phase Shift Keying (BPSK) modulation, where digital information is transmitted through changes in carrier phase while keeping carrier amplitude and frequency constant. The time domain expression of a BPSK modulation signal is shown in Eq. (2.11).

$$e_{BPSK}(t) = A\cos(2\pi f_c t + \varphi) \tag{2.11}$$

where f_c is the carrier frequency, φ is the absolute phase of the signal, and the value of φ is selected as shown in Eq. (2.12).

$$\varphi = \begin{cases} 0 & \text{Transmit } 0 \\ \pi & \text{Transmit } 1 \end{cases} \tag{2.12}$$

Combining Eqs. (2.11) and (2.12), we get Eq. (2.13).

$$e_{BPSK}(t) = \begin{cases} A\cos(2\pi f_c t) & \text{Transmit } 0 \\ -A\cos(2\pi f_c t) & \text{Transmit } 1 \end{cases} \tag{2.13}$$

From the above equation, it is known that the waveforms representing the two symbols are the same but with opposite polarities. Therefore, the BPSK modulation of the signal can be equated to the product of a bipolar non-return-to-zero rectangular pulse sequence and a sinusoidal carrier, as mathematically expressed in Eq. (2.14).

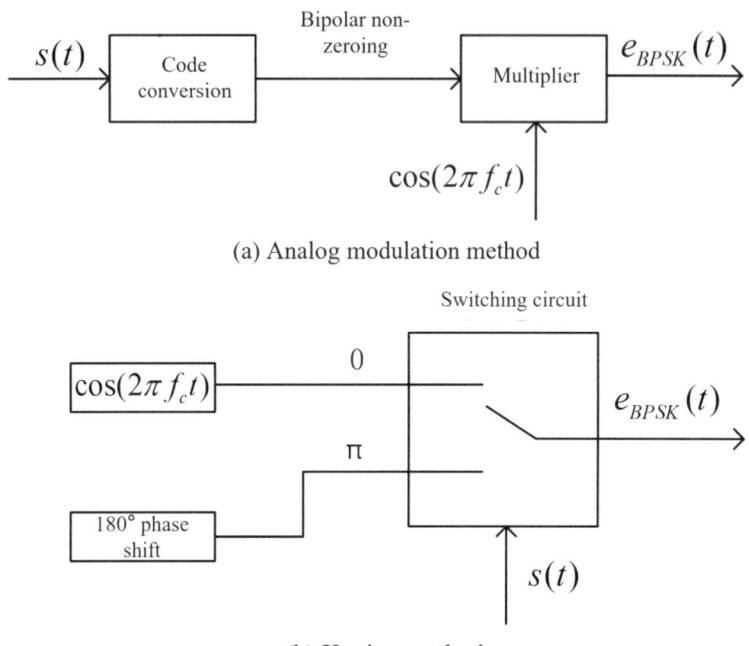

Fig. 2.10 BPSK modulation block diagram

From the above equation, it can be seen that the signal waveforms representing the two code elements are identical and of opposite polarity, thus the BPSK modulation of the signal can be equated to the product of a bipolar non-return-to-zero (NRZ) rectangular pulse sequence \(s(t)\) and a sinusoidal carrier. The mathematical expression is shown in Eq. (2.14) and the block diagrams in Figs. 2.10 and 2.11).

$$e_{BPSK}(t) = s(t)\cos(2\pi f_c t) \quad (2.14)$$

2.2.2 Quadrature Phase Shift Keying (QPSK)

Quadrature Phase Shift Keying (QPSK) is a four-phase modulation scheme where each symbol represents 2 bits of information, with four possible arrangements: 00, 01, 10, 11. The time domain expression of a QPSK modulation signal is shown in Eq. (2.15).

$$e_{QPSK}(t) = A\cos(2\pi f_c t + \theta_i) \quad i = 1,2,3,4 \quad (2.15)$$

2.2 Modulation Method

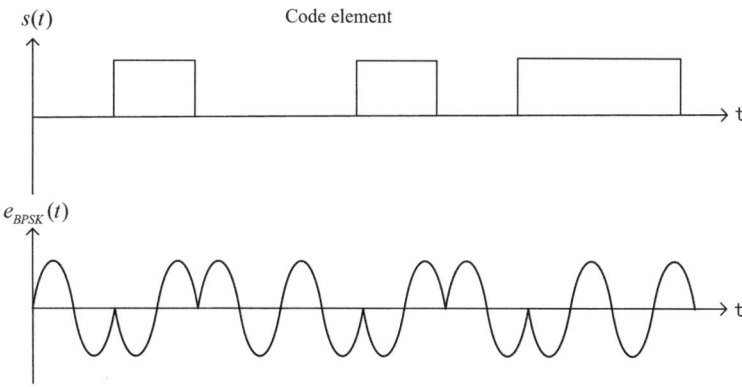

Fig. 2.11 BPSK signal waveform

Here, f_c is the carrier frequency and θ_i has four possible values, which are $\frac{\pi}{4}, \frac{3\pi}{4}, \frac{5\pi}{4}, \frac{7\pi}{4}$.

Using a single coherent carrier for QPSK demodulation leads to phase ambiguity. Therefore, the demodulation requires two orthogonal coherent carriers. Expanding Eq. (2.15) gives (Fig. 2.12 and 2.13):

$$e_{\text{QPSK}}(t) = I * \cos(2\pi f_c t) - Q * \sin(2\pi f_c t) \quad (2.16)$$

For simplicity, the amplitude of the output signal $e_{\text{QPSK}}(t)$ is set to 1, which requires the amplitude of the input signals to be adjusted to $\frac{1}{\sqrt{2}}$. By substituting $(+\frac{1}{\sqrt{2}}, +\frac{1}{\sqrt{2}})$ $(-\frac{1}{\sqrt{2}}, +\frac{1}{\sqrt{2}})$ $(-\frac{1}{\sqrt{2}}, -\frac{1}{\sqrt{2}})$ $(+\frac{1}{\sqrt{2}}, -\frac{1}{\sqrt{2}})$ as (I, Q) into Eqs. (2.15) and (2.16), respectively, the phase correspondence between the input and output signals can be obtained as shown in Table 2.2 (Fig. 2.14).

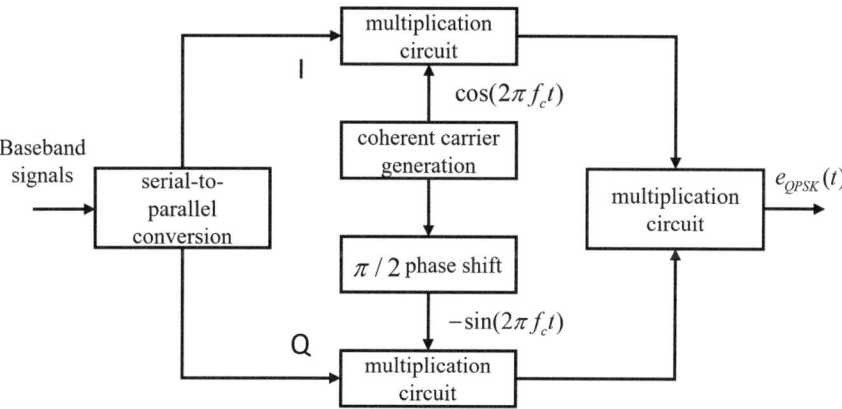

Fig. 2.12 QPSK multiplication circuit modulation

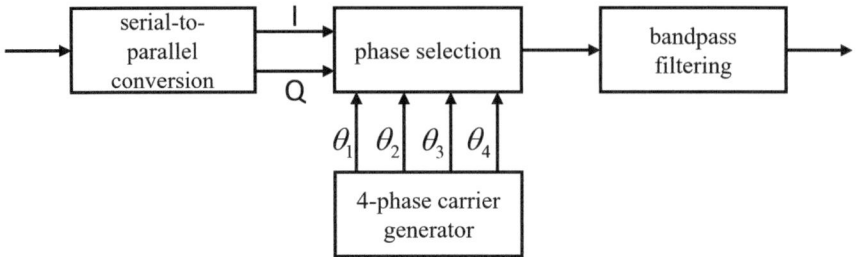

Fig. 2.13 QPSK selective modulation

Table 2.2 QPSK input and output signal phase correspondence

Input signal	I-signal	Q-signal	Output signal phase
00	$+\frac{1}{\sqrt{2}}$	$+\frac{1}{\sqrt{2}}$	$\frac{\pi}{4}$
01	$-\frac{1}{\sqrt{2}}$	$+\frac{1}{\sqrt{2}}$	$\frac{3\pi}{4}$
11	$-\frac{1}{\sqrt{2}}$	$-\frac{1}{\sqrt{2}}$	$\frac{5\pi}{4}$
10	$+\frac{1}{\sqrt{2}}$	$-\frac{1}{\sqrt{2}}$	$\frac{7\pi}{4}$

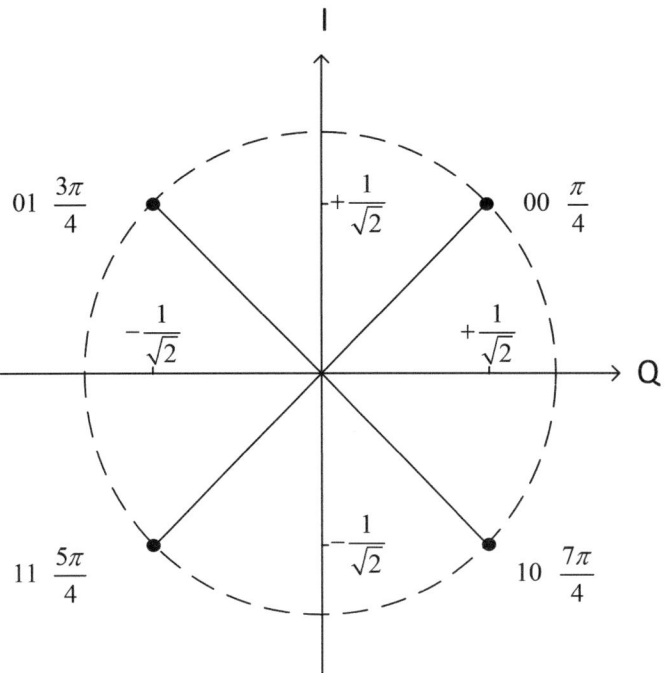

Fig. 2.14 QPSK modulation constellation diagram

2.2 Modulation Method

2.2.3 Binary-Offset-Carrier

BOC modulation was first proposed by Jhon W. Betz [1], a leading figure involved in the design of the Galileo signal, with the main purpose of solving the spectrum congestion issue. Currently, new signals in the BeiDou, GPS, and Galileo systems all adopt Binary Offset Carrier (BOC) modulation or its variants.

2.2.3.1 BOC Signals and Characteristics

In general, in GNSS signal processing, BOC is considered a kind of subcarrier modulation, equivalent to the symbol of a sine or cosine function. A BOC signal with a sine phase can be represented as the product of a code sequence and a subcarrier with a frequency of f_s,

$$x(t) = c(t) \cdot \text{sign}(\sin(2\pi f_s t)) \tag{2.17}$$

$$c(t) = \sum_k c_k h(t - kT_c) \tag{2.18}$$

where $h(t)$ is a code sequence equivalent to an inverse non-return-to-zero code with -1 or 1 in the interval $[0, T_c]$.

A BOC signal can usually be denoted as BOC(p, q), where the first parameter p indicates the subcarrier frequency and the second parameter q referring to the code rate: $f_s = p \cdot 1.023\text{MHz}$, $f_c = q \cdot 1.023\text{MHz}$, ratio $n = 2\frac{f_s}{f_c} = 2\frac{p}{q}$ is the number of half subcarrier periods in one chip, which can be either even or odd.

If the BOC signal has a sine phase, its normalized power spectral density can be represented as:

$$G_{BOC}(f) = \frac{1}{T_c} \left(\frac{\sin\left(\frac{\pi f T_c}{n}\right) \sin(\pi f T_c)}{\pi f \cos\left(\frac{\pi f T_c}{n}\right)} \right)^2 \quad n \text{ is even} \tag{2.19}$$

$$G_{BOC}(f) = \frac{1}{T_c} \left(\frac{\sin\left(\frac{\pi f T_c}{n}\right) \cos(\pi f T_c)}{\pi f \cos\left(\frac{\pi f T_c}{n}\right)} \right)^2 \quad n \text{ is odd} \tag{2.20}$$

If the BOC signal has a cosine phase, its normalized power spectral density can be expressed as

$$G_{BOC}(f) = \frac{1}{T_c} \left(\frac{\sin\left(\frac{\pi f T_c}{n}\right)\left(\cos\left(\pi f \frac{T_c}{n}\right) - 1\right)}{\pi f \cos\left(\frac{\pi f T_c}{n}\right)} \right)^2 \quad n \text{ is even} \tag{2.21}$$

$$G_{BOC}(f) = \frac{1}{T_c} \left(\frac{\cos\left(\frac{\pi f T_c}{n}\right)\left(\cos\left(\pi f \frac{T_c}{n}\right) - 1\right)}{\pi f \cos\left(\frac{\pi f T_c}{n}\right)} \right)^2 \quad n \text{ is odd} \tag{2.22}$$

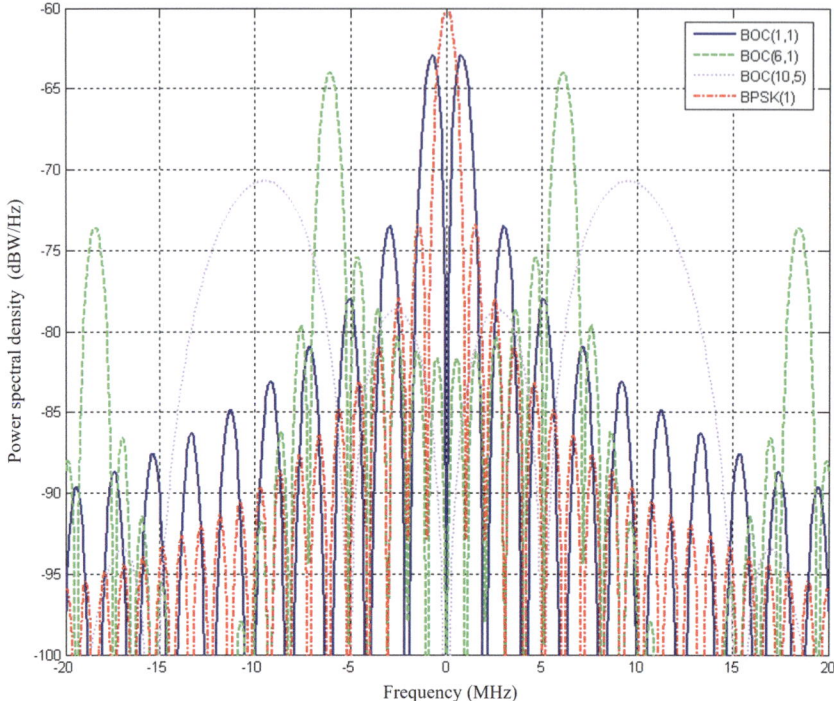

Fig. 2.15 Power spectral density of the BOC signal

Figure 2.15 shows the power spectral density plots of several typical BOC signals, and it can be seen that the power spectra of BOC signals have the following characteristics:

(1) The power spectral density of BOC signals is symmetrical on the left and right, with the main lobes distributed on both sides of the carrier frequency.
(2) The farther the ratio between the subcarrier and chip rate is from the center frequency, the more lobes the main lobe of the BOC signal has, with the sum of the main lobes and side lobes being n.
(3) More power can be provided on high frequencies offset from the center frequency, improving improving the system's resistance to narrowband interference and its multipath performance.

Corresponding to the power spectral density, the autocorrelation function of the BOC signal is also different from that of BPSK signals. Figure 2.16 shows the autocorrelation functions of several typical BOC signals. It can be seen that due to the influence of the subcarrier, the autocorrelation function of the BOC(p, q) signal has the following characteristics:

2.2 Modulation Method

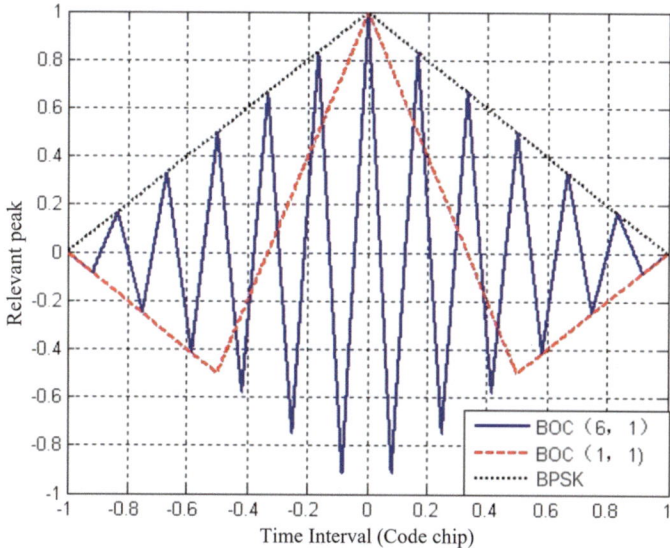

Fig. 2.16 Autocorrelation function of BOC signal

(1) It has multiple peaks, and as the value of p/q is larger, the number of side peaks is larger, and the side peaks adjacent to the main peak are also larger, and the width of the main peak decreases.
(2) The main peak is narrow, which can lead to high-precision code tracking and good multipath resolution, offering the potential for better navigation performance.
(3) The presence of multiple peaks may cause the receiver to lock onto the wrong peak, leading to positioning errors.

On one hand, the zero crossings in the auto-correlation function of BOC signals can cause signal acquisition failures, and the multi-peak nature of the auto-correlation function may lead to incorrect signal acquisition, resulting in tracking and calculation errors.

On the other hand, the autocorrelation function of BOC signals exhibits a multi-peak structure, which differ significantly from the correlation functions of conventional direct sequence spread spectrum communication. Among the multiple peaks of the BOC signal correlation function, only the central peak corresponds to a code phase error of 0. During signal tracking, tracking loop errors may lead to locking onto a side peak close to the central peak, causing incorrect lock.

Therefore, resolving or mitigating the ambiguity in BOC signal acquisition and tracking caused by these multiple peaks is a key focus of research on BOC signal reception processing.

2.2.3.2 MBOC Signals and Features

MBOC is the abbreviation of Multiplexed Binary Offset Carrier, and its Chinese name is Multiple Binary Offset Carrier. For example, MBOC (6,1,1/11) power spectral density is a mixture of BOC (1,1) spectrum and BOC (6,1) spectrum [3] which is defined as

$$G_{\text{MBOC}}(f) = \frac{10}{11} G_{\text{BOC}(1,1)}(f) + \frac{1}{11} G_{\text{BOC}(6,1)}(f) \tag{2.23}$$

The new-generation navigation system signals like GPS, BeiDou, Galileo different types of hybrid modulation based on MBOC are used. The main types include Time Multiplexed Binary Offset Carrier (TMBOC), Composite Binary Offset Carrier (CBOC), and Quadrature Multiplexed Binary Offset Carrier (QMBOC). For example, the GPS-L1C signal uses TMBOC modulation, where the data channel, occupying 1/4 of the total power, uses BOC(1,1) modulation. In the pilot channel, a time-multiplexed combination of BOC(6,1) and BOC(1,1) is used, where BOC(6,1) modulation in the 1st, 5th, 7th, and 30th chips out of every 33 pseudo-random code slices, and BOC(1,1) modulation in the rest of the code slices This results in a TMBOC(6,1,4/33) modulation format for the entire mixed signal. The spectrum of the GPS L1C signal can be represented as shown.

$$\begin{aligned}
G_{\text{L1C}}(f) &= \frac{1}{4} G_{\text{data}}(f) + \frac{3}{4} G_{\text{pilot}}(f) \\
&= \frac{1}{4} G_{\text{BOC}(1,1)}(f) + \frac{3}{4} \left(\frac{29}{33} G_{\text{BOC}(1,1)}(f) + \frac{4}{33} G_{\text{BOC}(6,1)}(f) \right) \\
&= \frac{10}{11} G_{\text{BOC}(1,1)}(f) + \frac{1}{11} G_{\text{BOC}(6,1)}(f)
\end{aligned} \tag{2.24}$$

Galileo E1 OS signal uses CBOC, where is added to the BOC(1,1) signal. The data channel component uses CBOC(+) modulation, while the pilot channel component uses CBOC(-) modulation. The power spectrum of the two components combined satisfies the power spectrum defined by MBOC.

Beidou B1C signal uses QMBOC modulation, where the BOC(1,1) and BOC(6,1) components are modulated on two orthogonal phases of the carrier. The power distribution between the data channel and pilot channel is 1/4 and 3/4, respectively. The power spectral density of QMBOC is the same as that of TMBOC.

2.2.3.3 AltBOC Signals and Features

The Galileo-E5 signal uses Alternate Binary Offset Carrier (AltBOC) modulation technology, which inherits the traditional BOC signal's ability to resist interference and multipath effects. It has superior spectrum compatibility and ranging performance, while also increasing spectrum utilization. The subcarrier spectrum of BOC

2.2 Modulation Method

modulation signals is symmetric with a double-sideband spectrum, whereas AltBOC modulation signals have a single-sideband spectrum. Therefore, the spectrum of BOC signals after modulation is symmetric and double-sideband, while AltBOC signals have a single-sideband spectrum. With traditional BOC modulation signals, both main lobes carry the same information, whereas with AltBOC modulation signals, different main lobes can carry different information. Like conventional BOC, AltBOC modulation is also denoted by AltBOC(m, n).

Taking the Galileo-E5 signal as an example, the E5 signal consists of upper and lower sidebands and modulates two signals, namely the E5a signal and the E5b signal, with carrier frequencies of 1176.45 MHz and 1207.14 MHz, respectively. E5a signal and E5b signal each contain two orthogonal channels, i.e., data channel and pilot channel. The data channel modulates navigational data code and pseudo-random code, while the pilot channel demodulates pseudo-random code. The Galileo-E5 signals are generated by AltBOC(15, 10) modulation, whose subcarrier frequency is $f_s = 15 \times 1.023$ MHz and the pseudo-random code rate is $f_c = 10 \times 1.023$ MHz. The Galileo-E5 signals can be represented as

$$s_{\text{AltBOC}}(t) = c_{a_I}(t)d_{a_I}(t)sc_{\text{boc}}(t) - c_{b_I}(t)d_{b_I}(t)sc_{\text{boc}}(t)$$
$$+ c_{a_Q}(t)sc_{\text{boc}}\left(t + \frac{\pi}{2}\right) - c_{a_Q}(t)sc_{\text{boc}}\left(t - \frac{\pi}{2}\right) \quad (2.25)$$

To keep the signal amplitude constant, the amplitude of signals in the horizontal, vertical and angular bisector directions are the same, and the zero phase point is eliminated. By performing constant envelope transformation on the signal, a signal with constant amplitude AltBOC signal as

$$s_{\text{AltBOC}}(t) = \frac{1}{2\sqrt{2}} \begin{bmatrix} (s_{a_I}(t) + j \cdot s_{a_Q}(t))(sc_s(t) - jsc_s(t - T_s/4)) + \\ (s_{b_I}(t) + j \cdot s_{b_Q}(t))(sc_s(t) + jsc_s(t - T_s/4)) + \\ (\bar{s}_{a_I}(t) + j \cdot \bar{s}_{a_Q}(t))(sc_p(t) - jsc_p(t - T_s/4)) + \\ (\bar{s}_{b_I}(t) + j \cdot \bar{s}_{b_Q}(t))(sc_p(t) + jsc_p(t - T_s/4)) \end{bmatrix} \quad (2.26)$$

where,

$$\begin{cases} s_{a_I}(t) = c_{a_I}(t) \cdot d_{a_I}(t) \\ s_{a_Q}(t) = c_{a_Q}(t) \\ s_{b_I}(t) = c_{b_I}(t) \cdot d_{b_I}(t) \\ s_{b_Q}(t) = c_{b_Q}(t) \end{cases} \quad (2.27)$$

$$\begin{cases} \bar{s}_{a_I}(t) = s_{a_Q}(t) \cdot s_{b_I}(t) \cdot s_{b_Q}(t) \\ \bar{s}_{a_Q}(t) = s_{a_I}(t) \cdot s_{b_I}(t) \cdot s_{b_Q}(t) \\ \bar{s}_{b_I}(t) = s_{a_I}(t) \cdot s_{a_Q}(t) \cdot s_{b_Q}(t) \\ \bar{s}_{b_Q}(t) = s_{a_I}(t) \cdot s_{a_Q}(t) \cdot s_{b_I}(t) \end{cases} \quad (2.28)$$

The newly introduced subcarrier functions are defined as follows

$$\begin{cases} sc_s(t) = \sum_{i=-\infty}^{+\infty} AS_{|i|_8} \text{rect}_{T_s/4}(t - i \cdot T_s/4) \\ sc_p(t) = \sum_{i=-\infty}^{+\infty} AP_{|i|_8} \text{rect}_{T_s/4}(t - i \cdot T_s/4) \end{cases} \quad (2.29)$$

where, the coefficients AS_i and AP_i are the four states on the AltBOC signal constellation diagram, as shown in Table 2.3 is shown.

After AltBOC modulation, the E5a and E5b signals can be regarded as two independent QPSK signals. The power spectral density function of AltBOC signal can be represented as Eqs. (2.30), (2.31) (Fig. 2.17).

$$G_{\text{AltBOC}}(f) = \frac{4f_c}{\pi^2 f^2} \frac{\sin^2\left(\frac{\pi f}{f_c}\right)}{\cos^2\left(\frac{\pi f}{2f_s}\right)} \cdot \left[\frac{\cos^2\left(\frac{\pi f}{2f_s}\right) - \cos\left(\frac{\pi f}{2f_s}\right)}{-2\cos\left(\frac{\pi f}{2f_s}\right)\cos\left(\frac{\pi f}{4f_s}\right) + 2} \right], \text{ 2 m/n odd}$$

(2.30)

Table 2.3 GALILEO-E5 signal AltBOC subcarrier mdulation coefficients

i	0	1	2	3	4	5	6	7
AS_i	$\frac{\sqrt{2}+1}{2}$	$\frac{1}{2}$	$-\frac{1}{2}$	$-\frac{\sqrt{2}+1}{2}$	$-\frac{\sqrt{2}+1}{2}$	$-\frac{1}{2}$	$\frac{1}{2}$	$\frac{\sqrt{2}+1}{2}$
AP_i	$-\frac{\sqrt{2}-1}{2}$	$\frac{1}{2}$	$-\frac{1}{2}$	$\frac{\sqrt{2}-1}{2}$	$\frac{\sqrt{2}-1}{2}$	$-\frac{1}{2}$	$\frac{1}{2}$	$-\frac{\sqrt{2}-1}{2}$

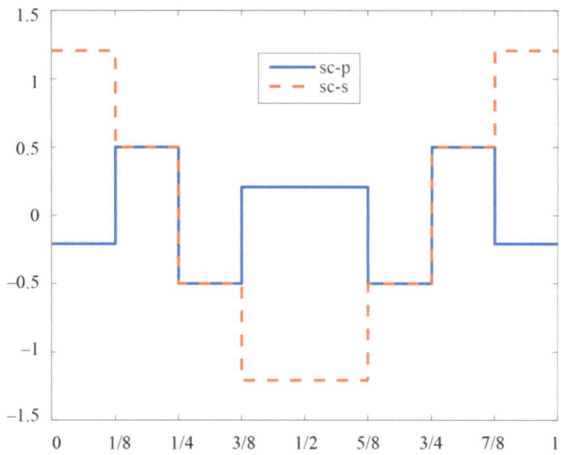

Fig. 2.17 Values of the subcarrier modulation coefficients of the GALILEO-E5 signal AltBOC over one cycle

2.3 Satellite Navigation Signal

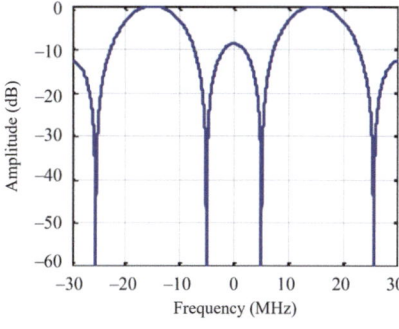
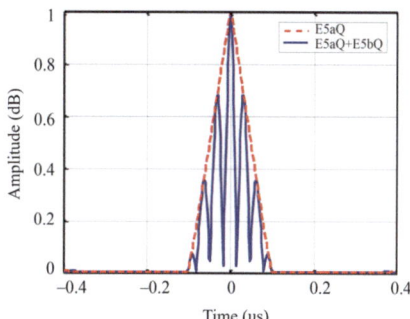

(a) Power Spectral Density of Galileo-E5 Pseudo-Random Codes

(b) Autocorrelation Function of Galileo-E5 Pseudo-Random Codes

Fig. 2.18 Power spectral density and sutocorrelation function of Galileo-E5 signal

$$G_{\text{AltBOC}}(f) = \frac{4f_c}{\pi^2 f^2} \frac{\cos^2\left(\frac{\pi f}{f_c}\right)}{\cos^2\left(\frac{\pi f}{2f_s}\right)} \cdot \left[\cos^2\left(\frac{\pi f}{2f_s}\right) - \cos\left(\frac{\pi f}{2f_s}\right) - 2\cos\left(\frac{\pi f}{2f_s}\right)\cos\left(\frac{\pi f}{4f_s}\right) + 2\right], \ 2 \text{ m/n even} \tag{2.31}$$

As shown in Fig. 18a, the spectrum of the Galileo-E5 signal modulated by AltBOC(15,10) exhibits a double-sideband, but it contains two independent channels, E5a and E5b, which can be regarded as two independent QPSK signals, and its autocorrelation function has a sharp central peak. As shown in Fig. 18b, the autocorrelation function of the full-channel E5 signal (E5aQ + E5bQ or E5aI + E5aI) is also a triangular wave modulated by a triangular wave.

2.3 Satellite Navigation Signal

2.3.1 Beidou Signal

2.3.1.1 Signaling System

The current BeiDou-3 system can provide five spatial signals for positioning and navigation services with, namely B1C, B2a, B2b, B1I and B3I. According to the "BeiDou Satellite Navigation System Public Service Performance Specification Version 3.0", which will be released on May 26th, 2021, the center frequencies of each signals are 1575.42, 1176.25, 1207.14, 1561.098 and 1268.52 MHz, and the bandwidth, symbol rate, modulation scheme, and polarization mode are shown in Table 2.4.

According to the BeiDou system ICD regulations, when the satellite elevation angle is greater than 5°, the minimum power level of the satellite's navigation signals

Table 2.4 BeiDou GNSS signaling regime

Signal component		Carrier frequency/ MHz	Bandwidths/ MHz	Symbol rate/ bps	Modulation method	Polarization mode
B1C	B1C_data	1575.42	32.736	100	BOC(1,1)	RHCP
	B1C_pilot			0	QMBOC(6,1,4/33)	
B2a	B2a_data	1176.45	20.46	200	BPSK(10)	RHCP
	B2a_pilot			0		
B2b	I-branc	1207.14	20.46	1000	BPSK(10)	RHCP
	Q-branch	/	/	/	/	/
B1I		1561.098	4.092	D1 Navigational messages 50 D2 Navigational messages 500	BPSK	RHCP
B3I		1268.52	20.46	D1 Navigational messages 50 D2 Navigational messages 500	BPSK	RHCP

reaching the receiver antenna output near the Earth's surface with a right-hand circularly polarized antenna of 0 dBi gain (or a linearly polarized antenna of 3 dBi gain) must meet the minimum received power levels shown in Table 2.5.

Table 2.5 Ground minimum received power level

Signal Name	Minimum user received power (greater than 5° elevation angle)/dBW
B1C	−159(MEO)
	−161(IGSO)
B2a	−156(MEO)
	−158(IGSO)
B2b	−160(MEO)
	−162(IGSO)
B1I	−163
B3I	−163

2.3.1.2 Signal Structure

The B1I and B3I signals are orthogonally modulated on the carrier by "ranging code + navigation message", with their signal expressions are as follows:

$$S_{B1I}^j(t) = A_{B1I} C_{B1I}^j(t) D_{B1I}^j(t) \cos\left(2\pi f_1 t + \varphi_{B1I}^j\right) \quad (2.32)$$

$$S_{B3I}^j(t) = A_{B3I} C_{B3I}^j(t) D_{B3I}^j(t) \cos\left(2\pi f_3 t + \varphi_{B3I}^j\right) \quad (2.33)$$

where the superscript j denotes the satellite number; A_{B1I} and A_{B3I} denote the amplitude of B1I and B3I signals, C_{B1I}^j and C_{B3I}^j denote the ranging codes of B1I and B3I signals, D_{B1I}^j and D_{B3I}^j denote the data codes modulated on the ranging codes of B1I and B3I signals, f_1 and f_3 denote the carrier frequencies of B1I and B3I signals, and φ_{B1I}^j and φ_{B3I}^j denote the initial phases of the carriers of B1I and B3I signals, respectively. The ranging codes of B1I and B3I signals are generated by two linear sequences G1, G2 modulo-2 addition and then truncated by 1 chip after generating the balanced Gold code. The code rate of the B1I signal Ranging code of is 2.046 Mcps, with a code length is 2046; and the code rate of the B3I signal ranging code is 10.23 Mcps, and a code length is 10230.

The two linear sequences G1 and G2 of the B1I signal ranging code are generated by an 11-stage linear shift register, with a generating polynomial as shown in Table 2.4

$$G1(X) = 1 + X + X^7 + X^8 + X^9 + X^{10} + X^{11} \quad (2.34)$$

$$G2(X) = 1 + X + X^2 + X^3 + X^4 + X^5 + X^8 + X^9 + X^{11} \quad (2.35)$$

The initial state of the G1 register is 01010101010; the initial state of the G2 register is 01010101010. ("State" is more precise than "phase" for shift registers).

The B1I signal ranging code generator is shown in Fig. 2.19 shows.

By different taps of the shift register generating the G2 sequence modulo 2, different phase offsets of the G2 sequence can be achieved, and different satellite pseudorange codes can be generated by modulo 2 with the G1 sequence. The phase allocation of the G2 sequence for different satellites is described in the relevant ICD document.

The G1 and G2 sequences of the B3I signal ranging code are generated by a 13-stage linear shift register with a period of 8191 code slices, and the generating polynomials are respectively

$$G1(X) = X^{13} + X^4 + X^3 + X + 1 \quad (2.36)$$

$$G2(X) = X^{13} + X^{12} + X^{10} + X^9 + X^7 + X^6 + X^5 + X + 1 \quad (2.37)$$

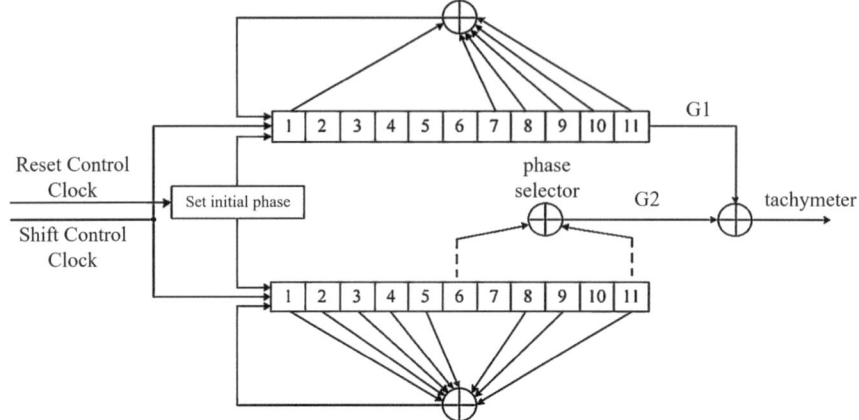

Fig. 2.19 B1I signal ranging code generator

The code generator is shown in Fig. 2.20.

The ranging code of 10230 chips is obtained by modulo 2 with CA and CB sequences with periods of 8190 and 8191 chips, respectively, where the CA sequence is obtained by truncating the code sequence generated by G1, and the CB sequence is obtained by G2. The initial phase of the G1 sequence is set to "11111111111100" at the beginning of each ranging code period (1 ms) or when the G1 sequence register phase is "11111111111100", and the initial phase of the G2 sequence is set at the beginning of each ranging code period (1 ms). The initial phase of the G1 sequence

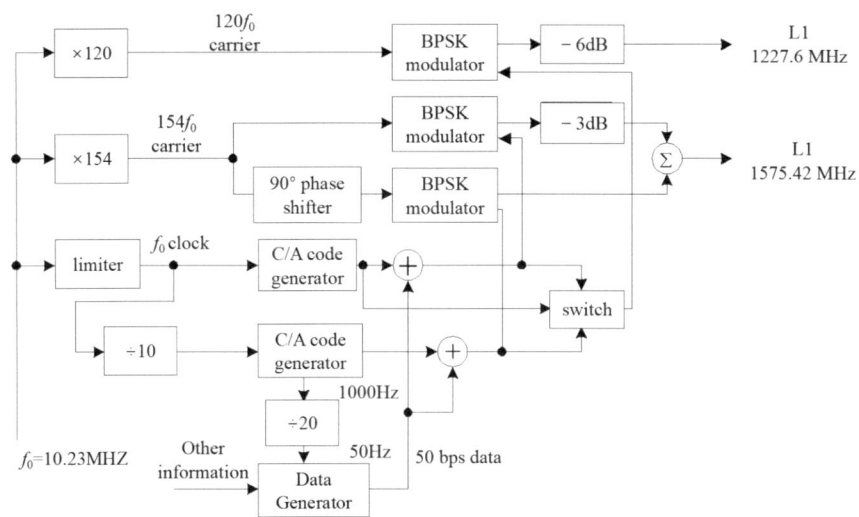

Fig. 2.20 B3I signal ranging code generator

2.3 Satellite Navigation Signal

is "1111111111111". The initial phase of the G2 sequence is formed by different numbers of shifts from "1111111111111", with different initial phases corresponding to different satellites. The phase allocation of the G2 sequence for different satellites is described in the relevant ICD document [4].

The I-branch component of the B2b signal is generated by the modulation of the navigation message data and the ranging code and can be expressed as follows

$$S_{B2b\_I}(t) = \frac{1}{\sqrt{2}} D_{B2b\_I}(t) C_{B2b\_I}(t) \tag{2.38}$$

$$D_{B2b\_I}(t) = \sum_{k=-\infty}^{\infty} d_{B2b\_I}(k) p_{T_{B2b\_I}}(t - kT_{B2b\_I}) \tag{2.39}$$

Where d_{B2b_I} is the data code of the navigation message of the B2b signal I branch, T_{B2b_I} is the chip width of the data code, and $p_{T_{B2b\_I}}$ is a rectangular pulse with a width of T_{B2b_I}.

The B2b signaling I-branch ranging code rate is 10.23 Mcps with a code length 10230, which is obtained from two 13-stage linear feedback shift registers by shifting and modulo-2-sum generating Gold code expansion. The generating polynomial is

$$G1(X) = 1 + X + X^9 + X^{10} + X^{13} \tag{2.40}$$

$$G2(X) = 1 + X^3 + X^4 + X^6 + X^9 + X^{12} + X^{13} \tag{2.41}$$

Detailed information on the structure of the B2b signal I-branch ranging code generator and the initial values of the G1 and G2 sequences are given in Reference.

The complex envelopes of both the B1C signal and the B2a signal can both be expressed as

$$s(t) = s_{\text{data}}(t) + js_{\text{pilot}}(t) \tag{2.42}$$

where, the data component of the B2a signal is generated by the modulation of the navigation message data and the ranging code; the guide frequency component includes only the ranging code; the data component of the B1C signal is generated by the subcarrier modulation of the navigation message data and the ranging code, and the the pilot component. Table 2.7 is the same revision is generated by the ranging code through subcarrier modulation. Tables 2.6 and 2.7 give the component compositions of the B1C and B2a signals, respectively, as well as the modulation mode, phase relationship, and power ratio of each component.

The ranging codes of both B1C and B2a signals use a hierarchical code structure, consisting of the main code and the sub-code . The width of the code piece of the sub-code is the same as the period of the main code, and the starting moment of the sub-code code piece is strictly aligned with the starting moment of the first code piece of the main code. Tables 2.8 and 2.9 give the ranging code parameters of B1C and B2a signals respectively, where the B1C data component does not contain

Table 2.6 B1C signal modulation characteristics

Component	Modulation method		Phase / °	Power ratio
$s_{B1C\_data}(t)$	Sin BOC(1,1)		0	1/4
$s_{B1C\_pilot\_a}(t)$	QMBOC(6,1,4/33)	Sin BOC(1,1)	90	29/44
$s_{B1C\_pilot\_b}(t)$		Sin BOC(6,1)	0	1/11

Table 2.7 B2a signal modulation characteristics

Component	Modulation method	Phase relation	Power ratio
$s_{B1C\_data}(t)$	BPSK(10)	0	1/2
$s_{B1C\_pilot\_a}(t)$	BPSK(10)	90	1/2

Table 2.8 B1C signal ranging code parameters

Signal component	Primary code type	Primary code length	Primary code period (ms)	Subcode type	Subcode length	Subcode period
B1C data component	Weilcode cut short	10230	10	(N/A)	(N/A)	(N/A)
B1C pilot component	Weilcode cut short	10230	10	Weilcode cut short	1800	18000 ms

Table 2.9 B2a signal ranging code parameters

Signal Component	Main Code Type	Main Code Length	Main Code Period (ms)	Sub-code Type	Sub-code Length	Sub-code Period (ms)
B2a data component	Gold	10230	1	fixed code	5	5
B2a data component	Gold	10230	1	Weil code truncation	100	100

subcode. For MEO and IGSO satellites, each satellite corresponds to a unique ranging code number (PRN number), and the B1C and B2a signals broadcasted by the same satellite adopt the same PRN number.

2.3.1.3 Navigation Messages

From the content of Sect. 2.3.1, it can be known that there are currently five signals with different frequency points in the BeiDou global satellite navigation system, and their navigation messages are also different. The corresponding relationship between the different types of satellites broadcasting signals and the types of navigation messages are as follows Table 2.10 shows.

(1) D1 Navigation Message and D2 Navigation Message

2.3 Satellite Navigation Signal

Table 2.10 Correspondence table for satellite type, broadcast signal and navigation message type

Signal type	Navigation message type	Satellite type
B1C	B-CNAV1	BDS-3I
B2a	B-CNAV2	BDS-3 M
B2b	B-CNAV3	
B1I B3I	D1	BDS-2I BDS-2 M BDS-3I BDS-3 M
	D2	BDS-2G BDS-3G

For B1I and B3I signals, due to differences in rate and structures, the navigation messages are divided into D1 navigation message (broadcast by MEO and IGSO satellites) and D2 navigation message (broadcast by GEO satellites). The D1 navigation message contains basic navigation information (basic navigation information of this satellite, all the satellite almanac information, and time synchronization information with other systems. The D2 navigation message contains basic navigation information and wide-area differential information (BeiDou system differential and integrity information and grid point ionospheric information).

The D1 navigation message rate is 50 bps and is further modulated by a secondary code, the Neumann-Hoffman code (NH code), at a rate of 1 kbps (referred to as NH code). The NH code is 20 bits long (0, 0, 0, 0, 0, 0, 1, 0, 0, 0, 1, 1, 0, 1, 0, 0, 0, 0, 1, 1, 1, 1, 0), with a period equal to the width of one navigation information bit, and the width of each 1-bit is the same as the period of the spread-spectrum code, which is modulated synchronously with spread-spectrum code and navigation information code in mode 2 sum form.

The D1 navigation message consists of a superframe, main frame and subframe. Each superframe is 36,000 bits long, consisting of 24 main frames (24 pages), taking 12 min. Each main frame consists of 5 subframes, each subframe has 10 words and each word includes 30 bits. Each word consists navigation message data and checksum. The first 15 bits of the first word in each subframe are coded without error correction, and the remaining 11 bits are encoded using a BCH(15,11,1) code plus interleaving for error correction, with a total of 26 bits; the other 9 words are encoded using a BCH(15,11,1) code plus interleaving for error correction, with a total of 22 bits. The simplified structure of the D1 navigation message frame is shown in Fig. 2.21 [5].

The D2 NAV message is also composed of superframe, main frame and subframe. Each superframe is 180,000 bits long, consisting of 120 main frames with the same structure oas that the D1 message, taking a total of 6 min. The simplified structure of the D2 Navigation Message Frame is shown in Fig. 2.22 [5]

(2) B-CNAV1 Message Format

52　　2 Overview of GNSS Satellite Navigation Signals

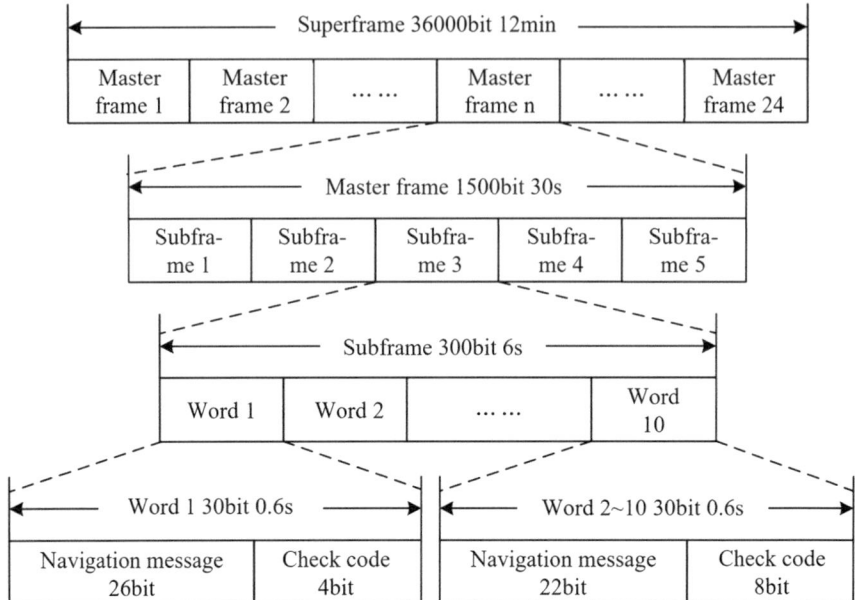

Fig. 2.21 D1 navigation message frame structure

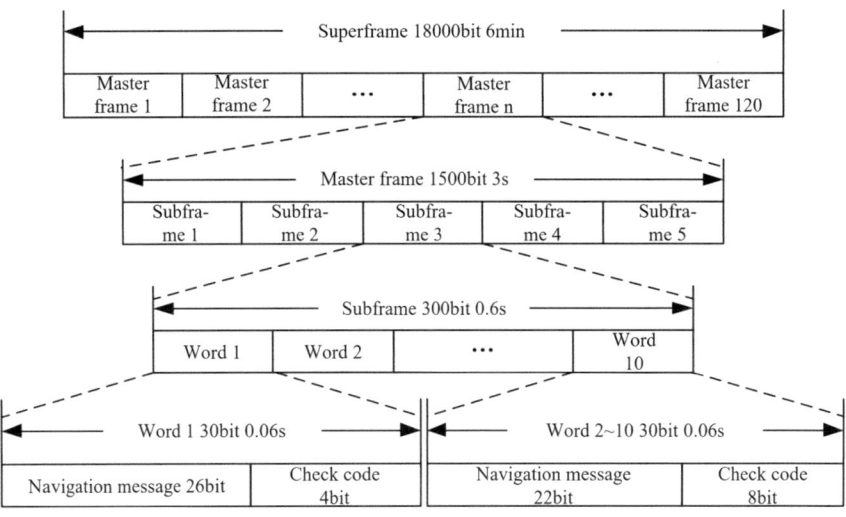

Fig. 2.22 D2 navigation message frame structure

2.3 Satellite Navigation Signal

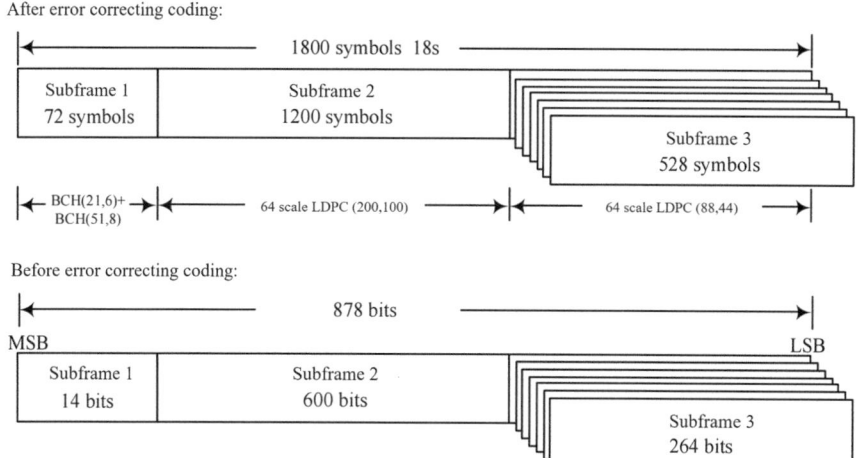

Fig. 2.23 B-CNAV1 frame structure

The B-CNAV1 navigation message is broadcast in the B1C signal, with each message frame consisting of 1800 symbol bits, a symbol rate of 100 sps (symbols per second), a period of 18 s, and composed of 3 subframes. The frame structure before and after error correction coding is shown in Fig. 2.23 [10].

Subframe 1 broadcasts the PRN number and the seconds of the hour (SOH), with a length of 14 bits before error correction coding, and a length of 72 symbol bits after BCH(21,6) + BCH(51,8) encoding. Subframe 2 broadcasts system time parameters, message data version number, ephemeris parameters, clock bias parameters, group delay correction parameters, and other system information. It has a length of 600 bits before error correction coding, and a length of 1200 symbol bits after LDPC(200,100) encoding. Subframe 3 broadcasts ionospheric delay correction model parameters, Earth orientation parameters, BDT-UTC time synchronization parameters, BDT-GNSS time synchronization parameters, medium-precision ephemeris, concise ephemeris, satellite health status, satellite integrity status indicators, space signal accuracy index, space signal monitoring accuracy index, and other information, divided into multiple pages. It has a length of 264 bits before error correction coding, and a length of 528 symbol bits after LDPC(88,44) encoding. Subframe 2 and Subframe 3 are each separately LDPC encoded and then interleaved.

(3) B-CNAV2 message format

The B-CNAV2 navigation message is broadcast in the B2a signal, with each message frame consisting of 600 symbols, a symbol rate of 200 sps, a period of 3 s, and the frame structure before and after error correction coding is shown in Fig. 2.24.

The first 24 symbols of each message frame are the frame synchronization header (fixed value 0xE24DE8), transmitted with the most significant bit first. Each message frame includes the PRN number, message type (MesType), Seconds of the Week (SOW), message data, and Cyclic Redundancy Check bits (CRC), totaling 288 bits

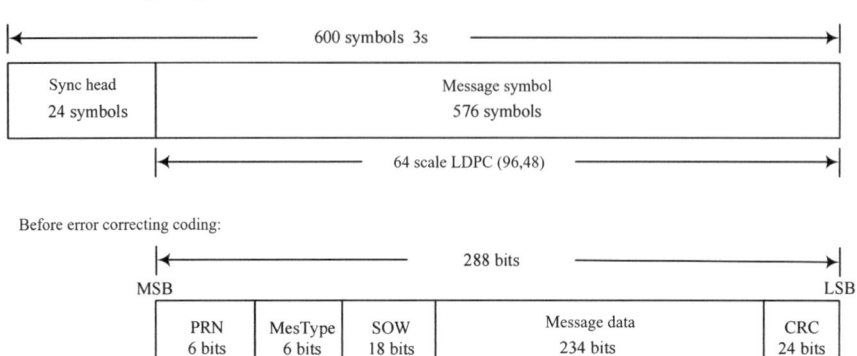

Fig. 2.24 B-CNAV2 Frame Structure

before error correction coding. After undergoing 64-ary LDPC(96,48) encoding, the length becomes 576 symbols.

(4) B-CNAV3 message format

The B-CNAV3 navigation message is broadcast in the B2b signal, including basic navigation information and basic integrity information. Each message frame has a length of 1000 symbols, a symbol rate of 1000 sps (symbols per second), a period of 1 s, and the frame structure before and after error correction coding is shown in Fig. 2.25 [9].

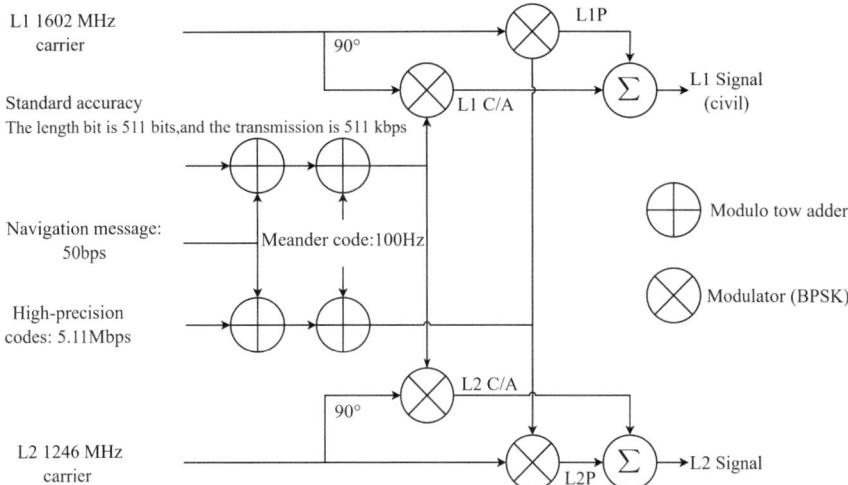

Fig. 2.25 B-CNAV3 frame structure

2.3 Satellite Navigation Signal 55

The first 16 symbols of each message frame are the frame synchronization header (fixed value of 0xEB90), sent with the most significant bit first. Each message frame includes the PRN number, message type (MesType), seconds of the week (SOW), message data, and cyclic redundancy check bits (CRC), totaling 486 bits before error correction coding. After 64-bit LDPC (162,81) encoding, the length becomes 972 symbols.

2.3.2 GPS Signal

2.3.2.1 GPS Signal Generation

GPS uses Code Division Multiple Access (CDMA) technology, modulating carriers with the same frequency. Different satellites use different pseudo-random codes to distinguish their signals. Each satellite signal uses direct sequence spread spectrum modulation with a pseudo-random code, as shown in Fig. 2.26.

As can be seen from Fig. 2.26, the signal sent by the GPS satellite to the user consists of two frequency components, L1 and L2, which have the same common frequency source and are generated by multiplying the fundamental frequency f_0 by 154 and 120, respectively.

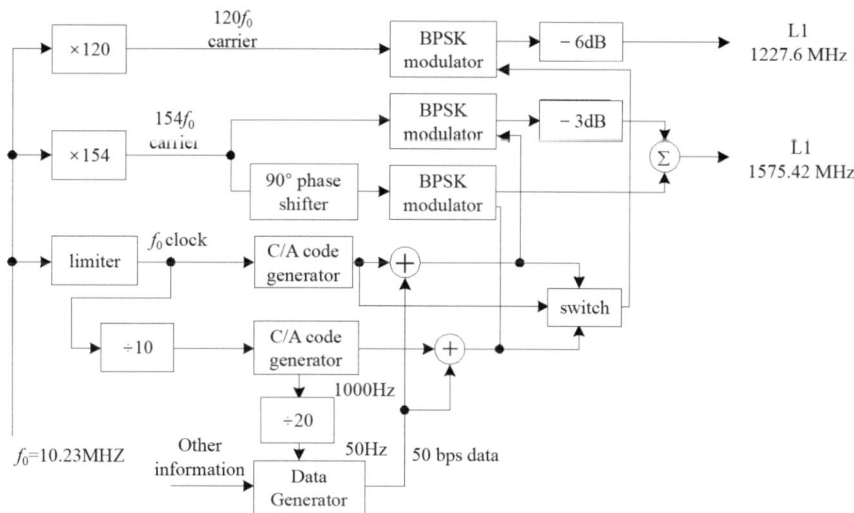

Fig. 2.26 GPS signal generation

$$f_0 = 10.23 \text{ MHz}$$
$$f_{L1} = 154f_0 = 1575.42 \text{ MHz} \qquad (2.43)$$
$$f_{L2} = 120f_0 = 1227.6 \text{ MHz}$$

where f_{L1} and f_{L2} are the center frequencies of the GPS signals.

The satellite navigation messages D-code, pseudo-random code P(Y)-code, and C/A code use BPSK modulation on carrier signals L1 and L2. The P(Y)-code and C/A code are pseudo-random sequences with values of 0 and 1. The P(Y)-code has a rate of 10.23 Mbps, a length of 235469592765×10^3 chips, a period of approximately 266 days and 9 h; the C/A code has a rate of 1.023 Mbps, a length of 1023 chips, and a period of 1 ms; the D-code is a navigation message code with a rate of 50 bps. The in-phase carrier component of the L1 carrier signal is modulated by the P(Y) $\oplus$ D composite code using BPSK, while the quadrature carrier component is modulated by the C/A $\oplus$ D composite code using BPSK. The L2 carrier signal can be modulated by the P(Y) $\oplus$ D composite code or the C/A $\oplus$ D composite code, and in some cases, only the P(Y) code is used. The specific choice depends on the control segment, but the composite code P(Y) $\oplus$ D is typically used.

The L1 signal emitted by the i th GPS satellite can be expressed as

$$\begin{aligned} S_{L1i}(t) = &A_{ci}C_i(t)D_i(t)\cos(2\pi f_{L1}\cdot t + \varphi_i) \\ &+ A_{pi}P_i(t)D_i(t)\sin(2\pi f_{L1}\cdot t + \varphi_i) \end{aligned} \qquad (2.44)$$

where A_{ci} and A_{pi} are the signal amplitudes, $P_i(t)$ and $C_i(t)$ are the P-code and C/A-code, respectively, $D_i(t)$ is the navigation message, and φ_i is the initial phase mainly due to phase noise and frequency drift.

The L2 signal emitted by the ith GPS satellite can be expressed as

$$S_{L2i}(t) = A_{L2i}P_i(t)D_i(t)\cos(2\pi f_{L2}t + \varphi_i) \qquad (2.45)$$

where A_{L2i} is the signal amplitude and φ_i is the corresponding initial phase.

Precise Positioning Services (PPS) is primarily military, and the user has access to all signals on L1 and L2 with full GPS accuracy. Since C/A codes are generally not available on L2, Standard Positioning Service (SPS, Standard Positioning Service) is primarily civilian and users cannot work dual-frequency. Single-frequency users cannot accurately measure the ionospheric delay and have lower positioning accuracy than dual-frequency users. Since the reception of P(Y) code requires a specialized auxiliary chip and key, SPS users generally only use C/A code on L1, and for reflection signal reception processing and applications, they also mostly use C/A code. Therefore, this book mainly focuses on introducing the C/A code signal on the L1. The C/A code, in addition to capturing satellite signals and providing pseudorange observations, the C/A code can also be used to assist in capturing the P-code.

The L1 frequency C/A code signal transmitted by GPS satellites reaches the ground with a maximum signal strength of -150 dBW and a minimum signal strength

2.3 Satellite Navigation Signal

of −158.5 dBW. According to the ICD IS-GPS-200 M, the minimum receiving power of the L1 C/A code for the user is as follows.

2.3.2.2 GPS Satellite Signal Navigation Message Structure

The navigation message of a GPS satellite is the data basis used by receivers to determine the user's position, speed, and time. It mainly includes: satellite ephemeris, clock corrections, ionospheric delay corrections, almanac and operational status information of other satellites in the GPS constellation, and information to assist in converting C/A code timing to P code timing. This information is transmitted in binary code format, framed according to specified formats, at a transmission rate of 50bps.

Each navigation message frame consists of 5 subframes, totaling 1500 bits (referred to as a long frame), taking 30 s. Each subframe is 300 bits long, taking 6 s, and includes 10 words of 30 bits each, with each bit taking 20 ms, corresponding to 20 C/A code cycles. Among them, The complete data for subframes 4 and 5 are transmitted over 25 consecutive frames; these 25 frames constitute a superframe. In these 25 frames, the 4th and 5th subframes of each main frame are also known as pages. The 1st, 2nd, and 3rd subframes broadcast the satellite's clock correction parameters and broadcast ephemeris, while the 4th and 5th subframes broadcast the almanac data of all satellites, ionospheric correction parameters, and other system information, with 25 pages constituting a complete set of system information. Each subframe/page should contain a telemetry word (TLM) and a handover word (HOW), with the TLM being transmitted first, followed by the HOW, and then eight data words. Each word in the frame contains 6 bits of parity for error checking and 24 bits of information. The navigation message structure is shown in Fig. 2.27 [1], with detailed navigation message frame structure and content available in literature [13].

The telemetry code is located at the beginning of each subframe, appearing once every 6 s, with the first 8 bits being a sync header that identifies the start of each subframe, making it easier for users to recognize each subframe, fixed at 10001011. Bits 9 to 22 are meaningful only to authorized users, followed by 6 bits of parity. Within the GPS satellite system, there is a 29-bit counter that records the number of X1 epochs (1.5 s cycles) from the GPS epoch (midnight on January 6, 1980), with the top 10 bits representing the week number since the GPS epoch (modulo 1024) and the lower 19 bits being the Time Of Week (TOW) count, indicating the number of X1 epochs since the end of the previous week. The top 17 bits of the TOW counter make up the first 17 bits of the handover word (HOW), with a counting range from 0 to 100799, increasing by 1 every 6 s (1.5 s × 4), indicating the start time of the week for the next subframe, adding a GPS time stamp to each subframe, eliminating the ambiguity in timing due to the multiple values of the C/A code 1 ms epoch cycle. The first subframe's message also includes the week number since the GPS epoch (modulo 1024, represented by 10 bits).

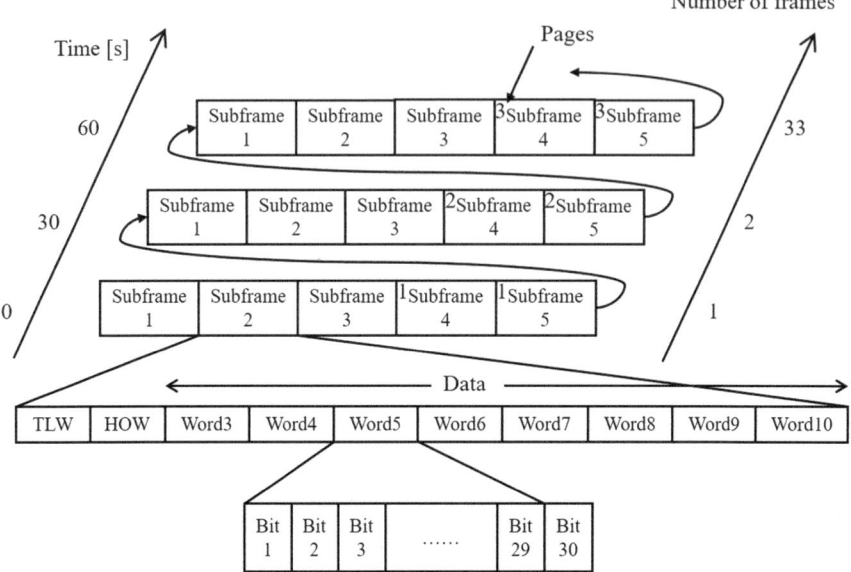

Fig. 2.27 GPS navigation message structure

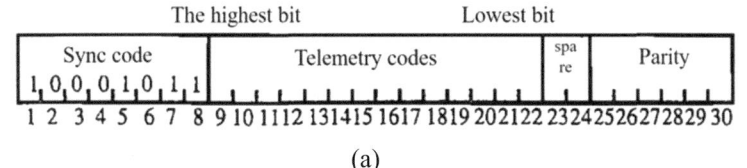

(a)

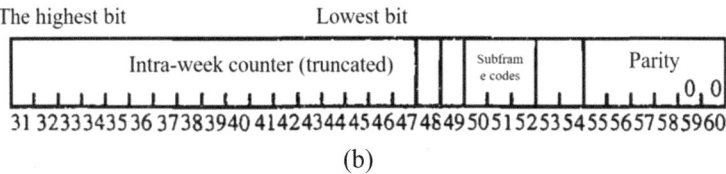

(b)

Figure 2.28 describes the relationship between the C/A code and GPS time. A C/A code receiver can use this relationship to determine the GPS signal transmission time of the received satellite signal, limited to within the base code time of a C/A code (977.5 ns), with the subdivision of the base code time achieved through the phase of the code NCO in the code tracking loop.

2.3 Satellite Navigation Signal 59

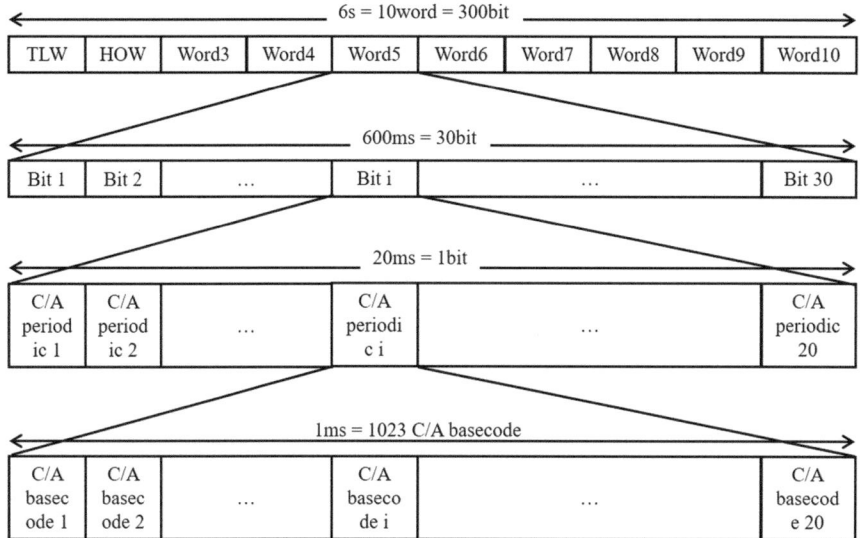

Fig. 2.28 The relationship between GPS C/A code and GPS time

2.3.2.3 Modernized GPS Signals

To maintain the technical leadership of the GPS system and sustain the momentum of the corresponding industries, the United States first proposed the GPS modernization plan in January 1999. Its main purposes include enhancing the competitiveness of GPS within the Global Navigation Satellite Systems (GNSS), and better protecting American interests; strengthening anti-jamming capabilities, etc.

The specifics of GPS modernization include the following three phases:

(1) Launch of improved GPS Block IIR satellites broadcasting L2C

Adding C/A code broadcast on the L2 frequency, which has stronger data recovery and signal tracking capabilities compared to the previous civilian code, further improving navigation and positioning accuracy. Broadcasting P(Y) code on L1 and L2 frequencies and testing the new military code (M code) signal, allowing the U.S. military the right to interfere with or temporarily shut down other signals in a region during wartime, depriving the opponent of the right to use GPS signals. The M code signal uses Binary Offset Carrier (BOC) modulation, with a subcarrier frequency of 10.23 MHz and a code rate of 5.115 Mcps.

(2) Launch of 6 GPS BLOCK IIF satellites

Based on the IIR satellite, further enhancing the power of M code transmission and adding a new civilian signal L5, with a frequency of 1176.45 MHz. To ensure global coverage of the M code, by 2008, at least 18 IIF satellites were operational in the air, and by 2016, the GPS satellite system was fully operational with these satellites, totaling $24 + 3$.

(3) Launch of GPS BLOCK III satellite

As of January 2022, The GPS Block III satellite launches are ongoing, with several satellites already in orbit. GPS BLOCK III satellites are a major upgrade to the existing constellation, providing three times the accuracy and eight times the immunity to interference of previous satellites. In addition to new civil signal, the five GPS III satellites in orbit have completed space testing of M-code and further enhanced their anti-jamming capabilities.

P-code and M-code are used by the U.S. military and special users. Civilian GPS signal resources include three carriers (L1, L2, L5) and five signals (L1 C/A, L1C, L2C, L5, and the semi-codeless L2 P(Y) for authorized users).

L2C signal modulation has two types of pseudocodes: CM code and CL code. The CM code is 10230 bits long, with a code rate of 511.5 kbps and a code period of 20 ms. CL code is a ranging code, 767,250 bits long, with a code rate of 511.5 kbps and a code period of 1.5 s [13]. The message data rate modulated on the L2C is 25bps, using a convolutional code with a coding efficiency of 1/2. CNAV (Civil Navigation) allows the control section to specify the data sequence, allocating time for each message content.

Carrier L5, at the frequency $f_{L5} = 115f_0 = 1176.45$ MHz, is used only on satellites IIF, GPS III, GPS IIIF, and subsequent satellites, and is modulated by a composite sequence generated from pseudo-random noise ranging codes, synchronization sequences, and downlink system data (referred to as L5 CNAV data), and their quadrature-phase sequences. Two PRN ranging codes are transmitted on L5: the I5 code and the Q5 code, and the I path modulates the navigation data at a pseudocode rate of 10.23 MHz and a navigation data rate of 50 bps, which is converted to a 100 bps stream after using a convolutional code with a coding efficiency of 1/2. The Q path does not modulate the navigation data. The L5C signal power for Block IIF and Block III/IIIF is -157.9 dBW and -157.0 dBW, respectively [7] (Table 2.11).

The L1C signal is mainly composed of two parts, the guide signal L1C$_P$ without any data modulation and L1C$_D$ with data modulation. The data on L1C$_D$ is represented as D_{L1C} (t) and consists of the satellite ephemeris, system time, system time offsets, satellite clock behavior, status information, and other data information modulated by (BOC) (1,1). The L1C$_P$ signal is modulated as TMBOC and consists of BOC (1,1) and BOC (6,1) [8] The nominal frequency of L1C is 1575.42 MHz, with an L1C code rate is 1.0229999999954326 MHz.

The GPS signals emitted by each type of satellite under the GPS modernization program are shown in Table 2.12.

Table 2.11 The minimum signal power that the user receives

Satellite	Signal	
	I5	Q5
Block IIF	-157.9 dBW	-157.9 dBW
GPS III/IIIF	-157.0 dBW	-157.0 dBW

2.3 Satellite Navigation Signal

Table 2.12 Signals emitted by various types of satellites

Satellite Signal	IIA	IIR	IIR-M	IIF	GPS III	GPS IIIF
L1 C/A	√	√	√	√	√	√
L1 P(Y)	√	√	√	√	√	√
L1 M			√	√	√	√
L1 C					√	√
L2 C	√	√	√	√	√	√
L2 P(Y)	√	√	√	√	√	√
L2 M			√	√	√	√
L2 C/A	√	√	√	√	√	√
L5 C				√	√	√

The modernized L1 signal includes C/A code, P(Y) code, L1C code and L1M code. The L2 signal includes C/A code, P(Y) code, L2C code and L2M code, and the L5 signal includes L5C code. Taking L1 as an example, its signal power spectrum is shown in Fig. 2.29.

The M code and P code are for military use, and civilian users cannot utilize their signal characteristics to enhance navigation and positioning service performance. However, in the application of reflective signal remote sensing, it is possible to utilize their bandwidth advantage for high-performance remote sensing detection without knowing their code structure.

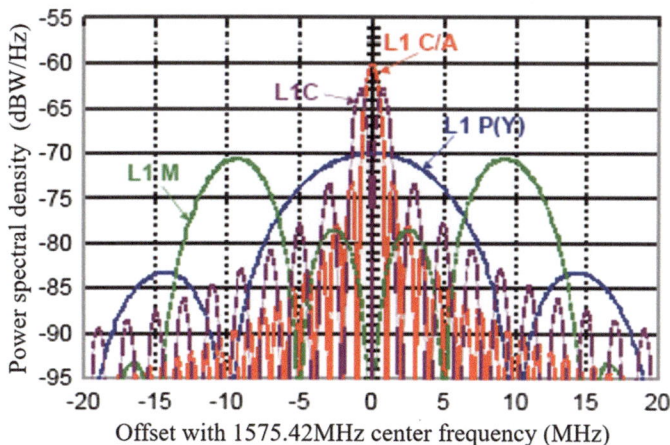

Fig. 2.29 Power spectrum of the modernized L1 signal

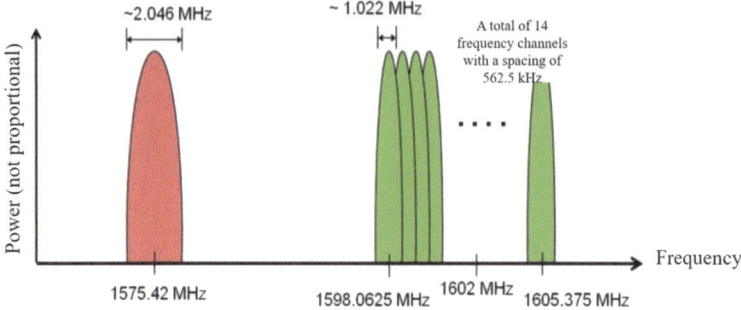

Fig. 2.30 GLONASS satellite signal spectrum versus GPS

2.3.3 Other Satellite Navigation Signals

2.3.3.1 GLONASS

The GLONASS system uses Frequency Division Multiple Access (FDMA) technology to differentiate between each satellite, with the L1 signal centered at 1602 MHz and adjacent channels spaced 562.5 kHz apart (~1597–1606 MHz), and the L2 signal centered at 1246 MHz and adjacent channels spaced 437.5 kHz apart, with the values of the individual carrier frequencies being

$$f_{L1} = 1602 + K \times 0.5625 \text{MHz}$$
$$f_{L2} = 1246 + K \times 0.4375 \text{MHz}$$
(2.46)

where $K = -7, -6, \ldots, 5, 6$ are the band numbers (Fig. 2.30).

Both L1 and L2 carriers in the GLONASS system are modulated with high-precision P code, while the low-precision C/A code appears only on L1. The C/A code rate is 511 kHz, with a code length of 511 and a code period of 1 ms, symbol width of about 2us (equivalent distance of approximately 586 m), and data modulation rate of 50bps. The P code rate is 5.11 MHz, for military use and encrypted. Figure 2.31 shows the generation process of the GLONASS signal.

2.3.3.2 GALILEO System

As shown in Fig. 2.32, Galileo navigation signals occupy four frequency bands, E5a, E5b, E6 and E1, with their frequency at 1176.45 MHz, 1207.14 MHz, 1278.75 MHz and 1575.42 MHz, respectively [5]. The E5a, E5b and E1 bands partially overlap with the L2 and L1 bands of GPS, which is beneficial for compatibility between the two systems. The signal power flux density proposed by the Galileo system protects aeronautical navigation services and allows for Radio Navigation Satellite Services (RNSS) in the lower L-band.

2.3 Satellite Navigation Signal

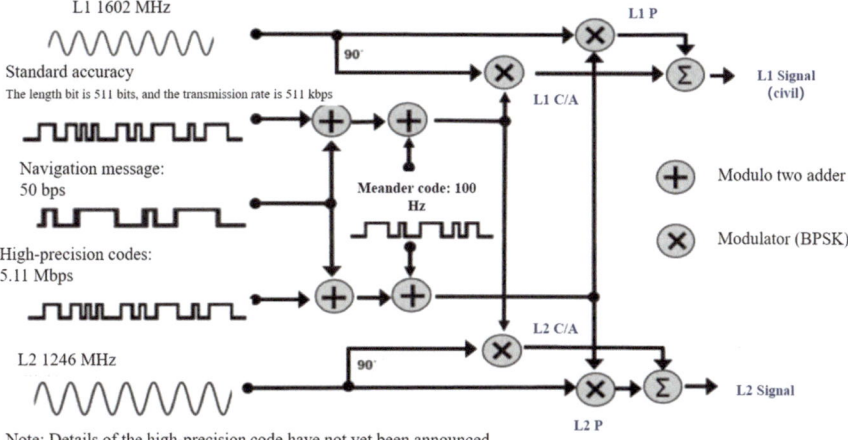

Fig. 2.31 Generation of the GLONASS signal

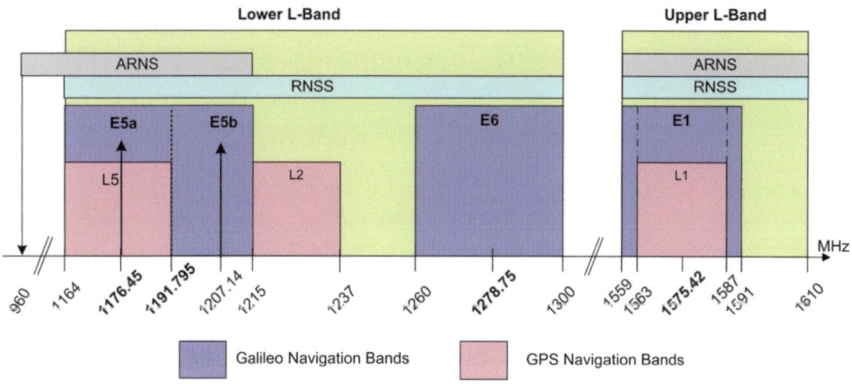

Fig. 2.32 Galileo frequency design

Galileo satellites transmit signals on four different carrier frequencies within the 1.1–1.6 GHz band. Figure 2.33 shows the modulation methods and spectrum structures of signals in each band [4].

(1) E1 is centered at 1575.42 MHz. The E1 Open Service (OS) signal is modulated using CBOC(6,1,1/11), including the E1-B signal that carries data components (error-corrected to 250sps) and the E1-C signal that carries no data component. The data rate of the OS signal is 125bps.

(2) E5a and E5b are centered at 1176.45 MHz and 1207.14 MHz, respectively, and transmitted at a center frequency of 1191.795 MHz after AltBOC multiplexing. The E5 signal consists of four parts modulated by AltBOC(15,10): E5a-I carrying F/NAV navigation data (data rate of 25bps, symbol rate of 50sps)

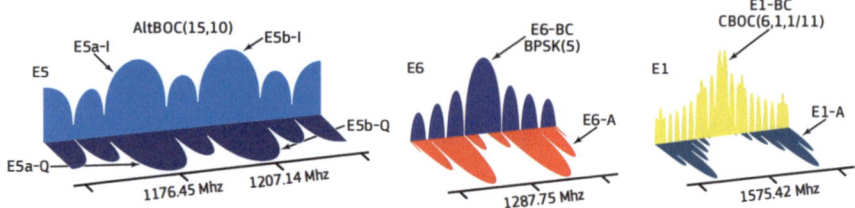

Fig. 2.33 Galileo signal spectrum structure and modulation mode

and the data-less pilot E5a-Q, E5b-I carrying I/NAV navigation data (data rate of 125bps, symbol rate of 250sps) and the data-less pilot E5b-Q.

(3) E6 is centered at 1278.75 MHz. The E6 signal includes a BOC(10,5) (pending) signal for channel A, a BPSK(5) signal for channel B, E6-B (carrying data with a data rate of 500bps), and a BPSK(5) signal for channel C, E6-C (data-less).

The minimum and maximum power levels of Galileo system signals received on the ground by users are as follows [19]:

$$-155.25\,\text{dBW} \sim -150\,\text{dBW}\,(\text{E5a/b, E6})$$

$$-157.25\,\text{dBW} \sim -152\,\text{dBW}\,(\text{E1})$$

2.4 Summary

This chapter has introduced the spread spectrum principle of GNSS navigation satellite signals and their various modulation methods. Since the most widely applied signals currently are those of GPS and BeiDou navigation satellite systems, this chapter has also provided detailed introductions to them. As satellite navigation systems continue to be built and upgraded, the applications of other navigation systems will become increasingly widespread. Readers are encouraged to refer to relevant materials for more information.

References

1. Betz JW. Binary offset carrier modulations for radio navigation. Navigation. 2001;48(4):227–46.
2. Kaplan E. Understanding GPS: principles and applications, 2nd ed. 2006.
3. Hein GW, et al. MBOC: the new optimized spreading modulation recommended for GALILEO L1 OS and GPS L1C//IEEE/ION position, location, & navigation symposium. IEEE;2006. p. 883–92.
4. Galileo Open Service Service Definition Document (OS SDD). Euro Space Agency.2019;(1.1).

References

5. Galileo Open Service Signal In Space Interface Control Document (OS SIS ICD). Euro Space Agency.2021;(2.0).
6. Tony Anthony RE. Interface control document IS-GPS-200M [EB/OL]. (2022-08-22) https://www.gps.gov/technical/icwg/.
7. Tony Anthony RE. Interface control document IS-GPS-705H [EB/OL]. (2022-08-01). https://www.gps.gov/technical/icwg/.
8. Tony Anthony RE. Interface control document IS-GPS-800H [EB/OL]. (2022-08-22). https://www.gps.gov/technical/icwg/.

Open Access This chapter is licensed under the terms of the Creative Commons Attribution-NonCommercial-NoDerivatives 4.0 International License (http://creativecommons.org/licenses/by-nc-nd/4.0/), which permits any noncommercial use, sharing, distribution and reproduction in any medium or format, as long as you give appropriate credit to the original author(s) and the source, provide a link to the Creative Commons license and indicate if you modified the licensed material. You do not have permission under this license to share adapted material derived from this chapter or parts of it.

The images or other third party material in this chapter are included in the chapter's Creative Commons license, unless indicated otherwise in a credit line to the material. If material is not included in the chapter's Creative Commons license and your intended use is not permitted by statutory regulation or exceeds the permitted use, you will need to obtain permission directly from the copyright holder.

Chapter 3
GNSS Signal Reception and Processing

Signals transmitted by navigation satellites reach the receiver antenna through the propagation path, are then amplified by the RF front-end, down-converted in frequency, and transformed from Analog to Digital (A/D) before entering the correlator for correlation processing. This process generates raw observational data from which the final positioning results can be calculated. This chapter uses the L1 C/A code signal of the GPS system as an example, starting from the signal model at various reference points of the receiver, to introduce the methods for receiving and processing GNSS signals.

3.1 Signal Model

A correct signal model is the basis and prerequisite for conducting in-depth theoretical analysis and for rational, efficient engineering design. This section establishes the signal model for key reference points of the GPS receiver from several aspects: the RF (Radio Frequency) signal received by the GPS receiver antenna, the digitized signal output from the RF front-end, and the correlation computation results output by the digital correlator.

3.1.1 Mathematical Description

This chapter focuses on the mathematical description of the direct signal; the description of the reflected signal model is given in the next chapter.

Since the GNSS signal can be viewed as a quasi-monochromatic, phase-modulated spherical wave signal, the direct signal at the receiving point R can be expressed as

$$E_d(\boldsymbol{R}, t) = A_{RF}(R_d) a\left(t - \frac{R_d}{c}\right) \exp(ikR_d - 2\pi i f_L t) \tag{3.1}$$

$A_{RF}(R_d)$ is the amplitude level of the received satellite RF signal, R_d denotes the distance from the transmitting point T to the receiving point R as a function of time. $a(t)$ is the GNSS modulated signal, c is the speed of light, and $i^2 = -1$. $k = k(f_L) = 2\pi f_L/c$ denotes the number of carriers between the transmitter (i.e., satellite) and the receiver, and f_L is the GNSS L-band carrier frequency. The signal amplitude is $A_{RF}(R_d) = \sqrt{P(R_d)}$ and $P(R_d)$ is the signal power at the distance from the satellite R_d, denoted as

$$P(R_d) = \frac{G_r \lambda^2}{(4\pi)^2} \frac{P_t G_t}{L_f R_d^2} \tag{3.2}$$

where $P_t G_t$ is the satellite transmit power, i.e., Effective Isotropic Radiated Power (EIRP), G_r is the receive antenna gain, L_f is the atmospheric loss, etc., and λ is the carrier wavelength, and so Eq. (3.1) is re-expressed as

$$E_d(\boldsymbol{R}, t) = A \frac{1}{R_d} a\left(t - \frac{R_d}{c}\right) \exp(ikR_d - 2\pi i f_L t) \tag{3.3}$$

where A is the magnitude factor.

$$A = \sqrt{\frac{P_t G_t G_r \lambda^2}{L_f (4\pi)^2}} \tag{3.4}$$

3.1.2 Reception and Digitization

The process of satellite signal reception and digitization is shown in Fig. 3.1.

Considering first the L1 RF signal modulated with C/A code transmitted by a single satellite, the LOS (Line Of Sight) signal arriving at the phase center of the receiver antenna at GPS time t can be expressed as [1].

$$S_{RF}(t) = A_{RF}(t) D[t - \tau(t)] C[t - \tau(t)] \cos[\phi(t)] \tag{3.5}$$

where $A_{RF}(t)$ denotes the received RF signal amplitude level of the satellite, $D[\cdot]$ denotes the broadcasted navigation message data of the satellite, $C[\cdot]$ denotes the civil PRN code (GPS C/A code) signal of the satellite, $\phi(t)$ denotes the received carrier phase, and $\tau(t)$ denotes the path of the PRN code signal modulated on the GPS L1 carrier that is emitted from the phase center of the satellite antenna to

3.1 Signal Model

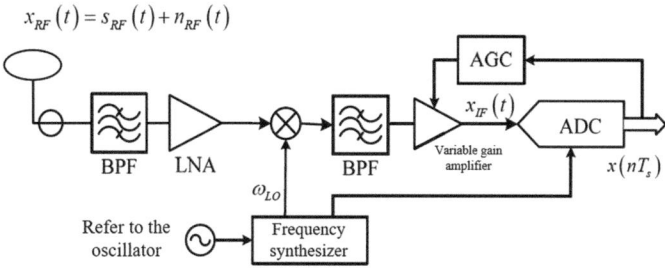

Fig. 3.1 Satellite signal reception and digitization

the phase center of the receiver antenna. Propagation group delay, i.e., the spatial propagation delay of the code phase.

According to ICD-GPS-200C, in the signals broadcast by GPS satellites, the time information in the navigation message and the phase state of the tuned signals are determined by the respective satellite clocks, and the difference between the satellite clocks and the GPS time (estimated by the ground monitoring system) is released in the first subframe of the satellite navigation message [2]. At the start of each GPS week epoch determined by GPS satellite time, the satellite signal resets to the initial phase state, i.e., at the beginning of the first page, the first frame, the first subframe, the first word, the first bit, the first PRN code cycle, and the start of the first chip. This enables the determination of the moments corresponding to D[·] and C[·] in the satellite signal. It is through the tracking and estimation of $\phi(t)$ and $\tau(t)$ in the received signals, and with the help of the time information in the demodulated D[·], the receiver calculates the code pseudo-distance and the carrier phase observation, and thus solves the user's position information.

The LOS signal spatial propagation group delay $\tau(t)$ consists of the vacuum propagation delay $\tau_{vacc}(t)$ and additional group delays in the atmosphere (mainly the ionospheric additional group delay $\tau_{iono}(t)$ and the tropospheric additional group delay $\tau_{trop}(t)$), namely

$$\tau(t) = \tau_{\text{vacc}}(t) + \tau_{\text{iono}}(t) + \tau_{\text{trop}}(t) \tag{3.6}$$

In practical applications, the electromagnetic waves received by GPS receiver antennas include thermal noise and various RF interferences, such as interference from the same satellite's P code signal, multipath interference from other satellites of the same constellation/system, interference from different frequency signals of the same constellation/system, interference from other satellite navigation systems, interference from other wireless systems, cosmic radiation, and various man-made interferences. The RF noise and interference, besides the useful signal received by the antenna, are usually modeled as Additive White Gaussian Noise (AWGN), selecting the output end of the passive antenna, the input end of the receiver's Low Noise Amplifier (LNA) (or the input end of the LNA in an active antenna) as a reference point, represented as the sum of an L1 C/A code signal from a certain GPS satellite and additive white Gaussian noise.

$$x_{RF}(t) = S_{RF}(t) + n_{RF}(t) \tag{3.7}$$

where $n_{RF}(t)$ denotes the additive Gaussian white noise with a bilateral power spectral density of $N_0/2$.

The carrier-to-noise ratio at the LNA input is

$$\mathrm{CNR}_{RF} = \frac{\overline{A_{RF}^2(t)}}{2N_0} \tag{3.8}$$

The analog Intermediate Frequency (IF) signal after downconversion, filtering, and amplification by the RF front-end is expressed as follows

$$x_{IF}(t) = A_{IF}(t)D[t - \tau(t)]C[t - \tau(t)]\cos[\phi_{IF}(t)] + n_{IF}(t) \tag{3.9}$$

where $A_{IF}(t)$ denotes the amplitude level of the analog IF signal and $\phi_{IF}(t)$ denotes the carrier phase of the analog IF signal, is given by

$$\phi_{IF}(t) = \int_0^t [\omega_{IF} + \omega_d(t)]dt + \phi_0 + \phi_{rhw} \tag{3.10}$$

$\omega_{IF} = |\omega_L - \omega_{LO}|$ denotes the downconverted analog IF, ω_d is the Doppler shift due to the relative motion between the satellite and the receiver, ϕ_{rhw} denotes the phase deviation caused by the receiver's local oscillator, filters, etc., and ϕ_0 is the initial phase.

The noise term is a narrowband Gaussian noise, which can be expressed as the sum of the in-phase and quadrature components.

$$n_{IF}(t) = n_{cIF}(t)\cos\omega_{IF}t + n_{sIF}(t)\sin\omega_{IF}t \tag{3.11}$$

Equation (3.9) is sampled by the A/D converter as

$$\begin{aligned}x_{IF}(nT_s) &= S_{IF}(nT_s) + n_{IF}(nT_s)\\ &= A_{IF}(nT_s)D[nT_s - \tau(nT_s)]C[nT_s - \tau(nT_s)]\cos[\phi_{IF}(nT_s)]\\ &\quad + n_{cIF}(nT_s)\cos(\omega_{IF}nT_s) + n_{sIF}(nT_s)\sin(\omega_{IF}nT_s)\end{aligned} \tag{3.12}$$

where $T_s = 1/f_s$ is the sampling period, f_s is the sampling frequency, and n is the serial number of the sampling moment. The reference epoch of the n th sampling period is taken as its midpoint, i.e.

$$t_{0n} = \left(n + \frac{1}{2}\right)T_s \tag{3.13}$$

3.1 Signal Model

The phase term can be expressed as

$$\phi_{IF}(nT_s) = [\omega_{IF} + \omega_d(nT_s)](nT_s - t_{0n}) + \phi_{IF}(t_{0n}) \tag{3.14}$$

Without considering the quantization error, let

$$A_n = C_D A_{IF}(nT_s)$$
$$\tau_n = \tau(nT_s)$$
$$\omega_{dn} = \omega_d(nT_s)$$
$$\phi_n = \phi_{IF}(t_{0n}) \tag{3.15}$$

where C_D is a scaling factor representing the conversion from the analog signal's amplitude level (in Volts) to the digital value after quantization and any amplification in the digital IF stage, then the A/D converter output by sampling and quantization of the digital IF signal can be expressed as follows

$$S_{IF}(n) = C_D S_{IF}(nT_s)$$
$$= A_n D[nT_s - \tau_n] C[nT_s - \tau_n] \cos[(\omega_{IF} + \omega_{dn})(nT_s - t_{0n}) + \phi_n] \tag{3.16}$$

Due to the frequency conversion effect of the sampling process itself, the center frequency of the carrier tracked by the baseband processing part of the receiver is not necessarily f_{IF}, but the center frequency of a mirror spectrum formed by sampling f_a. If this mirror is the original $S_{IF}(t)$ bisector spectrum of the positive frequency part of the translation, the digital IF signal output from the A/D converter can be expressed as

$$s(n) = A_n D[nT_s - \tau_n] C[nT_s - \tau_n] \cos[(\omega_a + \omega_{dn})(nT_s - t_{0n}) + \phi_n] \tag{3.17}$$

If the mirror image selected for tracking is a translation of the negative-frequency portion of the original $S_{IF}(t)$ bilateral spectrum, a Doppler shift and phase inversion occurs, i.e., the

$$s(n) = A_n D[nT_s - \tau_n] C[nT_s - \tau_n] \cos[(\omega_a - \omega_{dn})(nT_s - t_{0n}) - \phi_n] \tag{3.18}$$

Similarly, the noise at the useful mirror frequency at the output of the A/D converter can be expressed as

$$n(n) = n_{cn} \cos(\omega_a nT_s) + n_{sn} \sin(\omega_a nT_s) \tag{3.19}$$

In the formula, $n_{cn} = C_D n_{cIF}(nT_s)$, $n_{sn} = C_D n_{sIF}(nT_s)$.

Let the bilateral power spectral density of $n(n)$ be $n_0/2$, then the carrier-to-noise ratio of the digital IF $x(n) = s(n) + n(n)$ output from the RF front-end is

$$CNR_{\text{IF}} = \frac{\overline{A_n^2}}{2n_0} \qquad (3.20)$$

3.1.3 Correlation Operations

Due to GPS signals being submerged in noise, it is necessary to perform correlation operations to correctly extract the signal. In the digital reception channel of the receiver, the input digital signal is correlated with a locally generated signal using a correlator, which can be an active time-domain correlator composed of a multiplier and an integrator/accumulator (referred to as MAC correlator), or an equivalent structure of frequency domain multiplication or a matched filter. This section derives the output of correlation operations based on the MAC correlator, and the signal expression of its results is also applicable to correlators of other structures. Figure 3.2 shows the schematic diagram of a MAC correlator.

As shown in the figure, the digital signal input into the correlator can be a real signal, or a complex signal in the form of in-phase (I) and quadrature (Q). First, in the carrier mixer, it is multiplied by the local carrier generated by the Numerically Controlled Oscillator (NCO) (achieving carrier stripping), and the baseband I/Q sampling values generated are then correlated with the locally regenerated early,

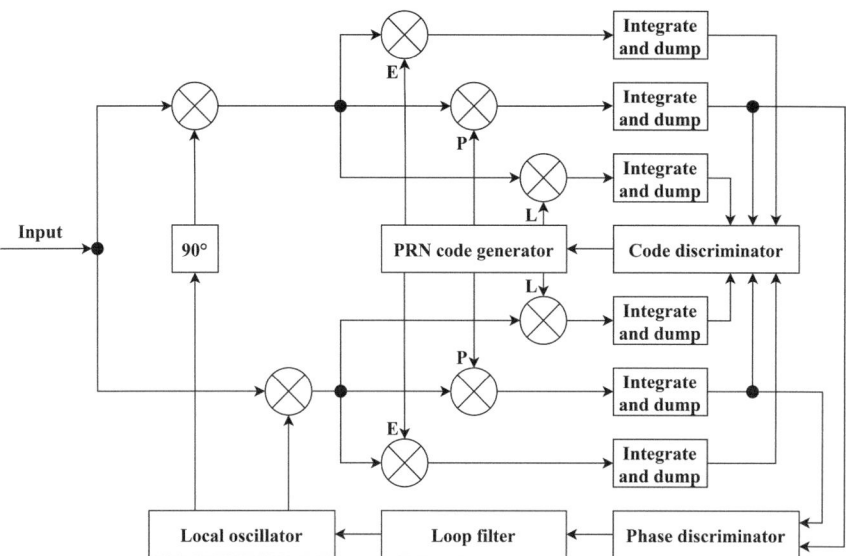

Fig. 3.2 MAC correlator structure

3.1 Signal Model

prompt, and late codes (achieving code stripping, i.e., despreading). Additionally, the locally regenerated codes used here are produced by a code generator and a 3-bit shift register under the control of the code NCO. When both the carrier loop and the code loop reach the tracking state, the output of the accumulator remains near its maximum value. The receiver demodulator output Signal-to-Noise Ratio (SNR) is the product of the input SNR and the PRN code despreading gain. In this state, the local PRN code phase and carrier phase (or phase difference) form the basis of the receiver's positioning measurements.

First, consider the signal in the I branch, where the locally generated carrier multiplied by the pseudocode signal is

$$s_l(n) = A_l C(nT_s - \hat{\tau} + \delta) \cos\left[(\omega_0 + \hat{\omega}_d)nT_s + \hat{\phi}_n\right] \quad (3.21)$$

where A_l is the local signal amplitude. For the ahead, immediate, and lag local code branches, respectively, $\delta = \frac{d}{2} \cdot T_c$, $\delta = 0$, $\delta = -\frac{d}{2} \cdot T_c$, and d are the correlation spacing, which is also the phase difference between the local ahead and lag codes, in units of chips (T_c). $\hat{\tau}$ is the local code delay estimate, $\hat{\omega}_d$ is the local Doppler estimate, and $\hat{\phi}_n$ is the carrier phase, which is generally taken to have an initial value of zero.

The product of the received signal and the local signal is

$$\begin{aligned} I(n) &= s(n)s_l(n) \\ &= A_n A_l D(nT_s - \tau_n)C(nT_s - \tau_n)C(nT_s - \hat{\tau} + \delta) \\ &\quad \cdot \cos[(\omega_0 + \omega_d)nT_s + \phi_n]\cos\left[(\omega_0 + \hat{\omega}_d)nT_s + \hat{\phi}_n\right] \end{aligned} \quad (3.22)$$

The sampling frequency within an integration period is fixed, and the clear control signal used at the end of integration is controlled by the code NCO. Due to the effect of Doppler, the values of each integration period may differ, with the actual integration period being

$$T = \frac{T_0}{1 + \hat{f}_d/f_{L1}} \quad (3.23)$$

where T_0 is the standard C/A code period. In order to complete the relevant operations, it is necessary to integrate the Eq. (3.22), and let the receiving code and the local code are basically aligned, i.e., $|\hat{\tau} - \tau_n| < T_c$, then omitting the high-frequency term can be obtained as follows

$$I(k) = \frac{A_n A_l}{2} \sum_{n=(k-1)T}^{kT} D(nT_s - \tau)C(nT_s - \tau)C(nT_s - \hat{\tau} + \delta) \cos(\Delta\omega_d nT_s + \Delta\phi_k) \quad (3.24)$$

where, $\Delta\omega_d = \hat{\omega}_d - \omega_d$ is the carrier Doppler angular frequency estimation error, $\Delta\phi_k = \hat{\phi}_k - \phi_k$ is the carrier phase estimation error, set the data bits do not change during the integration period, and the cumulative form is converted into the integral form to get

$$I(k) \approx \frac{A_n A_l D(k)}{2T_s} \int_{(k-1)T}^{kT} C(t-\tau)C(t-\hat{\tau}+\delta)\cos(\Delta\omega_d t + \Delta\phi_k)dt \quad (3.25)$$

Further collation yields

$$I(k) \approx \frac{A_n A_l D(k)}{2T_s} R(\varepsilon) \cdot \int_{(k-1)T}^{kT} \cos(\Delta\omega_d t + \Delta\phi_k)dt$$

$$= \frac{A_n A_l D(k)}{2T_s} \cdot R(\varepsilon) \cdot \frac{\sin(\Delta\omega_d t + \Delta\phi_k)\big|_{(k-1)T}^{kT}}{\Delta\omega_d}$$

$$= A \cdot D(k) \cdot R(\varepsilon) \cdot Sa(\Delta f_d \pi T) \cdot \cos\theta_k \quad (3.26)$$

where $A = \frac{A_n A_l T}{2T_s}$, $D(k)$ are the data bits, $R(\cdot)$ is the pseudocode autocorrelation function, $Sa(\cdot) = \frac{\sin(\cdot)}{\cdot}$, $\varepsilon = \Delta\tau - \delta$ is the code phase estimation error, $\Delta\tau = \hat{\tau} - \tau$ is the code delay estimation error corresponding to the instantaneous branch, $\theta_k = \Delta\omega_d(k-\frac{1}{2})T + \Delta\phi_k$, and identically

$$Q(k) \approx A \cdot D(k) \cdot R(\varepsilon) \cdot Sa(\Delta f_d \pi T) \cdot \sin\theta_k \quad (3.27)$$

For the noise term, it is analyzed mainly by its statistical properties and its magnitude relative to the signal power (i.e., signal-to-noise ratio or carrier-to-noise ratio). As can be seen in Fig. 3.2, the noise sample value I_{Nn} fed to the accumulator is the result of multiplying the IF noise sample value with the local carrier and local code. The bilateral power spectral density of the IF noise is known to be $n_0/2$, the precorrelation bandwidth 2B and sampling rate f_s are large enough with respect to the code rate, the equivalent noise bandwidth after sampling, carrier stripping and code correlation can be taken as f_s According to the literature [3, 4]. There are

$$E(I_{Ni}^2) = \frac{A_n^2}{4} n_0 f_s \quad (3.28)$$

Assuming that the noise samples are uncorrelated, they still obey a Gaussian distribution after being summed up, with a mean value of

$$E(I_{Ni}) = E\left[\sum_{n=n_{0i}}^{n_{0i}+N_i-1} I_{Nn}\right] = 0 \quad (3.29)$$

The variance is

3.2 Acquisition and Tracking

$$D(I_{Ni}^2) = D\left[\left(\sum_{n=n_{0i}}^{n_{0i}+N_i-1} I_{Nn}\right)^2\right] = \frac{A_n^2}{4}n_0 f_s \frac{T_i}{T_s} \tag{3.30}$$

Similarly, the noise power in the ith accumulated value output from the Q branch is

$$E(Q_{Ni}^2) = E(I_{Ni}^2) = \frac{A_n^2}{4}n_0 f_s \tag{3.31}$$

3.2 Acquisition and Tracking

Spread spectrum receivers generally complete despreading before demodulation to achieve a certain Signal-to-Noise Ratio (SNR) gain. The necessary condition for despreading is the synchronization of the received signal with the local code signal and carrier signal. The first step in the synchronization process, known as signal acquisition, refers to the process where the receiver searches for and captures the GPS satellite signal and brings the code phase and Doppler within the pull-in range of the tracking loops. Through a search process, the receiver identifies the satellite number of the received signal and makes a rough estimate of the phase of the satellite's PRN code and the carrier Doppler, then uses these estimates to initialize the tracking mode.

3.2.1 Acquisition

The acquisition system usually includes three parts: the correlator, signal detector, and search control logic. The various implementation options for these three components determine the different signal acquisition/detection methods and their characteristics.

The local GPS receiver needs to generate the captured satellite's PRN code, moving the phase of this regenerated code at a specific interval until it aligns with the satellite's PRN code. The correlation process between the satellite's transmitted PRN code and the receiver's local code has the same characteristics as the PRN code's autocorrelation process. The maximum correlation value occurs when the locally generated (or regenerated) code aligns in phase with the input satellite code. When the phase of the regenerated code shifts by more than one chip to either side from the phase of the input satellite code, the correlation value is minimal. This is the criterion for GPS receivers to detect satellite signals in the code phase domain while capturing satellite signals. Since the carrier frequency of the received signal includes Doppler shift, satellite signal detection also involves searching within the carrier phase domain. Therefore, the acquisition process of a certain GPS satellite signal can be viewed as a two-dimensional signal search process.

After the receiver determines which satellites each channel needs to search for according to a certain satellite selection logic, the synchronization acquisition process can be viewed as a binary hypothesis testing problem (also known as a binary decision detection problem) to verify/judge whether the signal from the satellite appears. According to the statistical detection theory of random signals, the optimal criteria for binary decision detection, such as the Bayes criterion, the minimum error probability criterion (including the maximum a posteriori probability criterion), the minimax criterion, and the Neyman-Pearson criterion, can all be used to form the decision logic of the signal detector. H_0 and H_1 are used to denote the signals "not present" and "present" respectively, and after n observations of the signal to obtain the observation vector $y = (y_1, y_2, \ldots y_n)$, the Bayesian judgment requires the "cost" and average risk also known as the Bayes risk to be extremely small, and thus requires knowing both the prior probabilities of H_0 and H_1 and the cost factors of the two types of errors (false alarms for misjudging H_0 as H_1 and missed alarms for misjudging H_1 as H_0). The minimum error probability criterion and the minimax criterion require knowledge of the a priori probability. The Neyman-Pearson judgment does not require these conditions and seeks to maximize the detection probability $P_D = P(H_1|H_1)$ given the false alarm probability $P_F = P(H_1|H_0)$. In GPS receivers, since false alarms cause false locks, it is important to limit P_F to be small enough; at the same time, it is desirable that P_D be as large as possible in order to detect signals as they come in and to reduce capture time. All of the above criteria can be summarized in the following likelihood ratio criterion:

$$\Lambda(y) = \frac{p(y|H_1)}{p(y|H_0)} = \frac{p_1(y)}{p_0(y)} \underset{H_0}{\overset{H_1}{\underset{<}{>}}} \psi \qquad (3.32)$$

Different decision criteria only affect the size of the corresponding decision threshold ψ.

As analyzed in Sect. 3.1, the signal received by the GPS receiver is a random parameter signal (its amplitude, phase, arrival time, and Doppler shift are either random or unknown non-random quantities). According to signal detection and estimation theory, by calculating the likelihood ratio, it can be inferred that the optimal detection system structure is an orthogonal receiver, or equivalently, a non-coherent matched filter connected in series with an envelope detector, where the envelope detector can also be replaced with a power detector (also called a square-law detector or energy detector). The test statistic (also called the detection quantity or decision variable) of the optimal detection receiver is a monotonic function of the likelihood ratio, obtained in an orthogonal receiver through the correlation (i.e., multiplication and integration) operations and square (or envelope) operations of the I and Q channel input signals with the local signal. The peak of the detection statistic (i.e., the correlation output) indicates the local code phase and Doppler frequency that best match the received signal, providing the receiver's initial estimates of these parameters.

3.2 Acquisition and Tracking

In GPS receivers, the detection quantities are constructed as the results of the correlation operations $I(k)$ and $Q(k)$, e.g., the squared detection quantities without considering noise are

$$I^2(k) + Q^2(k) = A^2 R^2(\varepsilon) Sa^2(\Delta f_d \pi T) \qquad (3.33)$$

(Note: $D^2(k) = 1$).

The maximum Doppler shift of GPS satellites relative to a stationary receiver on the Earth's surface is about ±5 kHz. Considering the dynamics of the receiver, the dynamic range of the Doppler shift can reach ±10 kHz. The maximum Doppler shift value (10 kHz) is much smaller than the code rate (C/A code 1.023 MHz). When the time delay difference is within one chip, (D(k)) is approximately 1, and the coupling of time delay and Doppler shift can be considered negligible. The joint effect of code delay and Doppler shift on the amplitude of the correlation value is shown in Fig. 3.3.

When there is no Doppler shift, the impact of the code time delay on the correlation value is shown in Fig. 3.4; when the pseudocode is fully aligned, the effect of the Doppler shift on the correlation value is shown in Fig. 3.5.

As can be seen in Fig. 3.3 through Fig. 3.5:

(1) When the Doppler shift is fixed, in the code phase direction, the amplitude of the signal correlation value presents the autocorrelation function pattern of the pseudocode sequence; when the delay is fixed, in the Doppler shift direction, the amplitude of the signal correlation value presents the sampling function pattern.

(2) When there is no Doppler shift and the phase offset between the received pseudocode signal and the local pseudocode signal is 1/2 chip, the magnitude of the correlation value decreases by 6 dB; whereas, when the pseudocode is perfectly aligned, a Doppler shift of ±500 Hz results in a 4 dB decrease in the energy of

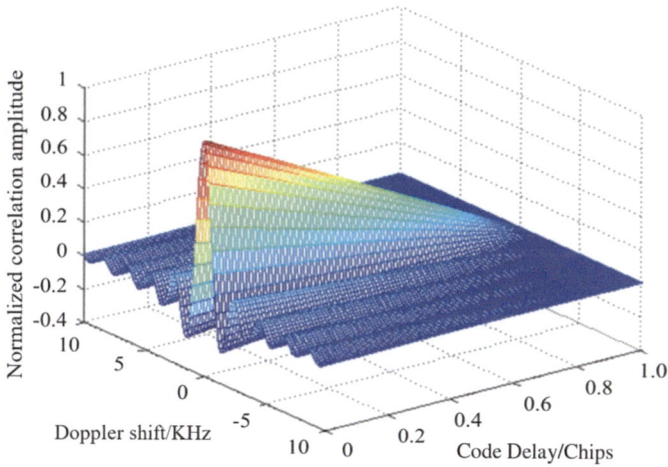

Fig. 3.3 Influence of delay and doppler shift on the amplitude of correlation values

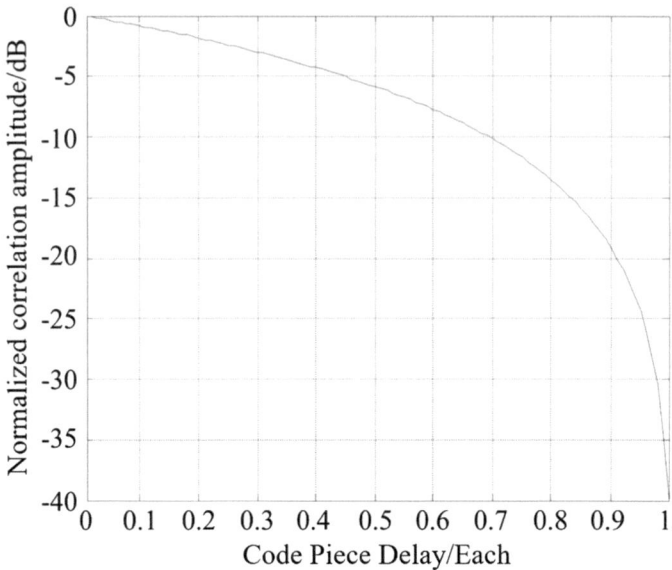

Fig. 3.4 No doppler shift

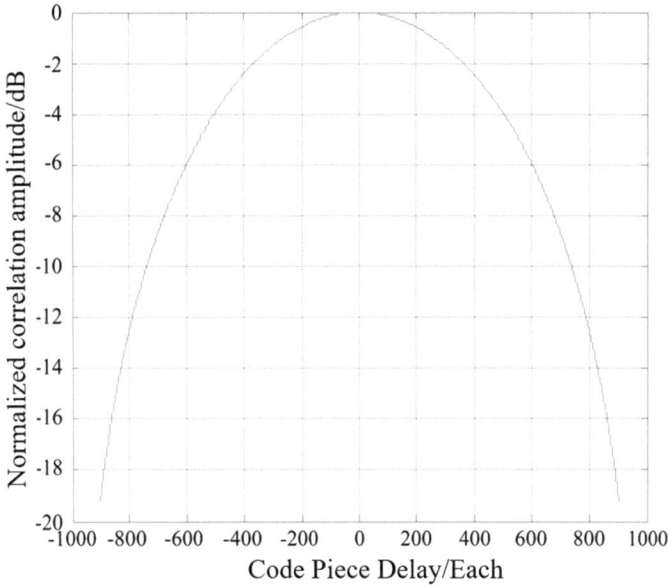

Fig. 3.5 Pseudocode fully aligned

the correlation value. This means that the signal correlation peak can only be effectively detected within a small range of Doppler shifts and code phase.
(3) The coupling effect of delay and Doppler shift on the correlation value is very small, and the Doppler effect on the code rate can also be neglected, so the capture of C/A code signals can consider delay and carrier Doppler shift separately.

GPS signal acquisition, as a two-dimensional search process, uses a specific algorithm to traverse all possible PRN code phase units (corresponding to the distance dimension) and carrier Doppler units (corresponding to the velocity dimension). Whether the signal appears is determined by whether the correlation peak exceeds a preset threshold. This exhaustive search strategy can be implemented as a serial sequential search, a parallel search using multiple correlators, or a hybrid approach.

For GPS applications, the false alarm probability and detection probability generated by a single decision generally do not meet performance requirements. Therefore, methods with verification logic, such as the dwell time method (e.g., N-out-of-M search detector) or the variable dwell time method (e.g., Tong search detector), are usually employed to improve performance [5].

Based on different acquisition methods in the time and frequency domains, GNSS signal acquisition can be divided into the following four strategies:

(1) Code serial, carrier serial search method

This method uses a single correlator to perform serial searches on code phase and carrier Doppler. The process is as follows: first, preset a roughly estimated carrier Doppler frequency, and at this Doppler frequency point, move the local code by one code phase unit (usually half a chip) each time. After shifting, perform a correlation operation with the input signal. If the result exceeds the set acquisition threshold, the acquisition is successful, and the corresponding code phase and carrier frequency are recorded for subsequent signal tracking. If the threshold is not exceeded, the local Doppler preset remains unchanged, and the process of moving the local code by half a chip is repeated. If the signal is not captured after moving the local code through one code period, change the local Doppler value to the next unit and repeat the above process with code phase unit movement until successful. Figure 3.6 shows the schematic diagram of the code serial, carrier serial search principle.

The advantage of this method is its simple hardware circuitry and ease of implementation, but the downside is the long acquisition time. When the code phase unit

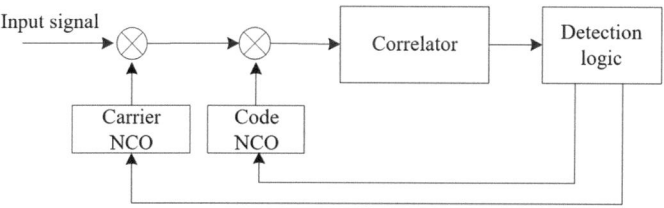

Fig. 3.6 Principle schematic diagram of code serial, carrier serial search

is set to half a chip (as in the following search methods), the time required to search the entire pseudocode and carrier Doppler range is (Tp × N × 2) × (F/Δf), where Tp is the pseudocode period, N is the code length, F is the carrier frequency range, and Δf is the Doppler frequency unit during the search process.

this method is suitable for situations where hardware resources are limited and acquisition time is not critical. In addition, it is more commonly used for occasions where accurate a priori information can be provided.

(2) Search method for code-serial carrier parallelism

This method employs a single code generator for serial code search, while utilizing Nf carrier correlators for parallel search across the carrier Doppler spectrum. Nf carrier correlators generate carriers of different frequencies, the number of which is related to the capture range. The code movement process is the same as in the code serial, carrier serial method. At each code phase, the correlation values for Nf frequency points are checked simultaneously. If any exceeds the threshold, the current code phase and the Doppler value corresponding to this correlation value are taken as the pseudocode phase and carrier Doppler estimates; otherwise, the code phase is moved by half a chip and the process is repeated. The principle schematic diagram is shown in Fig. 3.7.

The advantage of this method is its fast acquisition speed, which can be achieved without any prior information, typically taking Tp × N × 2, proportional to the code length; however, its hardware complexity is largely independent of the code length, allowing compatibility with various code lengths under the same hardware conditions.

(3) Pseudocode parallel, carrier serial search method

This method uses Nc independent code correlators, where, under normal circumstances, the code phase of each correlator is sequentially offset by half a chip from each other. All Nc code correlators share a single carrier Numerically Controlled Oscillator (NCO), with the carrier Doppler being searched in a serial scanning

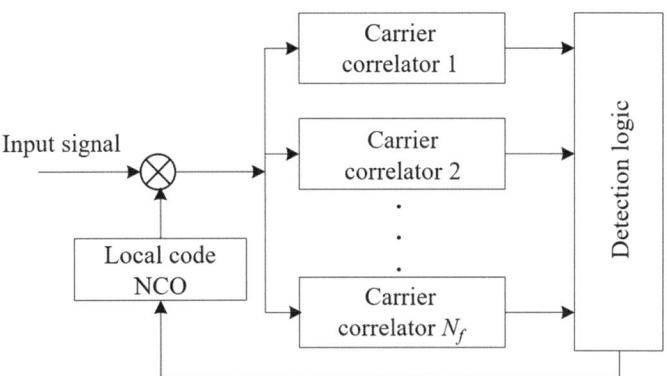

Fig. 3.7 Principle schematic diagram of code serial, carrier parallel search

3.2 Acquisition and Tracking

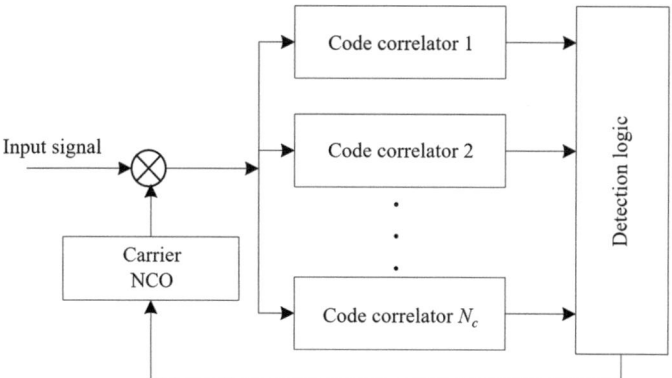

Fig. 3.8 Principle schematic diagram of pseudocode parallel, carrier serial search

manner. Assuming that the Doppler frequency unit during the search is $\Delta f = 1/T_c$, the Doppler range of ± 10 kHz is divided into 20 kHz/Δf intervals. After scanning each interval, the maximum value of N_c correlators is taken and compared with the capture threshold. If it is below the threshold, the search moves to the next interval; if it exceeds the threshold, the acquisition is deemed successful, and the carrier Doppler and code phase estimate are obtained based on the carrier interval and the location of the correlation peak (Fig. 3.8).

The advantage of this method is its fast acquisition speed, with only one code length period required to analyze each carrier interval. Without prior information, at most, 20 kHz/Δf intervals need to be scanned to complete acquisition. With prior Doppler frequency information, only one interval needs to be scanned. The drawback is that the computational load and hardware resource consumption are significantly higher.

(4) Code parallel, carrier parallel search method

This method also employs Nc independent code correlators, with the code phase of each correlator sequentially offset by half a chip, and uses Nf carrier Numerically Controlled Oscillators (NCOs). Each code correlator corresponds to Nf possible carrier frequencies, and each carrier correlator corresponds to Nc possible code phases. For parallel acquisition of code and carrier, the maximum correlation values corresponding to Nc code correlators and Nf carrier correlators are compared with the acquisition threshold. If the threshold is exceeded, the acquisition is deemed successful; otherwise, the search is restarted (Fig. 3.9).

This method offers fast acquisition speed, capable of signal acquisition within one code length period without prior information; however, its processing volume significantly increases, consuming a substantial amount of hardware resources.

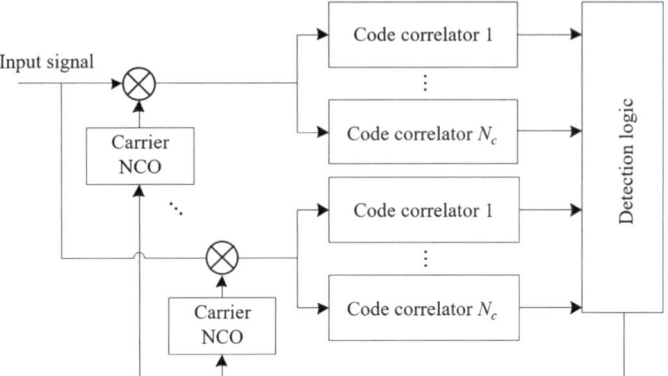

Fig. 3.9 Principle schematic diagram of code parallel, carrier parallel search

3.2.2 Code Tracking

Figure 3.10 shows the general block diagram of a C/A code tracking loop. It is mainly composed of correlator, code loop discriminator and code loop filter. This type of loop tracks the code phase by generating early and late replicas of the code and is commonly called a Delay Locked Loop (DLL).

In this code tracking loop block diagram, a code that is in phase with the C/A code in the received signal is first replicated, and the received digital IF signal carrier is stripped, and a correlation operation is done with the locally replicated code to strip the C/A code from the received signal.

Table 3.1 lists several types of discriminators commonly used in GPS receivers and their characteristics. In the table, the fourth type of discriminator is the third type of normalized form, can eliminate the amplitude sensitivity of the discriminator, so that it applies to different signal-to-noise ratios and different signal strengths of the signal, thereby improving the performance of anti-pulse RF interference. The other two discriminators can also be normalized in a similar way.

Figure 3.11 shows the outputs of the four discriminators, where the spacing between the early, prompt, and late correlators is half a chip.

3.2.3 Carrier Tracking

The influence of carrier Doppler frequency shifts caused by changes in the vehicle dynamics on the code tracking loop can be eliminated through carrier aiding. Thus, the dynamic performance of the receiver mainly depends on carrier tracking techniques. Generally, there are two types of loops used for carrier tracking: one is the coherent Phase Locked Loop (PLL), where the receiver needs to generate a coherent

3.2 Acquisition and Tracking

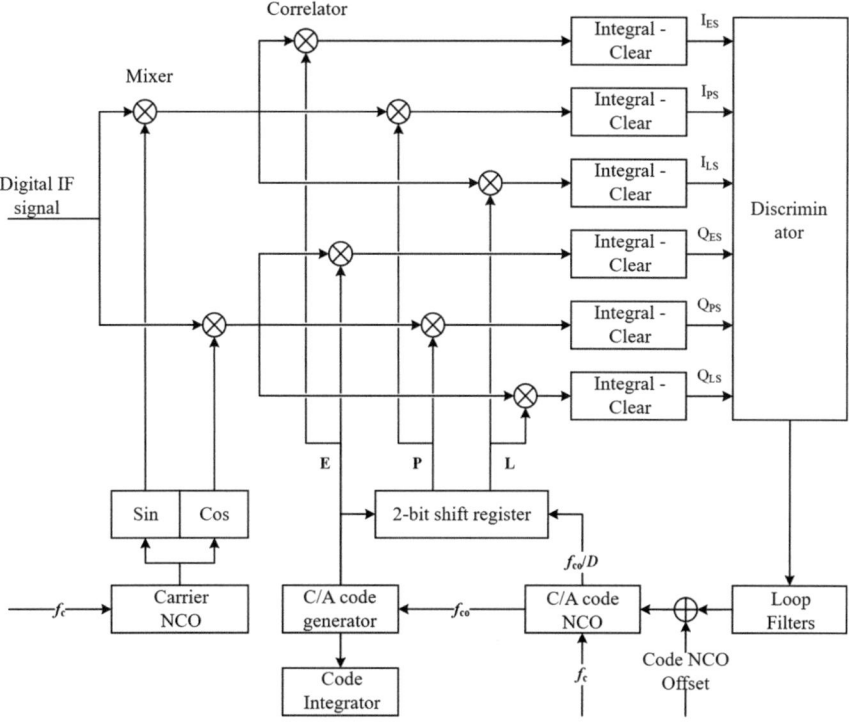

Fig. 3.10 General block diagram of code tracking loop

Table 3.1 Common discriminators for delay locked rings

	Discriminator	Characteristics
1	$-\sum (I_E - I_L)Q_P - \sum (Q_E - Q_L)Q_P$	Dot product power: Minimal arithmetic using 3 correlators. For 1/2 chip correlation spacing, produces near-true error output within ±1/2 chip input error
2	$\sum (I_L^2 + Q_L^2) - \sum (I_E^2 + Q_E^2)$	Early minus Late Power: Moderate arithmetic. Essentially the same error performance as the overrun minus hysteresis envelope within ±1/2 yardstick input error
3	$\sum \sqrt{(I_L^2 + Q_L^2)} - \sum \sqrt{(I_E^2 + Q_E^2)}$	Early minus Late Envelope: higher arithmetic. For correlation spacing of 1/2 code slice, has small tracking error within ±1/2 code slice input error
4	$\dfrac{\sum \sqrt{(I_L^2 + Q_L^2)} - \sum \sqrt{(I_E^2 + Q_E^2)}}{\sum \sqrt{(I_L^2 + Q_L^2)} + \sum \sqrt{(I_E^2 + Q_E^2)}}$	Normalized envelope: has small tracking error within ±1.5 yard input error for 1/2 yard correlation spacing. Becomes unstable when the input error is ±1.5 chips due to division by zero

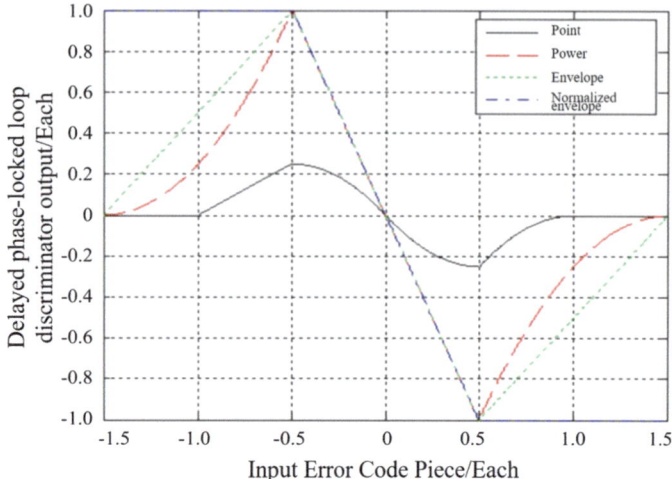

Fig. 3.11 Output of delay locked ring discriminator [6]

carrier that is in frequency and phase with the input carrier; the other is the non-coherent Frequency Locked Loop (FLL), where the receiver needs to generate a carrier that is in frequency with the input carrier but does not require phase coherence.

Figure 3.12 shows a block diagram of a GPS receiver carrier tracking loop, where the choice of discriminator algorithm directly determines the type (PLL/FLL) and performance of the tracking loop.

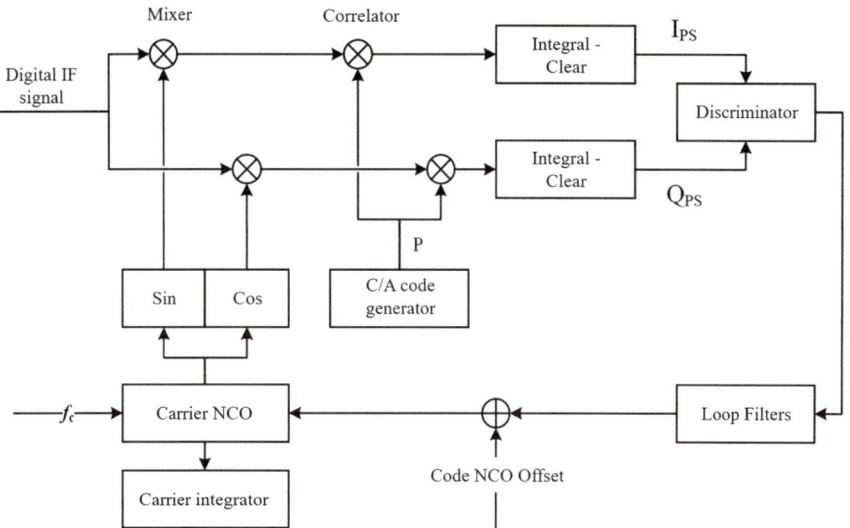

Fig. 3.12 General block diagram of the carrier tracking loop

3.2 Acquisition and Tracking

(1) Phase Locked Loop (PLL)

Generally, the PLL used in GPS receivers was invented in the 1950s by American engineer John P. Costas, also known as the Costas loop. The characteristic of this loop is that the phase discriminator uses the method of multiplying the in-phase and quadrature branches to obtain the phase error signal, making the phase detection process insensitive to data bit transitions "+1" and "−1". Since the Costas loop uses both in-phase and quadrature branches, it is also known as the I/Q loop.

When phase-locked (i.e., phase error is zero), the in-phase branch will provide the data bits, while the quadrature branch contains only noise components. At the same time, the phase error, after high-frequency components are filtered out by the loop filter, adjusts the carrier NCO, locking the carrier frequency along with the phase.

Table 3.2 presents the phase error output and characteristics of several Costas PLL discriminators.

(2) Frequency Locked Loop (FLL)

While the PLL reproduces the exact frequency of the input satellite signal for carrier stripping, the FLL reproduces an approximate frequency for the carrier stripping process. The FLL in a GPS receiver requires that there be no data bit transitions during a single frequency discrimination process, meaning the discriminator algorithm's t_1 and t_2 components are within the same data bit interval. Table 3.3 lists several commonly used FLL discriminators in GPS receivers and their output frequency error characteristics.

Table 3.2 Discriminators for costas loop

Discriminator	Output phase error	Character
$sgn(I_{PS}) \cdot Q_{PS}$	$\sin \phi$	Close to optimal at high signal-to-noise ratios. Slope is proportional to signal amplitude, less arithmetic
$I_{PS} \cdot Q_{PS}$	$\sin 2\phi$	Close to optimal at low signal-to-noise ratios. Slope is proportional to the square of the signal amplitude, and the operation is medium
Q_{PS}/I_{PS}	$\tan \phi$	Suboptimal, but good at both high and low signal-to-noise ratios. Slope is independent of signal amplitude, is more arithmetic, and must be verified to distinguish between 0 error when in the vicinity of ±90°
$ATAN(Q_{PS}, I_{PS})$	ϕ	Two Quadrant Arctangent. Optimal at both high and low signal-to-noise ratios, the slope is independent of signal amplitude, and the arithmetic is maximized

Table 3.3 Common discriminators for FLL

Discriminator	Output phase error	Character
$\frac{sgn(dot)\cdot cross}{t_2-t_1}$ $dot = I_{PS1}\cdot I_{PS2} + Q_{PS1}\cdot Q_{PS2}$ $cross = I_{PS1}\cdot Q_{PS2} - I_{PS2}\cdot Q_{PS1}$	$\frac{\sin[2(\phi_2-\phi_1)]}{t_2-t_1}$	Close to optimal at high signal-to-noise ratios. The slope is proportional to the signal amplitude A and the operation is moderate
$\frac{cross}{t_2-t_1}$	$\frac{\sin[(\phi_2-\phi_1)]}{t_2-t_1}$	Close to optimal at low signal-to-noise ratios. The slope is proportional to the square of the signal amplitude and the operation is low
$\frac{ATAN2(cross,dot)}{360(t_2-t_1)}$	$\frac{\phi_2-\phi_1}{(t_2-t_1)360}$	Four-quadrant Arctangent. It is a maximum likelihood estimator, optimal at both high and low signal-to-noise ratios, with slopes independent of signal amplitude and high arithmetic

3.3 Positioning Solving

3.3.1 Data Synchronization

The tracking of carriers and codes provides a basis for data demodulation. To correctly extract navigation messages from the received data stream, it is necessary first to recover the bit synchronization clock and the frame synchronization clock. Typically, the signal processing module has two epoch counters in each channel: a 1 ms epoch counter (counting range 0–19) and a 20 ms epoch counter (counting range 0–49), generating the corresponding clock signals.

(1) Bit synchronization

The bit synchronization process timing diagram is shown in Fig. 3.13. When the receiver's locally generated bit synchronization clock is not synchronized with the received data, there is an error between the zero moment of the receiver's 1 ms epoch counter and the starting moment of the data, known as bit synchronization clock error. Eliminating this error achieves bit synchronization.

In digital communication, bit synchronization is often achieved using a lead-lag Phase Locked Loop (PLL), subtracting a pulse when the bit synchronization clock is ahead, and adding a pulse when it is behind, continuing until the bit synchronization clock error is zero. Systems using a lead-lag PLL have strong anti-interference capabilities, but correcting errors takes a long time, especially at lower data rates.

GPS navigation data bits are 20 ms long, and correlators output data every 1 ms as samples of navigation data. At each epoch, the sampling values of 20 ms (one data bit) are accumulated and averaged, and the value at that moment is compared with the previous value. If it is greater than the previous value, that moment is considered the optimal decision moment for the code element.

3.3 Positioning Solving

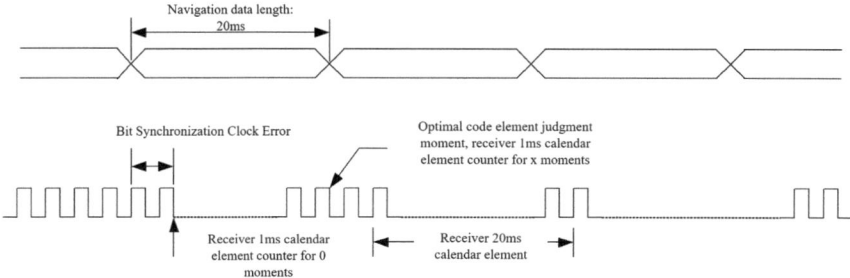

Fig. 3.13 Bit synchronization process timing diagram

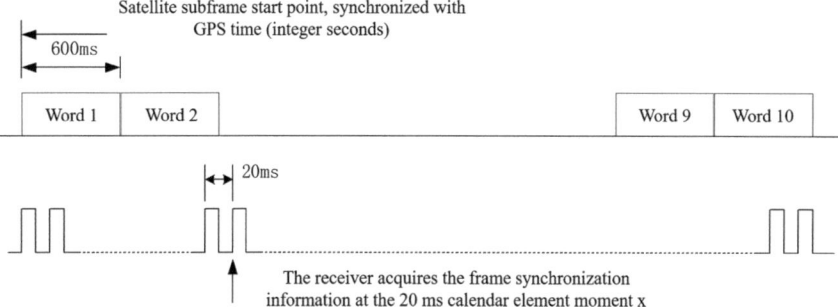

Fig. 3.14 Frame synchronization process timing

(2) Frame synchronization

Frame synchronization is based on bit synchronization, and Fig. 3.14 represents the timing of the frame synchronization process.

GPS satellites send navigation message at the time of integer seconds of GPS time, and each subframe of navigation message also starts at the time of integer seconds. Therefore, the frame synchronization time of the receiver should also be synchronized with integer seconds, meaning that the value of the receiver's 20 ms epoch counter should be 0 when it receives the first word of the navigation message subframe. Usually, when deciding whether the received signal is frame synchronized or not, the content of the first and second word code of the navigation message subframe is first obtained, which contains synchronization information such as frame synchronization telemetry word. If the value of the 20 ms epoch counter at the starting moment of the first word code is 0, the value of the 20 ms calendar counter at the ending moment of the second word code is 10 (the time of the two word codes is 1.2 s). If the value of the 20 ms epoch counter is not 10 at the frame synchronization extraction moment, it indicates that the frame synchronization time has not been recovered. The error read from the 20 ms epoch counter is [50−(40+X) mod 50]. Using this error to change the initial value of the 20 ms epoch counter can achieve the recovery of frame synchronization time.

It should be noted that the process of frame synchronization includes synchronization code, subframe number, zero data bit extraction verification and parity check. A phase ambiguity problem exists in Costas loops used for data demodulation, where a 180° phase shift in the carrier tracking can invert the data bits. For this, the navigation message sets the 30th bit in the previous word code of each word code as a flag bit to determine if the phase 180° blur problem exists. When phase 180° ambiguity exists, inverting the demodulated data bits can recover the correct data sequence. The first 1–8 bits of the telemetry word are a fixed synchronization code, the subframe number, and zero data bits are given by the handover word, and the parity of the telemetry word and handover word is checked following ICD-GPS-200 [2].

3.3.2 Positioning Principles and Methods

The GPS system adopts a high-orbit ranging system, with the distance between the observation station and the GPS satellite as the basic observation quantity. To obtain the distance observation quantity, two observation methods are mainly used: one is to measure the propagation time of the ranging code signal (C/A code or P code) transmitted by GPS satellites arriving at the user's receiver; and the other is to measure the phase difference between the carrier signal with Doppler frequency shift received by the receiver and the reference carrier signal generated by the receiver. Carrier phase observation is currently the most accurate observation method, which is of great significance for precision positioning. However, due to phenomena such as the integer ambiguity, cycle slips, and half-cycle slips in carrier phase observations, its data processing is relatively complex [6–8].

In navigation and dynamic positioning, there are two main types of methods: single-point positioning and differential positioning. Single-point positioning refers to independently determining the absolute position of the point to be determined in geocentric coordinates. The advantage of this method is that only one receiver is required. However, due to the fact that the results of single-point positioning are significantly affected by the satellite ephemeris error and the atmospheric delay error during the satellite signal propagation process, the positioning accuracy is lower. Differential positioning, on the other hand, involves placing a receiver at a known point on the ground as a reference point and conducting synchronized observations with receivers at all points to be measured. The reference point according to its accurately known coordinates can be derived from the positioning results of corrections to the coordinates (position-domain differential) or corrections to the pseudorange observations (observation-domain differential), and then through the data chain these corrections are transmitted in real time to the relevant users. Users use these corrections to correct their own positioning results or pseudo-distance observations, thus improving the accuracy of positioning results. Although the basic observation quantities are divided into ranging code and carrier phase, and the positioning methods include single-point positioning and differential positioning, the basic principles are consistent no matter what observation quantities and positioning method used. This

3.3 Positioning Solving

section takes the single-point positioning using the ranging code observation method as an example to illustrate the basic principles of GPS positioning.

Assuming that t_G^j and t^j denote the ideal GPS moment and satellite clock moment when the satellite j transmits signals, and T_G and T denote the GPS moment and receiver clock moment when the receiver receives the satellite signals, the satellite signal propagation time is

$$\tau^j = T_G - t_G^j \tag{3.34}$$

However, the observed quantities are t^j and T, so the actual measured satellite signal propagation times are

$$\tau^j = T - t^j \tag{3.35}$$

The deviation of the satellite clock and the receiver clock are represented by Δt^j and Δt, respectively, i.e.

$$\Delta t^j = t^j - t_G^j \tag{3.36}$$

$$\Delta t = T - T_G \tag{3.37}$$

follow

$$\tau^j = \left(T_G - t_G^j\right) + \Delta t - \Delta t^j \tag{3.38}$$

Factoring in the errors in the ionosphere, troposphere and multipath delays experienced during signal propagation yields

$$\tau^j = \left(T_G - t_G^j\right) + \Delta t - \Delta t^j + \Delta t_{ion}^j + \Delta t_{tro}^j + \Delta t_{mp}^j \tag{3.39}$$

(Multiply both sides of the above equation by the speed of electromagnetic wave propagation (i.e., the speed of light c), and you get

$$\rho^j = R^j + c\Delta t - c\Delta t^j + c\Delta t_{ion}^j + c\Delta t_{tro}^j + c\Delta t_{mp}^j \tag{3.40}$$

where $R^j = \sqrt{(x - x^j)^2 + (y - y^j)^2 + (z - z^j)^2}$ is the geometric distance from the receiver to the satellite; (x, y, z), (x^j, y^j, z^j) for the receiver and the satellite j in the geocentric space coordinate system in three-dimensional coordinates; ρ^j is actually available distance observation, due to the difference from the real distance R^j, so often referred to as "pseudo-distance". Equation (3.40) is the basic observation equation for GPS positioning. As the satellite clock difference Δt^j can be determined by the ground monitoring system and provided to the user through the satellite broadcast navigation message, the ionosphere and troposphere propagation delay $c\Delta t_{ion}^j$ and $c\Delta t_{tro}^j$ can also be used to correct the ionosphere and troposphere parameter model provided by the navigation message (of course, there is still a residual error after correction), multipath delay error Δt_{mp}^j through the antenna, siting and other methods of correction. Therefore, there are four unknowns in the Eq. (3.40), $x, y, z, \Delta t$ need to be solved. In order to find these 4 unknowns, at least 4 satellites need to be observed. This is the basic principle of the four-star positioning system proposed in the 1970s. Assuming that the number of satellites observed at a certain moment is N ($N \geq 4$), a system of observation equations can be obtained

$$\rho^j = R^j + c\Delta t - c\Delta t^j + c\Delta t_{ion}^j + c\Delta t_{tro}^j + c\Delta t_{mp}^j, j = 1, 2, \ldots, N \quad (3.41)$$

Given the approximate location of the receiver $r_0 = (x_0, y_0, z_0)$, applying a Taylor expansion to each equation in the above system of equations and omitting the higher terms yields

$$\frac{\partial R^j}{\partial x}(r_0)\Delta x + \frac{\partial R^j}{\partial y}(r_0)\Delta y + \frac{\partial R^j}{\partial z}(r_0)\Delta z + b \\ = \rho^j + c\Delta t^j - c\Delta t_{ion}^j - c\Delta t_{tro}^j - c\Delta t_{mp}^j - R^j(r_0) \quad j = 1, 2, \ldots, N \quad (3.42)$$

where $b = c\Delta t$; $\Delta x = x_u - x_0$; $\Delta y = y_u - y_0$; $\Delta z = z_u - z_0$. After finding the partial derivative of R^j in the above equation, Eq. (3.42) can be written as

$$e_1^j \Delta x + e_2^j \Delta y + e_3^j \Delta z - b \\ = R^j(r_0) - \rho^j - c\Delta t^j + c\Delta t_{ion}^j + c\Delta t_{tro}^j + c\Delta t_{mp}^j, \quad j = 1, 2, \ldots, N \quad (3.43)$$

where e_1^j, e_2^j, e_3^j are the components of the unit vector (direction cosines) from the receiver to the jth satellite. Writing the above system of equations in matrix form, we get formula

$$\mathbf{GX = L} \quad (3.44)$$

$$\mathbf{G} = \begin{bmatrix} e_1^1 & e_2^1 & e_3^1 & -1 \\ e_1^2 & e_2^2 & e_3^2 & -1 \\ \cdots & \cdots & \cdots & \cdots \\ e_1^N & e_2^N & e_3^N & -1 \end{bmatrix} \quad (3.45)$$

$$\mathbf{X} = \begin{bmatrix} \Delta x & \Delta y & \Delta z & b \end{bmatrix}^T \quad (3.46)$$

$$\mathbf{L} = \begin{bmatrix} l^1 & l^2 & \cdots & l^N \end{bmatrix}^T \quad (3.47)$$

3.3 Positioning Solving

$$l^j = R^j(r_0) - \rho^j - c\Delta t^j + c\Delta t^j_{ion} + c\Delta t^j_{tro} + c\Delta t^j_{mp} \quad j = 1, 2, \ldots, N \quad (3.48)$$

Solving by least squares gives the normal equation

$$\mathbf{G}^T\mathbf{G}\mathbf{X} = \mathbf{G}^T\mathbf{L} \quad (3.49)$$

When $\mathbf{G}^T\mathbf{G}$ is non-singular, its solution is

$$\mathbf{X} = (\mathbf{G}^T\mathbf{G})^{-1}\mathbf{G}^T\mathbf{L} \quad (3.50)$$

Since the linearization process of Eq. (3.42) inevitably introduces linearization error, Eq. (3.50) is usually iterated to achieve the receiver position solution. After analyzing the practical examples, we can see that under the condition of initial value setting within 15 km near the true value, good positioning accuracy can be obtained without iteration; only when the initial value setting is farther away from the true value (e.g., more than 150 km), the accuracy improvement obtained by iterative solving is more obvious.

Using $\delta\mathbf{X}$ to denote the error of $\mathbf{X}$ and $\delta\mathbf{L}$ to denote the error of $\mathbf{L}$, we have

$$\text{cov}(\delta\mathbf{X}) = (\mathbf{G}^T\mathbf{G})^{-1}\mathbf{G}^T\text{cov}(\delta\mathbf{L})\left[(\mathbf{G}^T\mathbf{G})^{-1}\mathbf{G}^T\right]^T \quad (3.51)$$

Assuming that each pseudo-ranging measurement is an independent equal-error measurement with an error variance of σ_0^2, and that the ranging error sequence is a normal (Gaussian) white noise sequence, then

$$\text{cov}(\delta\mathbf{L}) = \sigma_0^2 \mathbf{I} \quad (3.52)$$

where I is the $N \times N$ unit array. Thus

$$\text{cov}(\delta\mathbf{X}) = \sigma_0^2 (\mathbf{G}^T\mathbf{G})^{-1} = \sigma_0^2 \mathbf{Q} \quad (3.53)$$

Assuming

$$\mathbf{Q} = (\mathbf{G}^T\mathbf{G})^{-1} = \begin{bmatrix} q_{11} & q_{12} & q_{13} & q_{14} \\ q_{21} & q_{22} & q_{23} & q_{24} \\ q_{31} & q_{32} & q_{33} & q_{34} \\ q_{41} & q_{42} & q_{43} & q_{44} \end{bmatrix} \quad (3.54)$$

Then the error estimates for user position and clock bias are

$$\sigma_x^2 = q_{11}\sigma_0^2 \quad (3.55)$$

$$\sigma_y^2 = q_{22}\sigma_0^2 \quad (3.56)$$

$$\sigma_z^2 = q_{33}\sigma_0^2 \tag{3.57}$$

$$\sigma_t^2 = q_{44}\sigma_0^2 \tag{3.58}$$

Since **Q** is only related to the relative geometric position of the satellite and the receiver, the errors of the user 3D position and the receiver clock deviation are only related to two factors, which will be the relative geometric position of the satellite and the receiver and the error of the pseudo-range observation measurement. Defining the relative geometric position of the satellite and receiver into the form of various precision factors (Dilution of Precision (DOP)), we can obtain the relationship between the user position error, the clock deviation error and the pseudo-range observation measurement error as

$$\begin{aligned}
\sigma_g &= \sqrt{\sigma_x^2 + \sigma_y^2 + \sigma_z^2 + \sigma_t^2} = \text{DDOP} \times \sigma_0 \\
\sigma_p &= \sqrt{\sigma_x^2 + \sigma_y^2 + \sigma_z^2} = \text{PDOP} \times \sigma_0 \\
\sigma_h &= \sqrt{\sigma_x^2 + \sigma_y^2} = \text{HDOP} \times \sigma_0 \\
\sigma_v &= \sigma_z = \text{VDOP} \times \sigma_0 \\
\sigma_t &= \text{TDOP} \times \sigma_0
\end{aligned} \tag{3.59}$$

where GDOP (Geometric Dilution of Precision), PDOP (Position Dilution of Precision), HDOP (Horizontal Dilution of Precision), VDOP (Vertical Dilution of Precision) and TDOP (Time Dilution of Precision) are referred to as Geometric Dilution of Precision, Position Dilution of Precision, Horizontal Dilution of Precision, Elevation Dilution of Precision and Time Dilution of Precision factors, respectively, which depend only on the relative geometry between the satellites and the receiver.

$$\begin{aligned}
\text{GDOP} &= \sqrt{q_{11} + q_{22} + q_{33} + q_{44}} = \left\{ \text{tr}\left[\left(\mathbf{G}^T\mathbf{G}\right)^{-1}\right] \right\}^{1/2} \\
\text{PDOP} &= \sqrt{q_{11} + q_{22} + q_{33}} \\
\text{HDOP} &= \sqrt{q_{11} + q_{22}} \\
\text{VDOP} &= \sqrt{q_{33}} \\
\text{TDOP} &= \sqrt{q_{44}}
\end{aligned} \tag{3.60}$$

geometric dilution of precision (DOP) can also be regarded as the amplification factor of the pseudo-range view measurement error, where a larger value indicates worse positioning accuracy, hence it is often called the geometric dilution of precision (DOP). To reduce the user's positioning error, firstly, the geometric accuracy factor (or position accuracy factor) should be reduced; secondly, the pseudorange observation error for each satellite should be minimized.

3.3 Positioning Solving

When the receiver can observe more than four satellites simultaneously, choosing an appropriate combination of four satellites to minimize the geometric accuracy factor is crucial. This is known as the satellite selection problem. There are generally two methods of satellite selection: the best satellite selection method and the quasi-optimal satellite selection method. When the satellite elevation angle is too low, the atmospheric propagation error increases, significantly reducing the observation accuracy, so a minimum elevation angle limit, usually 5 degrees, should be specified for satellite selection. The best satellite selection method involves selecting various possible combinations of four satellites from all satellites with elevation angles greater than 5 degrees, to calculate the corresponding GDOP (or PDOP), and select the group of satellites with the smallest GDOP as the best choice. Since C_N^4 GDOP operations are required, and each GDOP operation involves matrix multiplication and inversion, it is more computationally intensive and time-consuming. Moreover, due to the continuous movement of satellites, satellite selection usually needs to be redone every 15 min. To reduce the computation for satellite selection, the quasi-optimal satellite selection method can be used. This method selects four satellites that maximize the volume of the tetrahedron formed with the receiver as the criterion [9]. Assuming that the volume of the hexahedron formed by the receiver and the four observing satellites is V, the analysis shows that the geometric accuracy factor GDOP is inversely proportional to the volume of this hexahedron, V, i.e.

$$\text{GDOP} \sim 1/V \tag{3.61}$$

In general the larger the volume of the hexahedron, the smaller the GDOP value.

As the number of channels for tracking satellite signals in receivers increases, the issue of satellite selection has become less critical, with a growing preference for using all visible satellites for positioning. The geometric dilution of precision (GDOP) decreases with an increasing number of satellites, and typically, positioning using all visible satellites provides higher accuracy than selecting just four satellites.

In solving the pseudorange localization equation described above, the instantaneous position of the GPS satellite in space is calculated using the orbital parameters provided by the navigation message. The coordinates of the satellite in the orbital plane rectangular coordinate system (with the X-axis pointing towards the ascending node) are

$$\begin{cases} x_k = r_k \cos u_k \\ y_k = r_k \sin u_k \end{cases} \tag{3.62}$$

where r_k is the satellite vector diameter, and u_k is the ascending intersection angle corrected by satellite uptake. The satellite in the orbit plane right-angle coordinate system for coordinate rotation transformation, can get the satellite in the center of the earth fixed coordinate system in three-dimensional coordinates as follows

$$\begin{bmatrix} X_k \\ Y_k \\ Z_k \end{bmatrix} = \begin{bmatrix} x_k \cos \Omega_k - y_k \cos i_k \sin \Omega_k \\ x_k \sin \Omega_k - y_k \cos i_k \cos \Omega_k \\ y_k \sin i_k \end{bmatrix} \quad (3.63)$$

where i_k is the satellite orbital inclination and Ω_k is the longitude of the ascending node at the observation time. Considering the effect of polar shift, the coordinates of the satellite in the protocol coordinate system are

$$\begin{bmatrix} X \\ Y \\ Z \end{bmatrix}_{CTS} = \begin{bmatrix} 1 & 0 & X_p \\ 0 & 1 & -Y_p \\ -X_p & Y_p & 1 \end{bmatrix} \begin{bmatrix} X_k \\ Y_k \\ Z_k \end{bmatrix} \quad (3.64)$$

where X_p, Y_p are the coordinate conversion factors.

In the above equations, r_k, u_k and Ω_k are calculated using the satellite parameters provided in the navigation message.

3.4 Summary

This chapter first summarizes the general model of signals at various key reference points in the L1 C/A code receiver based on the architecture of the GPS receiver. Based on a thorough analysis of the working principle of the receiver, a detailed derivation process is presented. On this basis, the methods for processing GNSS navigation satellite signals are introduced, including the signal acquisition, tracking, and positioning solution processes.

References

1. Misra P, Enge P. Global positioning system: Signals, measurements and performance, (revised second Edition). Ganga-Jamuna Press;2008. 年1月.
2. Interface Control Document ICD-GPS-200. Navstar GPS spacesegment/navigation user interfaces (Public Release Version C). ARINC Research Corporation, 10 October 1993.
3. Van Dierendonck FP, Ford T. Theory and performance of narrow correlator spacing in a GPS receiver. Navigation. 1992;39(3):265–83.
4. Parkinson BW, Spilker JJ. Global positioning system: Theory and applications, vol. I. Washington DC: American Institute of Aeronautics and Astronautics Inc.; 1996. p. 329–407.
5. Ward PW. GPS receiver search techniques. In: Proceedings of position, location and navigation symposium. Atlanta;1996. p. 604–11.
6. Kaplan ED. Understanding GPS: Principles and anplications. Artech House. MA;1996.

References

7. Teunissen PJG. The least-squares Am Biguity decorrelation adjustment: A method for fast GPS integer ambiguity estimation. J Geodesy. 1995;70(1):65–82.
8. Melbourne WG. The case for ranging in GPS-based geodetic systems. In: Proceedings of the first international symposium on precise positioning with the global positioning system. Rockville;1985. p. 373–86.
9. Noe PS, Mayer KA, Wu TK. A navigation algorithm for the low-cost GPS receiver. Navigation. 1978;25:258–64.

Open Access This chapter is licensed under the terms of the Creative Commons Attribution-NonCommercial-NoDerivatives 4.0 International License (http://creativecommons.org/licenses/by-nc-nd/4.0/), which permits any noncommercial use, sharing, distribution and reproduction in any medium or format, as long as you give appropriate credit to the original author(s) and the source, provide a link to the Creative Commons license and indicate if you modified the licensed material. You do not have permission under this license to share adapted material derived from this chapter or parts of it.

The images or other third party material in this chapter are included in the chapter's Creative Commons license, unless indicated otherwise in a credit line to the material. If material is not included in the chapter's Creative Commons license and your intended use is not permitted by statutory regulation or exceeds the permitted use, you will need to obtain permission directly from the copyright holder.

Chapter 4
Reflected GNSS Signal Fundamentals

This chapter focuses on the basic theories of Reflected GNSS Signals, including the geometric relations, polarization characteristics, scattering models, signal models and correlation function models.

4.1 GNSS-R Geometric Relations

4.1.1 Macroscopic Geometric Relations

In discussing the geometric relationship between direct and reflected signals, the concept of the specular reflection point is introduced, which is the point in the reflection area where the path delay between the reflective and direct paths is shortest. According to the geometric relationships between the transmitter, receiver and specular reflection point, the following local coordinate system is established: the specular reflection point is the coordinate origin, the z-axis as the normal direction to the Earth's surface, with the transmitter, specular point, and receiver in the yz plane, and the y-axis positive towards the transmitter direction, the x-axis determined by the right-hand rule, xy being the local tangent plane. The geometric relationship of GNSS-R is shown in Fig. 4.1 [1].

Where h_r and h_t are the heights of the receiver and transmitter to the reference ellipsoid, respectively; R_e is the radius of the Earth; G and L are the distances of the GNSS satellite (transmitter T in the figure) and the receiving platform (receiver R in the figure) to the Earth's center, respectively; γ_t is the angle between the GNSS satellite, the specular point of reflection, and the Earth's center line; Θ is the angle between the GNSS satellite, the receiving platform and the geocentric line; φ is the receiver viewing angle (the angle between the receiver mirror point line and the receiving platform's zenith direction); and θ is the elevation angle of the reflected

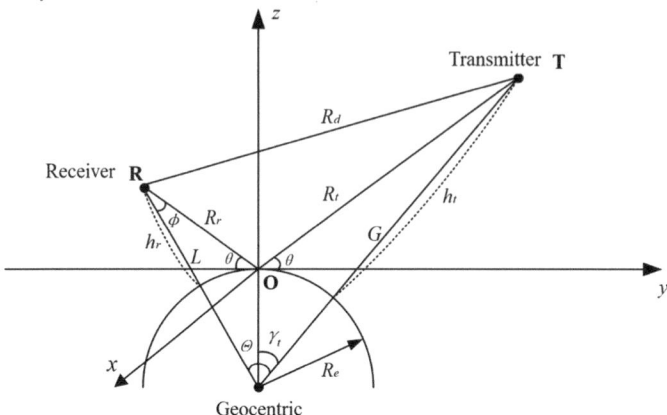

Fig. 4.1 GNSS-R geometric relations (T for transmitter)

signal with respect to the local tangent plane, whose value is equal to the satellite altitude angle; R_r and R_t are the distances from the receiver and transmitter to the specular reflection point, respectively; $\mathbf{T}$ and $\mathbf{R}$ are the position vectors of the transmitter and receiver, respectively. Given the receiver and transmitter heights h_r, h_t and the satellite altitude angle θ, the following order can be obtained R_r, R_t, γ_t, φ and Θ.

$$L = R_e + h_r \quad G = R_e + h_t \tag{4.1}$$

$$R_t^2 + R_e^2 + 2R_t R_e \sin\theta = G^2 \Rightarrow R_t = -R_e \sin\theta + \sqrt{G^2 - R_e^2 \cos^2\theta} \tag{4.2}$$

$$R_e^2 + G^2 - 2R_e G \cos\gamma_t = R_t^2 \Rightarrow \gamma_t = \arccos\left(\frac{R_t^2 - G^2 - R_e^2}{-2R_e G}\right) \tag{4.3}$$

$$R_r^2 + R_e^2 + 2R_r R_e \sin\theta = L^2 \Rightarrow R_r = -R_e \sin\theta + \sqrt{L^2 - R_e^2 \cos^2\theta} \tag{4.4}$$

$$R_r^2 + L^2 - 2R_r L \cos\phi = R_e^2 \Rightarrow \phi = \arccos\left(\frac{R_e^2 - R_r^2 - L^2}{-2R_r L}\right) \tag{4.5}$$

$$\Theta = \frac{\pi}{2} + \gamma_t - \phi - \theta \tag{4.6}$$

The positions of R, T in the local coordinate system are $(0, -R_r \cos\theta, R_r \sin\theta)$, $(0, R_t \cos\theta, R_t \sin\theta)$, respectively, and there are

$$R_d = |\mathbf{R}_d| = |\mathbf{R} - \mathbf{T}| = \sqrt{(R_t \cos\theta + R_r \cos\theta)^2 + (R_t \sin\theta - R_r \sin\theta)^2} \tag{4.7}$$

4.1.2 Antenna Coverage Area

The size of the antenna beam and the height and geometric relationship of the receiving platform determine the size and shape of the antenna footprint, as illustrated in Fig. 4.2 for an antenna footprint schematic. Low gain antennas are small and lightweight with low gain, but have a wide coverage area; high gain antennas are larger and heavier, but have high gain and a small coverage area. The size and shape of the antenna footprint vary with different antenna field of view angles. For different receiving platforms, the settings of antenna configuration and field of view angles are also different.

For low-gain antennas, it can be assumed that the antenna has the same gain in all directions, but for high-gain antennas, this assumption will introduce certain errors. Therefore, in GNSS-R remote sensing applications, precise antenna modeling is required. For the actual antenna, the antenna's directional coefficient D and gain G are generally used to express the relationship between the two as follows

$$G = kD \tag{4.8}$$

where k ($0 \leq k \leq 1$) is a dimensionless efficiency factor, and for well-designed antennas, the value of k can be close to 1 with $G \approx D$.

Assuming that $P(W \cdot m^{-2})$ denotes the power directional flap diagram of the antenna in a three-dimensional spherical coordinate system as a function of the meridian angle α and the azimuthal angle φ, the antenna's directionality coefficient D, which is the ratio of the maximum radiated power density to its average value on a certain sphere in the far-field region, is a dimensionless ratio greater than or equal to 1, and is denoted as

$$D = \frac{P_{max}}{P_{av}} \tag{4.9}$$

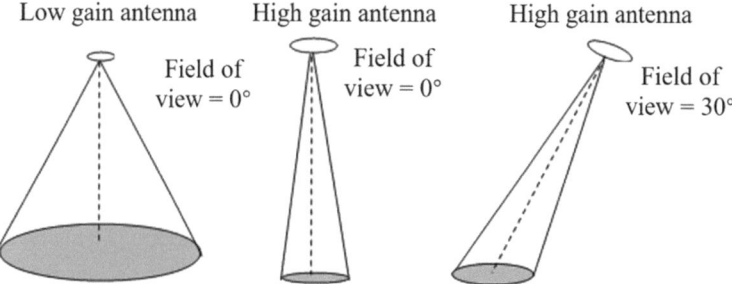

Fig. 4.2 Schematic diagram of the relationship between antenna footprint and antenna gain, field of view angle

Using $d\Omega = \sin\alpha \, d\alpha \, d\varphi$ to denote the steradian angle in steradian radians (sr) or square degrees, the average power density on the sphere is

$$P_{av} = \frac{1}{4\pi} \int_{\varphi=0}^{\varphi=2\pi} \int_{\alpha=0}^{\alpha=2\pi} P \sin\alpha \, d\alpha \, d\varphi$$

$$= \frac{1}{4\pi} \iint_{4\pi} P d\Omega \quad W \cdot sr^{-1} \tag{4.10}$$

Therefore, the direction coefficient can again be written as the following expression:

$$D = \frac{P_{max}}{\frac{1}{4\pi}\iint_{4\pi} P d\Omega} = \frac{1}{\frac{1}{4\pi}\iint_{4\pi} [P/P_{max}] d\Omega}$$

$$= \frac{4\pi}{\iint_{4\pi} P_n d\Omega} = \frac{4\pi}{\Omega_A} \tag{4.11}$$

where Ω_A denotes the beam stereo angle range, and P_n denotes the normalized power flap map, expressed as

$$P_n = P/P_{max} \tag{4.12}$$

The beam range of an antenna can usually be approximated as the product of the main flap half-power beam widths α_{HP} and φ_{HP} in the two main planes of the antenna, i.e.

$$\Omega_A \approx \alpha_{HP}\varphi_{HP} \quad (sr) \tag{4.13}$$

Half-Power Beam Width (HPBW) indicates the beam width (or -3 dB beam width) defined according to the half-power level point. In addition, Eq. (4.11), if the half-power width of an antenna is known, its directional coefficient can also be expressed as

$$D = \frac{41253}{\alpha_{HP}^\circ \varphi_{HP}^\circ} \tag{4.14}$$

where $41253 = 4\pi(180/\pi)^2$ square degrees, is the number of square degrees tensioned in the sphere. α_{HP}° 和 φ_{HP}° denote the half-power beam widths (expressed in degrees) in the two principal planes, respectively. Since the above equation ignores the effect of the sidelobes, it can be expressed in an approximate form as follows:

4.1 GNSS-R Geometric Relations

$$D = \frac{40000}{\alpha_{HP}^\circ \varphi_{HP}^\circ} \qquad (4.15)$$

If the beam width of an antenna is 20° in both principal planes, then there is $D = 100$ or 20 dB. Of course, the product of the beam widths 40,000 squared is a rough approximation, and for a particular type of antenna, there can be a more accurate value for each. It is also common to use the effective aperture of an antenna A_e as an indicator of the antenna's performance in relation to the beam width as follows

$$\lambda^2 = A_e \Omega_A \; (m^2) \qquad (4.16)$$

Thus, given the wavelength λ, Ω_A can be determined from the known A_e and vice versa. From Equation (4.11) and (4.16) another expression for the directivity index D of the antenna can be obtained as follows:

$$D = 4\pi \frac{A_e}{\lambda^2} \qquad (4.17)$$

Assuming that the height of the receiving platform is 10 km and the antenna beam widths (set $\alpha_{HP} = \varphi_{HP}$) are 4°, 20°, 45°, and 90°, the size of the long axis of the coverage area at an antenna boresight angle (FOV = 0°) is 0.7 km, 3.5 km, 8.2 km, and 20 km, respectively; the size of the long axis of the same antenna coverage area at the field-of-view angle of 30° is 0.9 km, 4.7 km, 11.7 km, and 40 km.

4.1.3 Isochronous Zone

When GNSS direct signal illuminates the sea surface, countless scattering points are formed within the glitter area. However, when the scattering signals from different scattering points are received by the receiver, the time of arrival at the receiver is different. According to the principle of spread spectrum signal reception, different propagation times of the scattering signals manifest as different code delays. An isochronous line is a curve composed of points with the same time delay compared with the reflected signals from the specular points. An iso-delay zone is an elliptical annular area defined by two isochronous lines, related to the receiver's sampling frequency, i.e., the code chip resolution. The higher the code chip resolution, the denser the isochronou zone under the same conditions.

The shape and size of the isochronou zone are related to the altitude of the receiver, the elevation angle and azimuth angle of the transmitter (GNSS satellite), etc., with specific calculation methods referred to in the literature [2]. When the receiver altitude is 5 km and the satellite elevation angle = 30°, the size, shape, and projection on the sea surface of the isochronous zone are shown in Fig. 4.3. The numbers on the isochronous linez in the left figure indicate the value of the delay code chip τ_c relative to the specular reflection point.

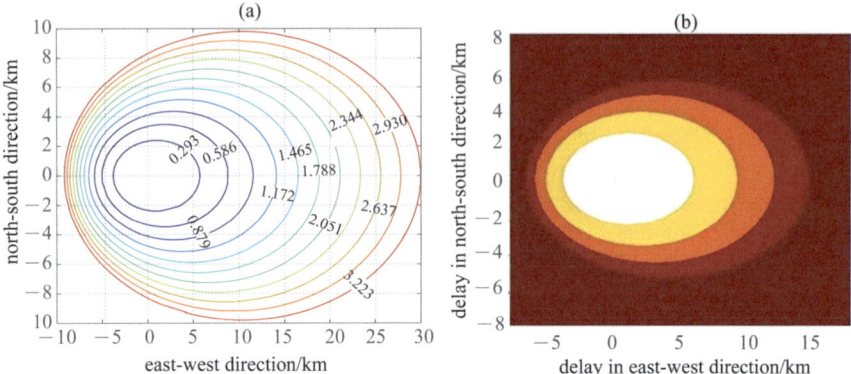

Fig. 4.3 Size and shape of the iso-delay zone (Size and shape of the isodelay zone and its projection on the sea surface $\Delta \tau = [1, 2, 3 \ldots 10] \cdot \tau_c$

(1) Impact of Satellite Elevation Angle

At a receiver altitude of 5 km and satellite altitude angles of 30°, 45°, 60° and 90°, the shape and size of the equal delay region are shown in Fig. 4.4.

It can be seen that the lower the satellite elevation angle, the flatter the ellipse and the greater the deviation from the specular reflection point in the transmitter direction. The isochronous line ellipse is symmetrical around the incident plane but centered on the specular point, exhibiting significant asymmetry in the direction perpendicular to the incident plane. And the lower the height angle, the more obvious the asymmetry. This phenomenon indicates that the scattering signals mainly come from the long axis direction of the isochronous zone, making wind direction detection feasible.

(2) Impact of Receiver Altitude

For the same satellite elevation angle (e.g., 60°), the size of the first isochronous zone (i.e., $\Delta \tau = [1] \cdot \tau_c$) under different receiver altitude conditions is shown in Fig. 4.5.

The higher the receiving altitude, the larger the isochronous zone, and the larger the corresponding scattering signal energy region is. For example, at a receiving altitude of 800 km, the long axis of the first isochronous zone already exceeds the length of a wind zone by 50 km. When solving for the power of scattering signals at different delays, this aspect must be fully considered to determine the specific code delay values.

(3) Impact of Sampling Rate

If the sampling rate is $f_s = 20.456$ MHz, the interval between sampled signals is $\tau_s = 1/f_s$, corresponding to a distance resolution of approximately 15 m. For the GPS C/A code is $\tau_c/20$, where the size and shape of the isochronous zone corresponding to the first 10 sampling points are shown in Figs. 4.6 and 4.7.

4.1 GNSS-R Geometric Relations

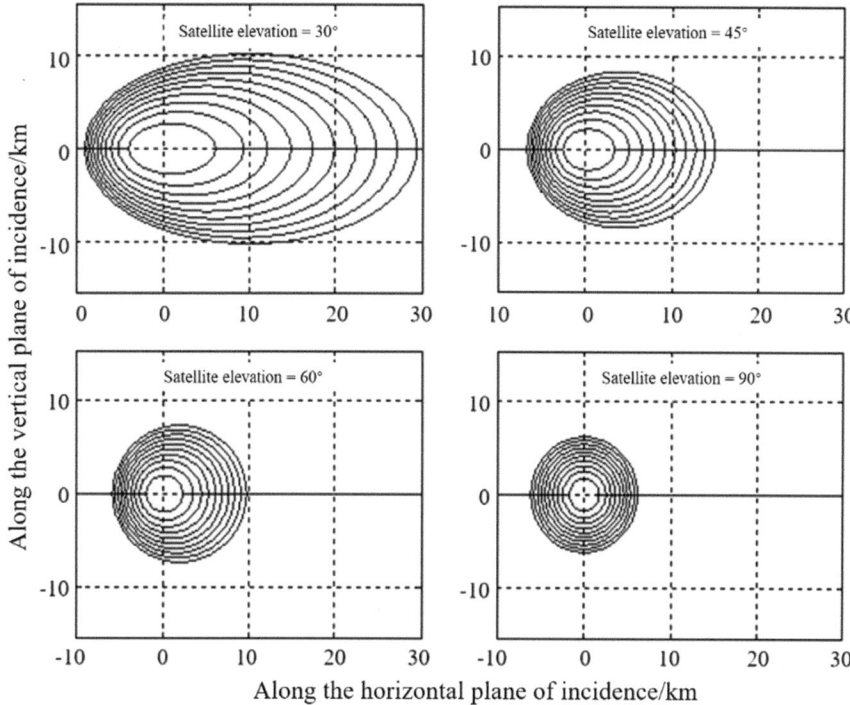

Fig. 4.4 Size and shape of the iso-delay zone ($\triangle \tau = [1, 2, 3 \ldots 10] \cdot \tau_c$)

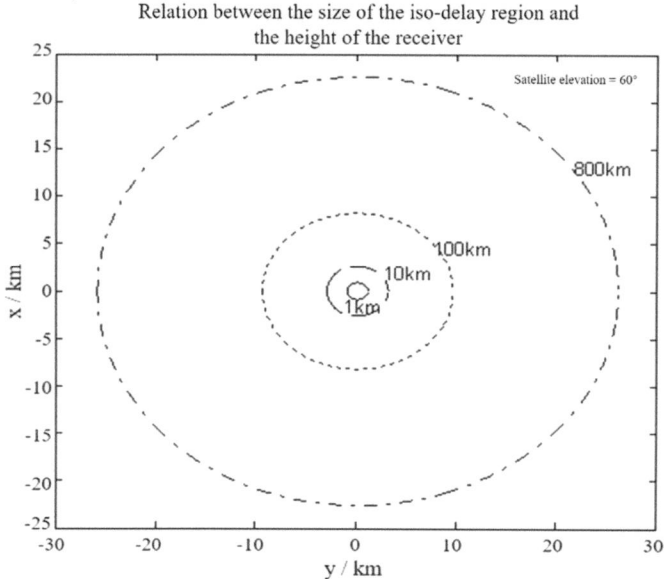

Fig. 4.5 Relationship between the size of the iso-delay zone and receiver altitude ($\triangle \tau = [1] \cdot \tau_c$)

Fig. 4.6 Size of the iso-delay zone ($h_r = 5km$, $\Delta \tau = [1, 2, 3 \ldots 10] \cdot \tau_c/20$)

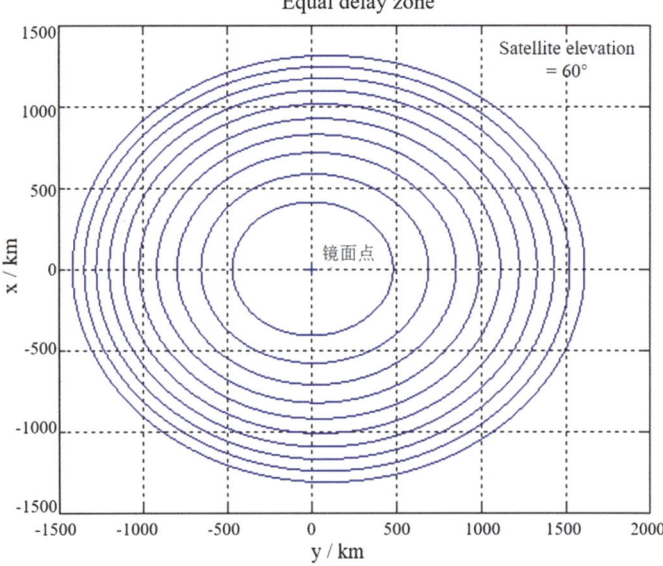

Fig. 4.7 Size of the iso-delay zone ($h_r = 5km$, $\Delta \tau = [1, 2, 3 \ldots 10] \cdot \tau_c/20$)

4.1.4 Iso-Doppler Lines

In GNSS bistatic radar systems for sea surface wind field detection, both the airborne or spaceborne receiver and the navigation satellite have different velocities, and the receiver's altitude may also change. Therefore, the scattering signals from different scattering points on the sea surface received by the receiver may have different Doppler frequencies, which change with the satellite parameters and the movement parameters of the receiving platform. The curve composed of sea surface scattering points with the same Doppler frequency is called an iso-Doppler line.

4.1 GNSS-R Geometric Relations

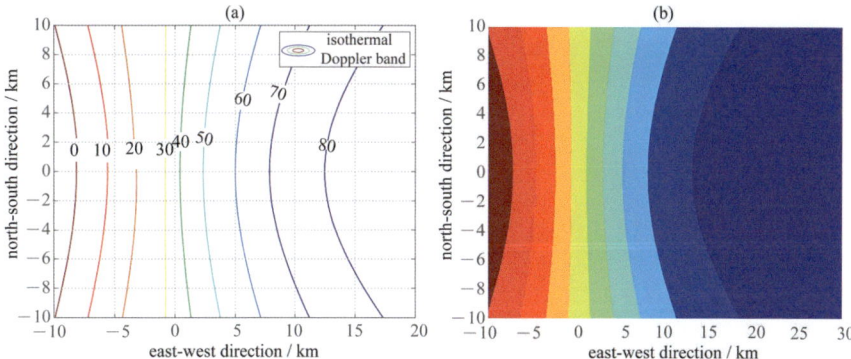

Fig. 4.8 Size and shape of the iso-Doppler regions and their projection on the sea surface

If the Doppler frequency at the mirror point $f_0 = f_D(0)$ is taken as the reference value, the points on the iso-Doppler line satisfy the following relation:

$$\Delta f = f_D(r) - f_0 = C \tag{4.18}$$

and C is a constant. The region that satisfies $\left|\Delta f - k\frac{1}{2T_i}\right| \leq \frac{1}{2T_i}$ is called the iso-Doppler region (k is a real number), and T_i is the reflection signal processing integration time. For GPS C/A code signals, the integration time (T_i) is usually taken as 1 ms.

The shape of an iso-Doppler region is typically a narrow strip, related to the receiver altitude, velocity magnitude and direction, and has little to do with parameters such as the velocity and azimuth of the navigation satellite. Figure 4.8 shows the receiver altitude $h_r = 10$ km, satellite altitude angle $\theta = 90°$, the size and shape of the iso-Doppler strip and the projection on the sea surface.

Isochronous lines and iso-Doppler lines are used to classify the sea surface scattering region in terms of time and frequency, respectively. In some cases, it may only be necessary to consider isochronous lines in the time dimension, while in other scenarios, both may need to be considered, depending on the specific application scenario.

For example, under airborne application conditions, with a typical flight speed of 0.17 km/s and GPS C/A code receiver integration time of 1 ms, the two-dimensional spatial division of the scattering region on the sea surface at different receiver altitudes $h_r = 1$ km, 10 km, 30 km is shown (Fig. 4.9).

The delay lines in the figure correspond to $\Delta \tau = [1, 2, 3 \ldots 18, 19] \cdot \tau_c$ from inside to outside, and the equal Doppler line intervals $f_B = 1/2T_i = 500$Hz and f_0 corresponding to the Doppler curves passing through the coordinate zeros in the figure. $\Delta f_c > 0$ is shown by a solid line. It is evident that under airborne altitude and speed, the iso-Doppler lines coarsely divide the scattering region, and in all three situations shown in the figure, most of the isochronous zone areas are contained within the Doppler range of -500 to $+500$ Hz. If the integration time is increased,

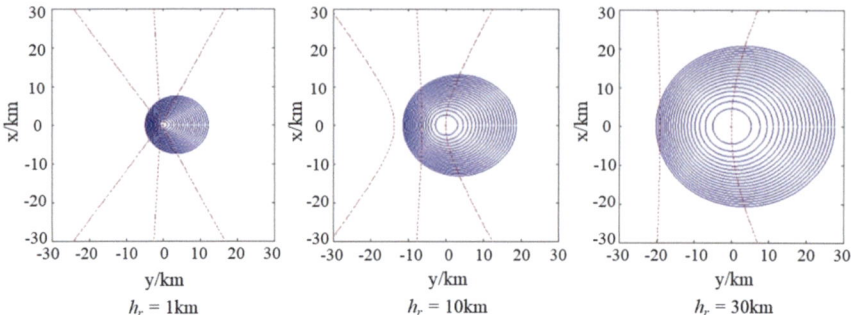

Fig. 4.9 Two-dimensional segmentation of iso-delay and iso-Doppler regions (satellite elevation angle of 60°)

e.g., $T_i = 10$ ms, the Doppler resolution is increased by tenfold and the spatial resolution strips will also shrink by tenfold.

On spaceborne platforms, due to their high velocity, also at $T_i = 1$ ms, the division of iso-Doppler lines will be very dense. For example, at an altitude of 800 km and a speed of 7.6 km/s, the division of isochronous lines and iso-Doppler lines is as shown in Fig. 4.10.

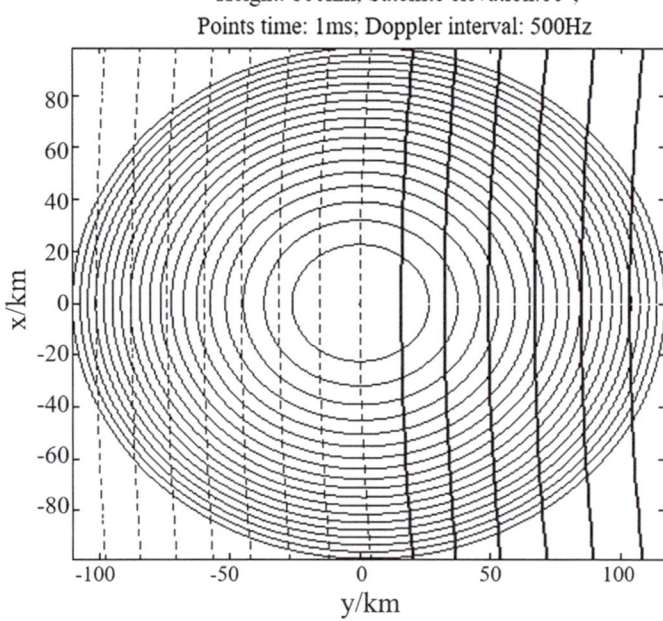

Fig. 4.10 Isochronous and iso-Doppler lines corresponding to a spaceborne receiving platform

4.2 Characterization of the Reflected Signal

In the ocean, the length of a wind zone is generally 50 km, and if only one-dimensional time-delay correlated power calculations are performed to divide the isodoppler zone, the spatial resolution requirement of wind field inversion cannot be met. In this case, it is necessary to calculate the iso-Doppler lines simultaneously, thereby finely dividing the scattering region to the scale of a single wind cell. Given the time delay and Doppler shift, the corresponding observational unit on the sea surface can be determined. By setting the range and interval of time delay and Doppler shift, the size of the observation area and resolution unit can be determined to meet different remote sensing task needs. Of course, the division of iso-Doppler lines differs when the receiver is moving in different directions, which needs to be considered comprehensively in practical applications.

From the above analysis, it can be seen that the delay and Doppler partitioning of the scattering region is symmetrically distributed around the specular reflection point, which means that a scattering signal with the same time-delay and Doppler may come from two different regions or be a superposition of signals from two regions. The division of the two areas utilizes the autocorrelation function of the ranging code (for GPS L1, the C/A code) and the Doppler sinc function. For example, Fig. 4.11a shows the scattering signals under the conditions of a receiving platform at an altitude of 5 km and a satellite elevation angle of 60°, with an integration time of 5 ms (corresponding to a Doppler interval of 100 Hz), where the shadow-filled annular band corresponds to the iso-delay zone for $\Delta \tau = [2\tau_c \ 4\tau_c]$, whose signal sampling is accomplished by spatial filtering using the autocorrelation function of the GPS C/A code, as shown in Fig. 4.11b. The shadow-filled stripe is the iso-Doppler zone of $\Delta f = \left[-1/T_i \ 1/T_i\right]$ signal sampling completed with a sinc function, as shown in Fig. 4.11c. The interaction of the isochronous and iso-Doppler regions results in two identical delay-Doppler regions corresponding to $\Delta \tau = 3\tau_c$, $\Delta f = 0$ Hz, which form a spatially resolved ambiguity, as shown in Fig. 4.11d, impacting wind field inversion.

4.2 Characterization of the Reflected Signal

4.2.1 Polarization Characteristics

4.2.1.1 Classification of Polarization

As mentioned earlier, the direction of the electric field (E) and magnetic field (H) of a uniform plane wave does not change within an equiphase surface. However, in practical engineering, the direction of field strength can change over time according to a certain pattern, a concept described as polarization. Since the relationship among the electric field strength E, magnetic field strength H, and propagation direction K is fixed, the polarization of electromagnetic waves is generally represented by the

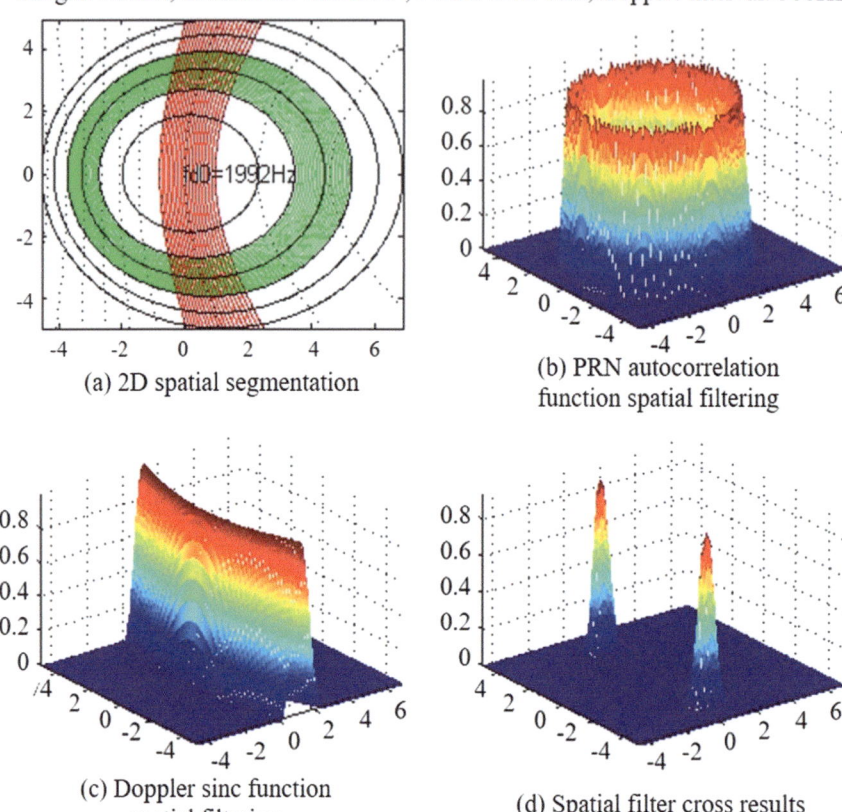

Fig. 4.11 Two-dimensional spatial sampling of iso-delay and iso-Doppler regions (x/y-axis in km, z-axis 0–1)

trajectory described by the vector endpoint of the electric field strength E changing over time at any fixed point in space.

Assuming a uniform plane wave propagates along the z-axis, with the electric field strength and magnetic field strength both in the plane perpendicular to the z-axis, let the electric field strength (E) be decomposed into two mutually orthogonal components Ex and Ey, with the same frequency and direction of propagation, denoted as:

$$\begin{cases} E_x = E_{x_o} \cos(\omega t + \varphi_x) \\ E_y = E_{y_o} \cos(\omega t + \varphi_y) \end{cases} \quad (4.19)$$

The trajectory equation of the E vector endpoint can be obtained through trigonometric operations, as:

$$\left(\frac{x}{E_{x_o}}\right)^2 + \left(\frac{y}{E_{y_o}}\right)^2 - 2\frac{x}{E_{x_o}} \cdot \frac{y}{E_{y_o}} \cos(\varphi_y - \varphi_x) = \sin^2(\varphi_y - \varphi_x) \quad (4.20)$$

Based on the amplitude and phase relationship of Ex and Ey, wave polarization is divided into three types.

(1) Linear polarization

The electric field E vibrates only in one direction, i.e., the trajectory of the vector E endpoint is a straight line.

If the two components have the same phrase, i.e., $\varphi_y - \varphi_x = 0$, and E_{x_o}, E_{y_o} are non zero, then there are:

$$y = \frac{E_{y_o}}{E_{x_o}} \cdot x \quad (4.21)$$

That is, the trajectory is a straight line passing through the point 0 and in one or three quadrants;

If the two components differ in phase by π, i.e., $\varphi_y - \varphi_x = \pm\pi$, and E_{x_o}, E_{y_o} are not zero, then:

$$y = -\frac{E_{y_o}}{E_{x_o}} \cdot x \quad (4.22)$$

That is, the trajectory is a straight line passing through the point 0 and in the second and fourth quadrants;

If $E_{x_o} = 0$, then:

$$\begin{aligned} x &= 0 \\ y &= E_{y_o} \cos(\omega t + \varphi_y) \end{aligned} \quad (4.23)$$

That is, the trajectory is a straight line varying along the y axis;

If $E_{y_o} = 0$, then we have

$$\begin{aligned} y &= 0 \\ x &= E_{x_o} \cos(\omega t + \varphi_x) \end{aligned} \quad (4.24)$$

The trajectory is a straight line varying along the x-axis. The four forms of linear polarization are shown in Fig. 4.12

(2) Circular polarization

When $E_{x_o} = E_{y_o} = E_0$, $\varphi_y - \varphi_x = \pm\pi/2$, the trajectory equation of the vector E endpoint of is

$$x^2 + y^2 = E_0^2 \quad (4.25)$$

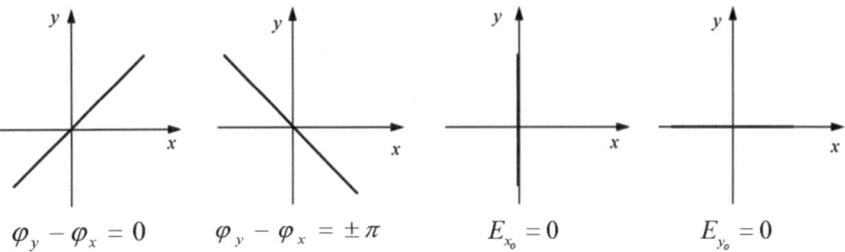

Fig. 4.12 Four cases of linear polarization

This is the equation of a circle with a radius of E_0, hence it is called circle polarization, as shown in Fig. 4.13. If E_y is $\pi/2$ behind E_x, the rotation of the electric field vector and the wave propagation direction satisfy the right-handed helix relationship, called the right-handed circular polarization; otherwise, it is called the left-handed circular polarization.

(3) Elliptical polarization

Generally, the two components of the electric field are not equal in amplitude and phase, and do not satisfy the condition that the phase difference is $\pi/2$ or an integer multiple of $\pi/2$, then the trajectory of the endpoints of the electric field vector is an ellipse, and so it is called elliptical polarization, as in Fig. 4.14 is shown.

Linear polarization and circular polarization are special cases of elliptical polarization. Waves of all three polarization forms can be decomposed into the superposition of orthogonally polarized waves in space. Any linearly polarized wave can also be decomposed into the superposition of two amplitude-equal, opposite-spin circularly polarized waves; similarly, any elliptically polarized wave can be decomposed into the superposition of two circularly polarized waves.

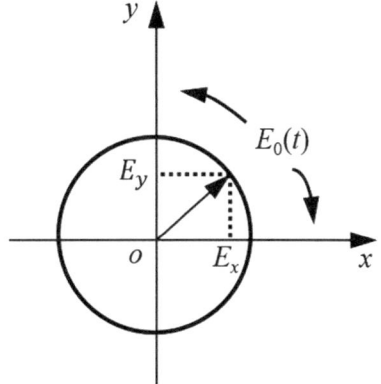

Fig. 4.13 Circular polarization

Fig. 4.14 Elliptical polarization

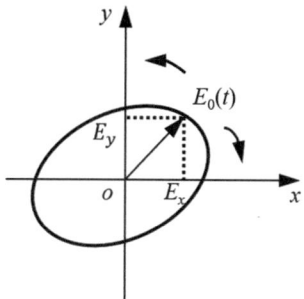

4.2.1.2 Application of Polarization

The polarization of electromagnetic waves is widely used in communication, broadcasting, electronic reconnaissance, aerospace, and other fields. For instance, the electromagnetic waves transmitted by amplitude modulation (AM) radio stations utilize vertical polarization, thus a radio antenna should be perpendicular to the ground to properly receive signals; television signals, on the other hand, often use horizontal polarization, requiring antennas to be parallel to the ground. During rocket and satellite operations, the constantly changing attitude leads to continuous changes in antenna orientation, hence circular polarization is often used for electromagnetic wave transmission to ensure unobstructed communication.

In radio communication, utilizing two orthogonal linear polarizations can double the capacity of a single-polarization system within the same allocated frequency band.

In addition, the polarization type of the electromagnetic wave can be used to identify the target. When an electromagnetic wave of a certain polarization type irradiates a target, the polarization type of its reflected wave may change. The change in polarization type depends on the shape, size, structure and material properties of the target. By studying the changes in polarization type of electromagnetic wave, characteristics of the target can be extracted, which is called polarization identification technology. This forms the physical basis for target detection and identification using reflected electromagnetic waves. GNSS-R technology leverages the polarization properties of electromagnetic waves upon reflection.

4.2.2 Reflection Coefficient

The direct signal from navigation satellites is incident on the Earth's surface and is reflected by the Earth's surface. At the air-Earth's surface interface, the energy relationship between the reflected and incident electromagnetic waves is determined by the Fresnel reflection coefficient. Take the sea surface as an example, the expression of Fresnel reflection coefficient is

$$\mathfrak{R}_{VV} = \frac{\varepsilon \sin\theta - \sqrt{\varepsilon - \cos^2\theta}}{\varepsilon \sin\theta + \sqrt{\varepsilon - \cos^2\theta}} \qquad (4.26)$$

$$\mathfrak{R}_{HH} = \frac{\sin\theta - \sqrt{\varepsilon - \cos^2\theta}}{\sin\theta + \sqrt{\varepsilon - \cos^2\theta}} \qquad (4.27)$$

$$\mathfrak{R}_{RR} = \mathfrak{R}_{LL} = \frac{1}{2}(\mathfrak{R}_{VV} + \mathfrak{R}_{HH}) \qquad (4.28)$$

$$\mathfrak{R}_{LR} = \mathfrak{R}_{RL} = \frac{1}{2}(\mathfrak{R}_{VV} - \mathfrak{R}_{HH}) \qquad (4.29)$$

where R, L, V and H represent the right-hand circular polarization, left-hand circular polarization, vertical and horizontal line polarization, respectively; ε is the complex dielectric constant of the sea surface. Taking GPS L1 band (1575.42 MHz) as an example, the sea water temperature is taken as room temperature 25 °C and salinity is taken as 35, $\varepsilon = 70.53 + 65.68i$. Substituting the complex permittivity of the sea surface into the Fresnel reflection coefficient formula, the magnitude of the Fresnel reflection coefficient and the time-delay Doppler partitioning of the scattering region is are as shown in **Fig. 4.16.** and **Fig. 4.15.** (Figs. 4.15 and 4.16).

As shown in **Fig. 4.15.** and **Fig. 4.16**, the left circular polarization component of the reflected signal $\mathfrak{R}_{RL}$ increases with the in elevation angle of the satellite e, and the right-circular polarization component of the reflected signal $\mathfrak{R}_{RR}$ decreases with the increase of the satellite altitude angle. The horizontal component of the reflected signal line polarization $\mathfrak{R}_{HH}$ decreases with the elevation angle, and the vertical component $\mathfrak{R}_{VV}$ increases after the satellite altitude angle exceeds about 6.8 degrees. This indicates that the right circularly polarized satellite signal undergoes a polarity conversion after scattering off the sea surface, transforming into left circularly polarized waves, with a significant proportion of energy conversion.

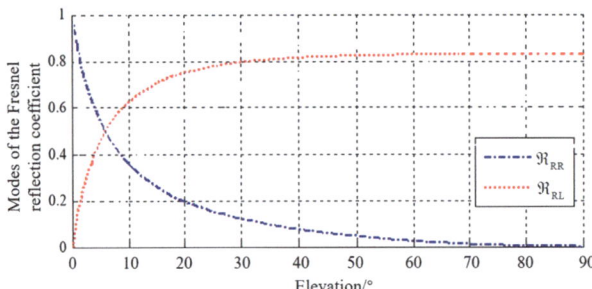

Fig. 4.15 Relationship between the reflected signal's right and left circular polarization components and the satellite elevation angle

4.2 Characterization of the Reflected Signal

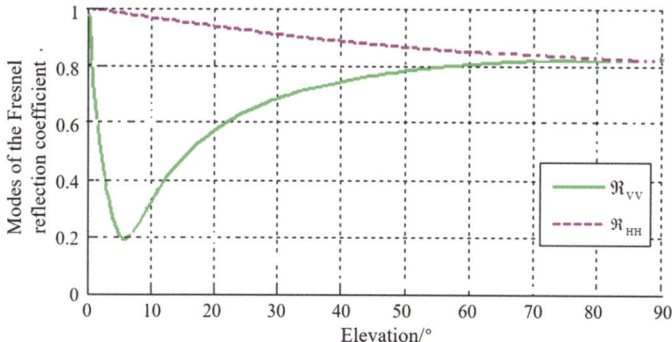

Fig. 4.16 Relationship between the reflected signal polarization characteristics' vertical and horizontal components and the satellite elevation angle

4.2.3 Description of Reflected Signals

Here, we analyze the model of reflected signals from navigation satellites on the ocean surface, using the reflection signal as an example. The ocean surface reflected signal results from the combined effects of different ocean surface reflection regions. Given that the reflection region area is relatively small, the impact of Earth's curvature can be neglected. The relationship diagram of the reflected signal is shown in Fig. 4.17 [3].

Assume that the coordinates of the reflection point S are (x, y, ζ), where $\zeta = \zeta(x, y)$ is a random variable representing the sea surface height, corresponding to the horizontal position vector $r = (x, y)$. m, n represent the unit vectors from the transmitter to the reflection point and from the reflection point to the receiver, respectively, we have

$$m = \frac{R_t}{R_t} = \frac{S - T}{|S - T|} \tag{4.30}$$

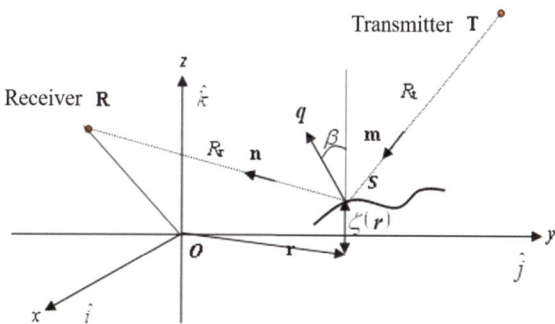

Fig. 4.17 Schematic diagram of reflected signal relationship

$$n = \frac{R_r}{R_r} = \frac{R - S}{|R - S|} \quad (4.31)$$

$R_r(t) = |R - S|$, $R_t(t) = |T - S|$ are the distances from the receiver and transmitter to the reflection point, respectively. q is defined as the reflection vector:

$$q = k(n - m) = (q_x, q_y, q_z) = (q_\perp, q_z) \quad (4.32)$$

where q_x, q_y and q_z are the components of the reflection vector in x, y and z directions respectively, $q_\perp = (q_x, q_y)$ denotes the horizontal component of the reflection vector, and β is the angle between the reflection vector and the z axis. The incident signal at the reflection point S can be expressed as

$$E(S, t) = A \frac{1}{R_t} a \left(t - \frac{R_t}{c} \right) \exp(ikR_t - 2\pi i f_L t) \quad (4.33)$$

According to the Kirchkov approximation model, the reflected field at the receiver R can be expressed as

$$\begin{aligned} E_s(R, t) &= \frac{1}{4\pi} \iint D(r, t) \Re \frac{\partial}{\partial N} [E(S) \frac{\exp(ikR_r)}{R_r}] d^2 r \\ &= \iint D(r, t) \left[\frac{\partial E(S)}{\partial N} + E(S) \frac{\partial R_r}{\partial N} \left(ik + \frac{1}{R_r} \right) \right] \frac{\Re}{4\pi} \frac{\exp(ikR_r)}{R_r} d^2 r \end{aligned} \quad (4.34)$$

$D(r, t)$ is the directivity function of the receiving antenna, $\Re$ is the reflection coefficient for different polarizations, and $\frac{\partial}{\partial N}$ is the normal derivative. Substituting Eq. (4.33) into Eq. (4.34) gives:

$$\begin{aligned} E(R, t) = A \iint D(r, t) a \left(t - \frac{R_t + R_r}{c} \right) \left[\left(k + \frac{i}{R_t} \right) \frac{\partial R_t}{\partial N} + \left(k + \frac{i}{R_r} \right) \frac{\partial R_r}{\partial N} \right] \\ \times \left(-\frac{\Re}{4\pi i} \frac{\exp[ik(R_r + R_t)]}{R_r R_t} \exp(-2\pi i f_L t) \right) d^2 r \end{aligned} \quad (4.35)$$

Since:

$$\frac{\partial R_t}{\partial N} = -\nabla R_t \cdot N = -\frac{R_t}{R_t} \cdot N = -m \cdot N \quad (4.36)$$

$$\frac{\partial R_r}{\partial N} = \nabla R_r \cdot N = \frac{R_r}{R_r} \cdot N = n \cdot N \quad (4.37)$$

where N is the normal unit vector, the $[\cdot]$ part can be expressed as

4.2 Characterization of the Reflected Signal

$$\left(k + \frac{i}{R_t}\right)(-\boldsymbol{m} \cdot \boldsymbol{N}) + \left(k + \frac{i}{R_r}\right)(\boldsymbol{n} \cdot \boldsymbol{N}) \approx \boldsymbol{q} \cdot \boldsymbol{N} \quad (4.38)$$

For rough sea surface, there are

$$\boldsymbol{q} \cdot \boldsymbol{N} \approx \frac{q^2}{q_z} \quad (4.39)$$

Thus, Eq.(4.35) can be expressed as

$$E(\boldsymbol{R}, t) = A \cdot \exp(-2\pi i f_L t) \cdot \iint D(\boldsymbol{r}, t) a[t - (R_t + R_r)/c] g(\boldsymbol{r}, t) d^2 r \quad (4.40)$$

where:

$$g(\boldsymbol{r}, t) = -\frac{\Re}{4\pi i R_t R_r} \exp[ik(R_t + R_r)] \frac{q^2}{q_z} \quad (4.41)$$

In fact, the receiver, the transmitter, and the sea surface reflector elements are all in motion, and thus R_t, R_r are functions of time, and a first-order Taylor series expansion of $R_t(t_0 + \Delta t)$ and $R_r(t_0 + \Delta t)$ at t_0 leads to

$$R_t(t_0 + \Delta t) \approx R_t(t_0) + \Delta t[\boldsymbol{v}_s - \boldsymbol{v}_t] \cdot \boldsymbol{m} \quad (4.42)$$

$$R_r(t_0 + \Delta t) \approx R_r(t_0) + \Delta t[\boldsymbol{v}_r - \boldsymbol{v}_s] \cdot \boldsymbol{n} \quad (4.43)$$

where $\boldsymbol{v}_t$, $\boldsymbol{v}_r$, $\boldsymbol{v}_s$ is the velocities of the transmitter, receiver and sea surface reflection elements, respectively. Substituting Eqs. (4.42) and (4.43) into (4.41) and considering only the variation in the exponential term gives:

$$g(\boldsymbol{r}, t_0 + \Delta t) = g(\boldsymbol{r}, t_0) \exp\{-2\pi i f_D(\boldsymbol{r}, t_0) \Delta t\} \quad (4.44)$$

where $f_D(\boldsymbol{r}, t_0)$ is the total Doppler shift, composed of Doppler shifts caused by the relative motion of the transmitter and receiver as well as by the relative motion of the reflection element, with:

$$f_D(\boldsymbol{r}, t_0) = f_{D0}(\boldsymbol{r}, t_0) + f_{rD}(\boldsymbol{r}, t_0) \quad (4.45)$$

$$f_{D0}(\boldsymbol{r}, t_0) = [\boldsymbol{v}_t \cdot \boldsymbol{m}(\boldsymbol{r}, t_0) - \boldsymbol{v}_r \cdot \boldsymbol{n}(\boldsymbol{r}, t_0)]/\lambda \quad (4.46)$$

$$f_{rD}(\boldsymbol{r}, t_0) = [\boldsymbol{m}(\boldsymbol{r}, t_0) - \boldsymbol{n}(\boldsymbol{r}, t_0)] \cdot \boldsymbol{v}_s/\lambda = -\boldsymbol{q}(\boldsymbol{r}, t_0) \cdot \boldsymbol{v}_s/2\pi \quad (4.47)$$

For the ocean surface, $q_\perp \ll q_z$, the main contribution to f_{rD} comes from the vertical velocity component of the ocean surface gravity waves v_{sz}, thus

$$f_{rD} \approx q_z v_{sz}/2\pi \tag{4.48}$$

Generally, v_{sz} is very small, its effect of f_{rD} can be neglected in Doppler shifts.

4.3 Correlation Function of the Reflected Signal

As mentioned in Chap. 2, GNSS signals (e.g., GPS, GALILEO, and BDS) are direct-sequence spread spectrum signals, where the signals transmitted by satellites are distributed over a wide frequency band. Due to the limitations of satellite transmission power and the free-space attenuation caused by long-distance spatial transmissions, the GNSS signals received on the ground are buried in noise. Their power is too low to measure directly; it can only be measured through correlation processing, which provides processing gain. It can only be captured and measured through correlation processing. The power of the reflected signal is lower than that of the direct signal, and likewise, can only be analyzed after obtaining higher gains through correlation processing.

In the processing of direct GNSS signals, the correlation function of the local PRN code copy, a at any t_0 moment and the signal output by the receiving antenna at $t_0 + \tau$ moment is defined as u_D. The correlation function is defined as:

$$Y_D(t_0, \tau) = \int_0^{T_i} u_D(t_0 + t' + \tau) a(t_0 + t') \exp[2\pi i(f_c + \hat{f}_d)(t_0 + t')] dt' \tag{4.49}$$

where T_i is the integration time, f_c is the center frequency of the received signal, and $\hat{f}_d$ is the local Doppler estimate to compensate for the Doppler shift of the received signal. For a direct signal, the difference between u_D and a differ by only one time delay, and by correlating with different delay moments τ The maximum value of the correlation function can be obtained by correlating the local code at different delay moments, when the local code and the received signal slice are aligned. The time delay information of the direct signal indicates the distance information from the transmitter to the receiver and can be used for navigation localization.

The correlation function of the reflected signal is defined in a similar way to that of the direct signal, as well as the correlation value between the received reflected signal and the locally generated standard signal. However, due to the roughness of the reflecting surface, the signal characteristics are more complex, which are characterized by the attenuation of the signal amplitude and the superposition of different time delays and different Doppler signals, which correspond to the different reflective units of the reflecting surface, [4] as shown in Fig. 4.18.

Therefore, the correlation value of the reflected signal needs to be considered in terms of both time delay and Doppler frequency. Given this characteristic of the reflected signal, the correlation function of the reflected signal is analyzed from three

Fig. 4.18 Correspondence between reflective surface units and time-delay-Doppler units

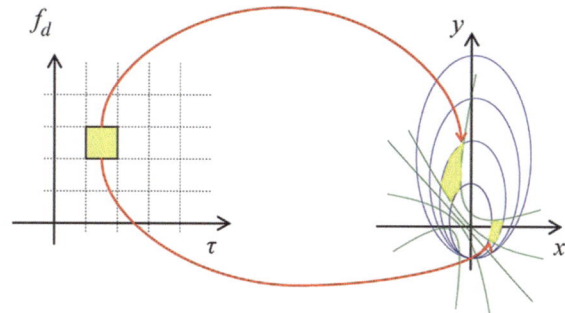

perspectives: one-dimensional time-delay correlation function, tone-dimensional Doppler correlation function, and two-dimensional time-delay-Doppler correlation function.

4.3.1 One-Dimensional Time Delay Correlation Function

The one-dimensional time delay correlation function of the reflected signal is defined in the same way as the correlation function of the direct signal, as shown in Eq. (4.50):

$$Y_{R-Delay}(t_0, \tau) = \int_0^{T_i} u_R(t_0 + t' + \tau) a(t_0 + t') \exp[2\pi i (f_c + \hat{f}_d + f_0)(t_0 + t')] dt' \tag{4.50}$$

It can be seen that the one-dimensional time-delay correlation function of the reflected signal is the correlation value between the received signal and the local pseudo-code signal under a specific Doppler shift f_0 with different time delays. It represents the one-dimensional variation trend of the reflected signal correlation value with time delay, reflecting the distribution of the reflected signals in specific equal Doppler regions on the reflection surface.

The relationship between the one-dimensional time delay correlation function of the reflected signal and the characteristics of the reflection surface is very close. For example, in the remote sensing of the sea surface wind field, the energy value of the one-dimensional time delay correlation function is closely related to physical parameters such as sea surface wind speed and wind direction. Figure 4.19 is the time-delay one-dimensional correlation power curve of the reflected signal under the conditions of the receiver height 5 km, satellite altitude angle 60°, wind direction angle to the incident plane of 0°, and the wind speed of 4-10 m/s with intervals of 2 m/s. It can be seen that as the wind speed increases, the peak power of the reflected signal decreases and is shifted towards greater time delays, and the slope of

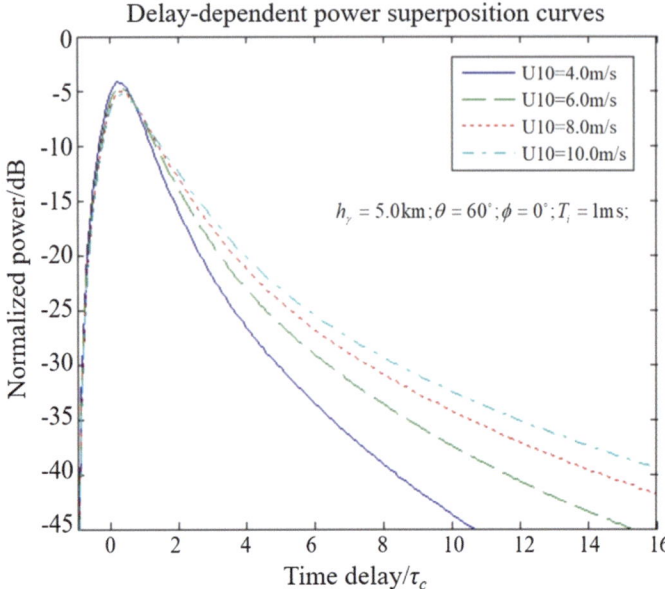

Fig. 4.19 One-dimensional time delay correlation power at different wind speeds

the reflected signal decreases along the falling edge. This indicates the distribution of the reflected signal power in the illuminatio area; as the wind speed increases, the sea surface roughness increases, and the signal power in the quasi-mirror direction decreases; the trailing of the reflected signal correlation power increases with time-delay spreading, which is due to the increase in sea surface roughness, the illumination area enlarges, and the reflected signals far from the specular point reach the receiver with greater time delays; moreover, at lower wind speeds, the spacing changes between the scattering signal power curves are significant, while at higher wind speeds the changes are smaller.

4.3.2 One-Dimensional Doppler Correlation Function

The one-dimensional Doppler correlation function refers to the correlation value between the received signal and the local carrier signal at different Doppler frequency shifts at a specific certain code delay τ_0, as shown in Eq. (4.51)

$$Y_{R-Doppler}(t_0, f) = \int_0^{T_i} u_R(t_0 + t' + \tau_0) a(t_0 + t') \exp[2\pi i (f_c + \hat{f}_d + f)(t_0 + t')] dt' \quad (4.51)$$

4.3 Correlation Function of the Reflected Signal

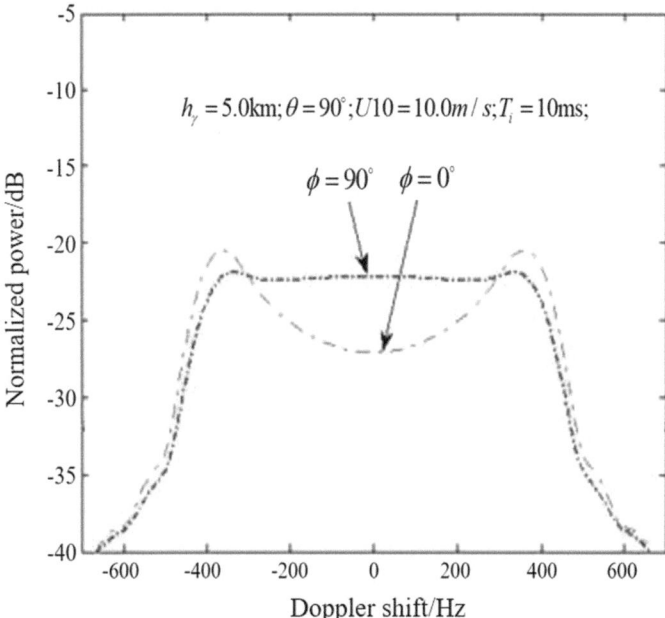

Fig. 4.20 One-dimensional Doppler correlation power under different wind directions

It can be seen that the one-dimensional Doppler correlation function characterizes the frequency domain properties of the reflected signal, reflecting the distribution of the reflected signal in different iso-Doppler zones within a specific iso-delay ring on the reflecting surface.

The one-dimensional Doppler correlation function of the reflected signal is also closely related to the characteristics of the reflecting surface. Figure 4.20 gives the one-dimensional Doppler correlation power values of the reflected signals under different wind direction conditions in the remote sensing application of the sea surface wind field, showing the sensitivity of the Doppler one-dimensional correlation value power curve to wind direction.

4.3.3 Time-Delay-Doppler Two-Dimensional Correlation Function

By integrating the time-delay one-dimensional correlation function and the Doppler one-dimensional correlation function, the time-delay-Doppler two-dimensional correlation function of the reflected signals can be obtained, as shown in Eq. (4.52)

$$Y_{R-Delay}(t_0, \tau, f)$$
$$= \int_0^{T_i} u_R(t_0 + t' + \tau) a(t_0 + t') \exp\left[2\pi i \left(f_c + \hat{f}_d + f\right)(t_0 + t')\right] dt' \quad (4.52)$$

It reflects the correlation values of the reflected signals at the cross regions of the delay lines and Doppler lines in the reflection zone, and is the most comprehensive way to describe the reflected signals. Figure 4.21 shows a typical time-delay-Doppler two-dimensional correlation power waveform diagram in ocean remote sensing.

The two-dimensional correlation power values can describe the strength of reflection signals from different reflection surface units, where its peak amplitude can indicate the reflectivity of the reflection medium to GNSS reflection signals. The time delay of two-dimensional correlation values can describe the path delay relationship of reflection signals compared to direct signals, and the phase of two-dimensional correlation values can describe the coherence characteristics of the reflection signals themselves. These physical parameters are crucial for remote sensing using GNSS reflection signals, making the effective acquisition of this two-dimensional correlation value matrix a key issue for feature extraction of reflection signals.

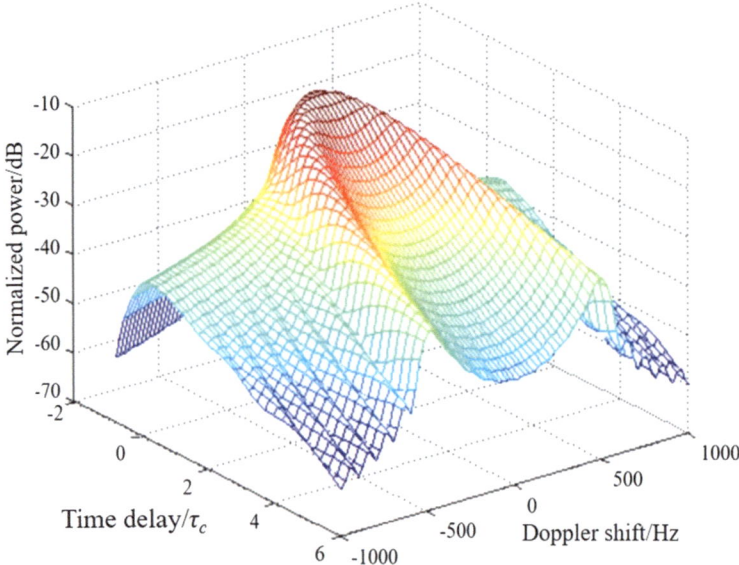

Fig. 4.21 Typical time-delay-Doppler two-dimensiona correlation power

4.4 Summary

Compared to direct signals, the essence of GNSS reflection signals is multipath. This chapter, based on the signal model from the previous chapter and starting from the reflection characteristics of electromagnetic waves, provides a detailed analysis of the polarization and reflection coefficients of electromagnetic waves. It discusses the geometric relationship between navigation satellites, receivers, and reflection surfaces and presents the mathematical expressions for reflection signals. Relying on classical processing methods for spread spectrum signals, the correlation function of reflected signals is elaborated in detail from three different perspectives: time dimension, frequency dimension and a two-dimensional combination of time and frequency. Using ocean remote sensing applications as examples, it demonstrates several reflection signal correlation power diagrams, laying the foundation for subsequent analyses.

References

1. Zavorotny VU, Voronovich AG. Scattering of GPS signals from the ocean with wind remote sensing application [J]. IEEE Trans Geosci Remote Sens. 2000;38(2):951–64.
2. Solat F. sea surface remote sensing with GNSS and sunlight relfections [D]. Universitat Politecnica De Catalunya;2003.
3. Shah R, Garrison JL. Application of the ICF coherence time method for ocean remote sensing using digital communication satellite signals [J]. IEEE J Sel Top Appl Earth Obs Remote Sens. 2014;7(5).
4. Huai-Tzu Y. Stochastic model for ocean surface reflected gps signals and satellite remote sensing applications [D]. Purdue University;2005.

Open Access This chapter is licensed under the terms of the Creative Commons Attribution-NonCommercial-NoDerivatives 4.0 International License (http://creativecommons.org/licenses/by-nc-nd/4.0/), which permits any noncommercial use, sharing, distribution and reproduction in any medium or format, as long as you give appropriate credit to the original author(s) and the source, provide a link to the Creative Commons license and indicate if you modified the licensed material. You do not have permission under this license to share adapted material derived from this chapter or parts of it.

The images or other third party material in this chapter are included in the chapter's Creative Commons license, unless indicated otherwise in a credit line to the material. If material is not included in the chapter's Creative Commons license and your intended use is not permitted by statutory regulation or exceeds the permitted use, you will need to obtain permission directly from the copyright holder.

Chapter 5
Reception and Processing of Reflected GNSS Signals

The signals reflected by navigation satellites in the target area contain rich physical characteristics information of the target, and the accurate extraction and metrics of this information to invert the physical state of the target is the premise and theoretical basis for remote sensing detection using reflected signals. This chapter, from the correlation value characteristics of the reflected signal, introduces the general method for receiving and processing GNSS reflection signals.

5.1 Generalized Model of a Reflected Signal Receiver

5.1.1 General Model

Referring to the GNSS receiver general structure, Fig. 5.1 presents the GNSS-R receiver general structure.

A GNSS-R receiver generally consists of two antennas: a Right Hand Circular Polarization (RHCP) antenna directed towards the zenith for receiving direct satellite signals and a Left Hand Circular Polarization (LHCP) antenna pointing towards the reflecting surface for receiving reflected signals. The receiver contains multiple parallel channels: the direct channel connected to the RHCP antenna acquire the pseudorange and Doppler observations and calculate positioning solutions; reflection channels connected to the LHCP antenna acquire correlation power values for different time delays and Doppler shifts through an open-loop approach.

As to GNSS-R remote sensing applications, the role of the direct signal is as follows: (1) The code phase and Doppler shift tracked from the direct signal tracking can be used as the essential auxiliary information for rapid processing of reflection signals; (2) The power of the direct signal can serve as a normalization reference for the reflected signal, correcting for power variations caused by satellite transmit

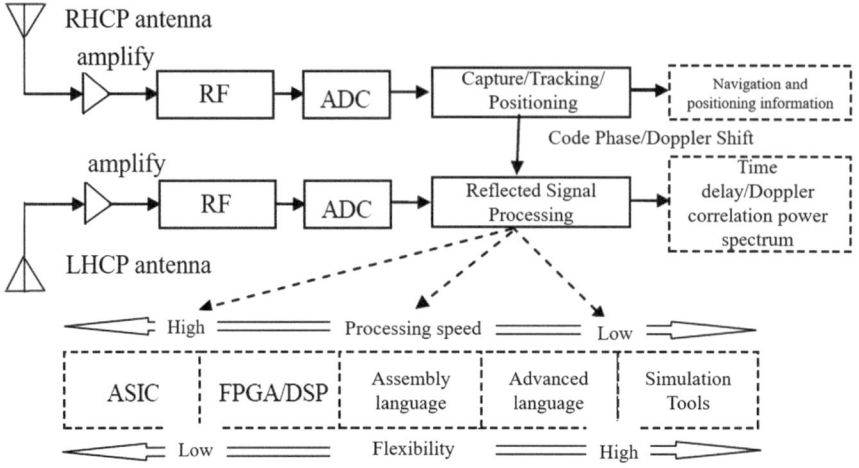

Fig. 5.1 General structure of a GNSS-R receiver

power changes, atmospheric attenuation, and changes in satellite elevation angle; (3) Precise positioning solutions and timing information provide information such as the moment of signal acquisition as well as the position and speed of the receiver.

5.1.2 Implementation

At the early stage, GPS reflected signal receivers were improved based on traditional navigation receivers by using a single RHCP antenna. By orienting the antenna horizontally during observation, both direct and low-elevation angle reflected signals could be received [1]. However, as the satellite elevation angle increased, the signal polarization changed, which could not be received by the ordinary GPS RHCP antenna, thus preventing the reception of all reflected signals. To solve the aforementioned problem, generally, two different polarized antennas were used: one standard GPS RGCP antenna directed towards the zenith for direct signals, and another high-gain LHCP antenna for reflected signal from the reflected area. Also, some research institutions had used two commercial receivers to carry out the test of sea surface wind field detection, one for receiving and processing direct signals with an RHCP antenna for positioning solutions, and another for receiving sea surface reflected signals with an LHCP antenna, with the results from the first receiver assisting in processing the reflected signals. The advantage of this approach was to realize the reception of the reflected signal; the disadvantage was that the time delay information of the direct signal and the reflected signal was complicated to compare, and the reflected signal processing process lacked real-time. Later, NASA's Jet Propulsion Laboratory developed a 16-channel reflection signal receiver based on conventional receivers with four RF front ends, allowing for different antenna configurations. The

5.1 Generalized Model of a Reflected Signal Receiver

initial reflection signal receiver, known as the Delay Mapping Receiver (DMR), was designed to map reflection signal correlation data from different reflection areas to different time delays as the receiver's output, producing one-dimensional time-delay-correlation power data. This has evolved into the Delay/Doppler Mapping Receiver (DDMR), outputting time-delay-correlation power data, Doppler-correlation power data, and two-dimensional time-delay-Doppler-correlation power data, representing the typical receiver for current GNSS-R applications [2]. Currently, the technical development and classification of GNSS-R receivers are shown in Fig. 5.2.

At present, GNSS-R receivers can be divided into two categories according to their implementation: software receivers and hardware receivers. Software receivers mainly consist of an RF front end and software, where the RF front end down-converts the signal to baseband, and after sampling and analog-to-digital conversion, the raw data is stored directly for software-based signal processing. Hardware receivers perform signal processing with correlator chips, directly outputting waveforms after correlation processing. Software receivers offer simplicity and flexibility, making it easier to change algorithms and parameters during signal processing. However, they produce a large volume of data and struggle to output waveforms in real-time,

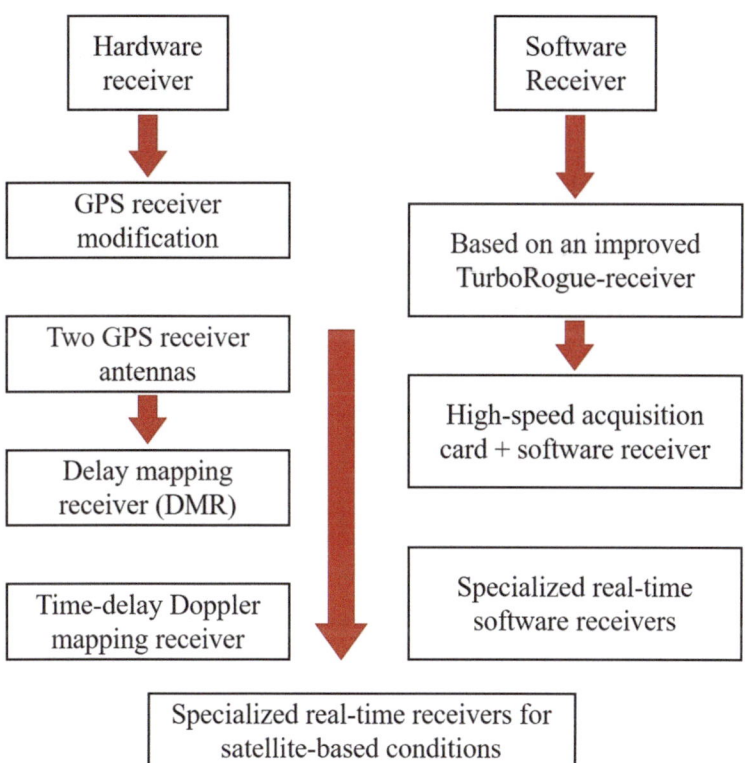

Fig. 5.2 GNSS-R receiver technology trends and classifications

which some applications, such as space borne conditions, cannot accept. Real-time capability is precisely the advantage of hardware receivers. The following summarizes the GNSS-R receivers developed and used by foreign entities reported in the literature.

Currently, there are several types of software receivers available:

(1) Two sets of equipment developed by ESA. The first set's downconversion and sampling are based on a GPS development toolkit from GEC-Plessey, with data storage completed by Vitrek's Signatec high-speed data acquisition system, allowing for the continuous collection of raw data for only 2.56 s. Experimental data analysis results are presented in reference [3]. The second set of equipment is based on the improved Turbo Rogue GPS receiver and Sony's SIR-1000 storage device, capable of collecting L1, L2 band I, Q dual-channel signals. However, ESA's equipment only used the L1 band I channel signal and conducted corresponding experiments [4–6].
(2) Equipment made by NASA, similar to ESA's second set of equipment but using L1, L2 band I, Q dual-channel signals for corresponding experiments [7, 8]
(3) A device developed by the Johns Hopkins University Applied Physics Laboratory, also utilizes a GEC-Plessey chip, with a sampling frequency of 5.714 MHz [9]. The related experiments completed can be found in the reference [10, 11].
(4) Equipment developed by the Institute of Space Studies of Catalonia, similar to ESA's second set of equipment but using L1 band I, Q dual-channel signals [12].
(5) Equipment developed by the University of Colorado, focuses on the advantages of software design. Reference [13] describes it in detail how to select a software receiver that is suitable for the current environment and anticipated future needs in embedded systems.
(6) Equipment developed by Starlab in Spanish, which can be deployed on multiple platforms for coastal monitoring operations. In references [14] the basic components and processing of the device are described.
(7) Equipment developed by the Surrey Satellite Space Center also uses the GEC-Plessey chip. Currently, the device has been tested on the UK-DMC satellite, successfully collecting GPS signals scattered from the surface of the Earth from outer space [15].
(8) Equipment developed by NAVSYS Corporation. This equipment consists of a digital front-end and a high-speed storage system, and its major advantage is that it can control the beam of the antenna for greater gain [16].

The main types of hardware receivers available are as follows:

(1) Equipment developed by NASA, using GEC-Plessey 2021 correlator chip. The receiver first performs 1 ms of coherent accumulation, then followed by 0.1 s of incoherent accumulation, and outputs a correlation waveform with 12 correlation function values, each delayed by 0.5 μs apart. The literature [17] describes its experimental setup.

(2) Equipment developed by the German Geological Research Center, based on open-source software from Reference [18], with a hardware environment similar to NASA developing equipment. Experiments using this equipment are described in the reference [19].
(3) Equipment developed by the Institute of Space Studies of Catalonia, consisting of 10 groups, each with 64 correlators. Each group of correlators can use different template signals, with delays of 50 ns between each correlator [20].

Hardware receivers can be configured in two different modes: serial operation mode and parallel operation mode. In serial mode, a scattering signal is input into a correlator, which computes the signal collected at different times with different parameter template signals to produce the output waveform. In parallel mode, a scattering signal is input into multiple correlators, each of which computes the signal collected at the same time with different parameter template signals to produce the required waveform. The parallel mode offers better real-time performance because the different sampling points of the correlated waveform are obtained by processing the signal collected at the same time, thus containing more accurate information. The disadvantage of the parallel mode is that it requires more correlators.

5.2 Reflected Signal Processing Methods

The reception and processing of GNSS direct signals maximize the correlation power by altering the time delay and Doppler frequency shift. The point of maximum correlation power contains precise pseudorange information, hence the correlation operation mainly targets a specific time delay-Doppler point. In contrast, the reception and processing of GNSS reflection signals are concerned with the one-dimensional correlation values of time delay near the maximum point, the one-dimensional correlation values of Doppler frequency, or the two-dimensional correlation surface of time delay-Doppler. Thus, the correlation operation expands from "point" to "line" and "surface". Since the one-dimensional correlation value of time delay and Doppler frequency of the reflection signal are special forms of the two-dimensional correlation value of time delay-Doppler, this section primarily introduces the calculation method for the two-dimensional correlation value of time delay-Doppler.

5.2.1 Discrete Forms of Reflected Signals

The reflected signal consist of components with different time delays and Doppler shifts, and their characteristics can be described by the correlation values of the reflected signal under different code delays and Doppler shifts. In practical reception and processing of reflected signals, the signals are given in discrete form. For this reason, Eq. (5.1) presents the discrete form of the time delay-Doppler correlation function of the reflected signal.

$$DDM_k(\tau_{N\_delay}, f_{N\_Doppler}) = \sum_{n=(k-1)T_i f_S}^{kT_i f_S} s_R(nT_S) \cdot C(nT_S - \tau_D - \tau_E - \tau_{N\_delay}) \cdot \exp[j2\pi(f_{IF} + f_D + f_E + f_{N\_Doppler})nT_S] \quad (5.1)$$

Here, DDM(·) is the complex two-dimensional correlation function with respect to time delay and Doppler shift. T_i is the coherence accumulation time, f_S is the sampling frequency of the received signal, T_S is the sampling interval of the received signal, $s_R(nT_S)$ is the reflected digital IF signal, C(·) is the pseudo-random code of the navigational satellite, f_{IF} is the center frequency value of the reflected digital IF signal, τ_D is the time delay of the direct signal, τ_E is the time delay of the reflected signal with respect to the direct signal, f_D is the carrier Doppler frequency of the direct signal, f_E is the average carrier Doppler shift of the reflected signal with respect to the direct signal, τ_{N_delay} is the time delay relative to the specular reflection point, and $f_{N\_Doppler}$ is the Doppler shift value relative to the center frequency of the reflected signal.

For the discrete form of the two-dimensional correlation function of reflection signals, the following basic definitions for the calculation of the two-dimensional correlation value of reflection signals are used to describe the process of reflection signal processing.

Reference point: in terms of specular reflection point time delay $\tau_0 (\tau_0 = \tau_D + \tau_E)$ and Doppler shift $f_0 (f_0 = f_D + f_E)$ as the time delay/Doppler coordinate zero point.

Time delay window: the time delay range of the reflection signal that needs to be processed, denoted as T_w.

Time Delay interval: the time delay interval during reflection signal collection, denoted as ΔT_w, whose value is controlled by the receiver resources and maximum original signal sampling rate.

Doppler Window: the Doppler range that needs to be processed for reflection signal collection, denoted as F_w.

Doppler Interval: The interval of Doppler during reflection signal collection, denoted as ΔF_w, the value of which is limited by the receiver resources.

Figure 5.3 represents the relationship among these definitions. Based on this, the two-dimensional correlation processing of reflection signals is as follows: (1) For each selected satellite, estimate the code delay and Doppler shift of the specular reflection point based on the direct channel information and set it as the reference zero point; (2) According to the requirements of the signal collection task, set the range and interval of the time delay window/Doppler window for reflection signal collection; (3) Perform coherent and incoherent accumulation of the reflection signal to calculate the correlation values under different time delays/Doppler shifts.

From the two-dimensional correlation process, it can be seen that to complete the calculation of the two-dimensional correlation value of reflection signals, it is

5.2 Reflected Signal Processing Methods

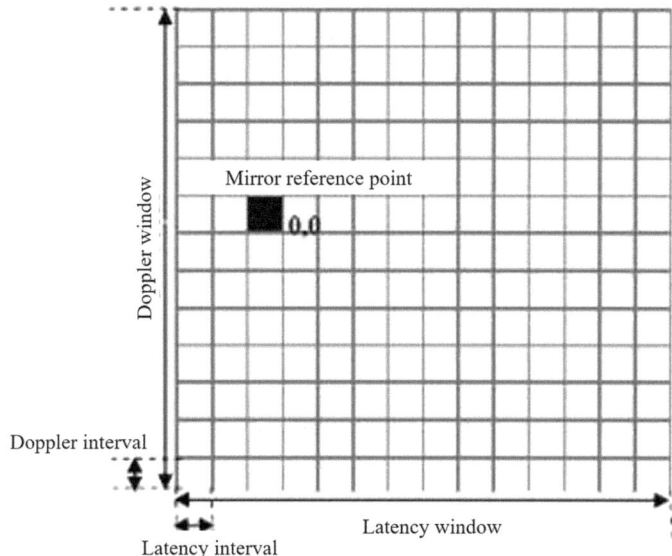

Fig. 5.3 Shows the definitions involved in the calculation of the two-dimensional correlation values of reflection signals

necessary to consider several aspects, such as the form of the calculation, the selection of the reference point, and the method of generating the local signal. In addition, due to the low signal-to-noise ratio of the reflection signals, it is necessary to improve the signal-to-noise ratio of the two-dimensional correlation values to enhance inversion accuracy.

5.2.2 Multi-channel Correlation Processing Algorithm

Similar to the processing of direct signals, the calculation of two-dimensional correlation values can be conducted in both serial and parallel modes, which can be further categorized into time-delay-serial-Doppler serial, time-delay-serial-Doppler parallel, time-delay-parallel-Doppler serial, and time-delay-parallel-Doppler parallel.

(1) Time-delay serial-Doppler serial

In this processing mode, one correlator is used for a single satellite, and the code phase and carrier Doppler are calculated serially. The process is as follows: first, a carrier Doppler frequency is preset within the specified Doppler range, and at this Doppler frequency point, the local code phase is moved one code phase unit at a time to perform correlation operation with the input signal. After completing all time delay units, set the local Doppler value to the next Doppler unit until all time delay/Doppler units are completed.

The advantage of this method is that the hardware circuit is simple and easy to implement, but the drawback the long calculation time. When the impact of the Doppler shift is minor, the Doppler can be set to a single fixed value for time delay correlation value calculation. Theoretically, this method can also serially be used to calculate correlation values under different Doppler shifts, but its efficiency is too low that it is generally not adopted.

(2) Time-delayed serial-Doppler parallel

For a single satellite, this method employs a single code generator to compute the code serially, while employing N_f carrier correlators toparallel process the carrier Doppler rangel. The N_f carrier correlators generate carriers at different frequencies, with the number of correlators related to the Doppler window range, and and Nf is at least equal to Fw/△Fw.

When GNSS-R receiver resources are limited but two-dimensional computation is needed, this method can be adopted. Taking a 12-channel receiver as an example, the method can be set up as follows: the first 6 channels are used to process the direct satellite signal, and the 6 reflection channels process the reflection signal of a single satellite in parallel, with each Doppler shift value processed by one reflection channel.

(3) Time-delay Parallel-Doppler Serial

This method uses Nc independent code correlators, where, under normal circumstances, the code phase difference between correlators is 1/2 chip. The Nc code correlators share a single carrier NCO, with carrier Doppler calculated in a serial scanning manner.

Still taking a 12-channel receiver as an example, only 1 channel is used to track the code phase and Doppler shift of a single satellite direct signal, and the remaining 11 channels are all processing the satellite signal, configured with different time delays, and the interval between channels is 1/2 a chip, allowing for 11 different time delay points to be set. The Doppler shift can be sequentially controlled or set to a fixed value. For a1ms pseudocode period, a time delay Doppler curve can be computed every 1 ms. Further, incoherent accumulation (e.g. 100 times) can also be performed before output.

(4) Time-delay parallel-Doppler parallel

This method is the most frequently used in the calculation of two-dimensional correlation values of reflection signals, and its typical structure is shown in Fig. 5.4. It employs $N_c * N_f$ independent correlators to parallel generate C/A code sequences and carrier sequences, enabling the acquisition of multiple time-delay-Doppler two-dimensional correlation values of a single satellite within an integration period.

To obtain effective two-dimensional correlation values of reflection signals, it is necessary to correctly select the time-delay-Doppler reference point for two-dimensional correlation values, and the time-delay and Doppler values. In the application of reflection signals, the time delay and Doppler values at the specular reflection point are generally used as reference points.

5.2 Reflected Signal Processing Methods

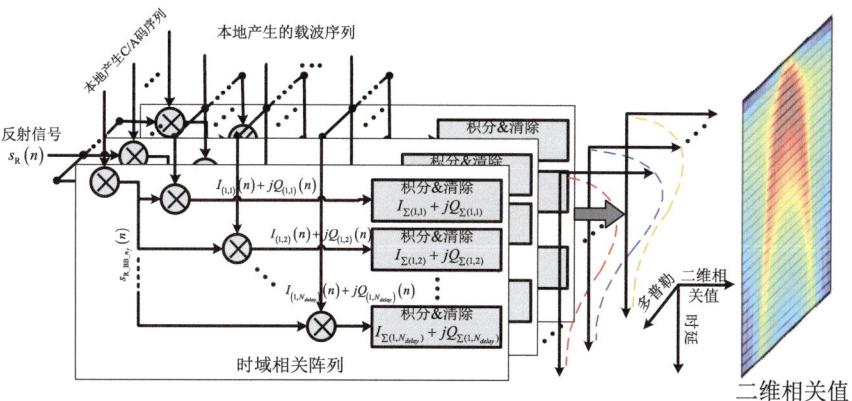

Fig. 5.4 Typical structure of the time-delay-Doppler two-dimensional correlation value solver

(1) Calculation of time delay values at specular reflection points

The path delay of the reflected signal with respect to the GNSS launch point is expressed in Eq. (5.2):

$$\rho_R = c \cdot \tau_R = \rho_D + \Delta\rho_E \tag{5.2}$$

where τ_R is the time delay of the reflected signal relative to the launch point, c is the speed of light, ρ_D is the path delay of the direct signal relative to the launch point, and $\Delta\rho_E$ is the path delay of the reflected signal relative to the direct signal. Since the path delay of the direct signal relative to the launch point can be obtained directly by processing the direct signal, it is only necessary to calculate the path delay of the reflection signal relative to the direct signal to establish the time-delay reference for the reflected signal's two-dimensional correlation function.

The receiver first determines the positions of the navigation satellites and itself from the received direct signals, then calculates the height of the receiver platform with respect to the reference horizontal plane h_R and the elevation angle of the visible navigation satellites θ. Using the geometric relationship of reflection events, a rough estimate of the time delay of the reflected signal is obtained as follows: [21]

$$\Delta\rho_E^{Coarse} = 2h_R \sin\theta \tag{5.3}$$

(2) Calculation of Doppler values at specular reflection point

The carrier Doppler frequency shift of the reflected signal is expressed in Eq. (5.4):

$$f_R = f_D + f_E \tag{5.4}$$

where f_D is the Doppler frequency of the direct signal, f_R is the Doppler frequency of the reflected signal, and f_E is the Doppler frequency shift of the reflected signal relative to the direct signal.

The Doppler frequency component of the direct signal can be obtained through the capture and carrier tracking process of the direct signal, and f_E can be estimated by Eq. (5.5):

$$f_E = [\mathbf{v_t} \cdot \mathbf{u_i} - \mathbf{v_r} \cdot \mathbf{u_r} - (\mathbf{v_t} - \mathbf{v_r})\mathbf{u_{rt}}]/\lambda \tag{5.5}$$

where $\mathbf{v}_t$ and $\mathbf{v}_r$ are the operating speeds of the navigation satellite and the receiving platform, respectively, $\mathbf{u_i}$ and $\mathbf{u}_r$ are the unit direction vectors of the incident and reflected signal paths and $\mathbf{u}_{rt}$ is the unit direction vector between the navigation satellite and the receiving platform, all of which are obtained by obtained from the navigation and positioning processing.

5.2.3 Calculation of Specular Reflection Point

The intersection of the plane formed by the satellite's vertical line and the receiving antenna with the Earth's surface creates a curve, and the reflection point lies on this curve. According to the principle of reflection, the angle of incidence should equal the angle of reflection. Assuming that the receiving antenna is stationary, the reflection point S is also a moving point with the movement of the satellite; at the same time, different satellites will produce different reflection points. The specular reflection point position is calculated by a two-step method, first assuming that the reference surface is an ellipsoid surface, and calculating it iteratively, then making corrections.

In the WGS84 coordinate system, let the center of the sphere be O. The coordinate vectors of the transmitter, receiver, and mirror points are $\boldsymbol{T}$, $\boldsymbol{R}$, and $\boldsymbol{S}$, respectively, and $\boldsymbol{T}$, $\boldsymbol{R}$, and the heights H_T and H_R relative to the ellipsoid are known so that $\gamma_t + \gamma_r$ can be calculated.

As shown in Fig. 5.5, $\boldsymbol{R}'$ is the mirror point of $\boldsymbol{R}$ along the line OM, and $\boldsymbol{C}$ is the mirror point of $\boldsymbol{M}$ along the line RR'. From the geometric relationship, it can be seen that $RS = R'S$, $RM = CR'$, which can be concluded as follows:

$$\begin{aligned}
& RM/MT = CR'/MT = SR'/ST = SR/ST = H_R/H_T \\
\Rightarrow\ & RM/RT = H_R/(H_R + H_T) \\
\Rightarrow\ & RM = H_R/(H_R + H_T)RT \\
\Rightarrow\ & \boldsymbol{M} = \boldsymbol{R} + H_R/[(H_R + H_T)](\boldsymbol{T} - \boldsymbol{R})
\end{aligned} \tag{5.6}$$

5.2 Reflected Signal Processing Methods

Fig. 5.5 Schematic diagram of the geometric relationship for calculating the specular reflection point

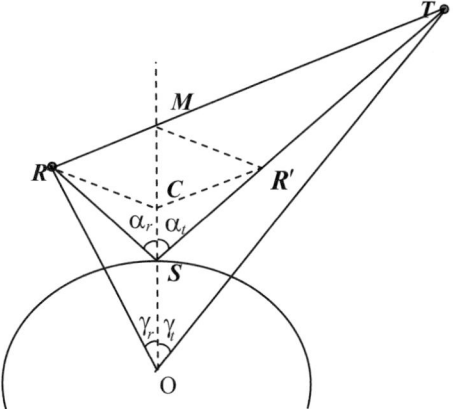

Thus, the position of vector M, as well as γ_t, γ_r can be calculated. According to the triangular relations OST and OSR, we can calculate α_t and α_r respectively. However, usually, α_t and α_r are not equal after the above calculation, so they can be re-estimated α_t and α_r and new angles αr' and αt' are calculated as:

$$\alpha'_r = \alpha'_t = (H_T \alpha_t + H_R \alpha_r)/(H_T + H_R) \tag{5.7}$$

The OSR and OST triangles are used to recalculate γ_r and γ_t, which are denoted as γ'_t and γ'_r, respectively, and the average value of γ_t is calculated according to Equation $(\gamma_t + \gamma_r + \gamma'_t - \gamma'_r)/2$. After solving for M based on the new value of γ_t, α_t and α_r are calculated. The above process is iterated until $\alpha_t = \alpha_r$. Actual simulations show that after 10 iterations, the difference between α_t and α_r can be less than 10^{-5} radians.

The above calculations assume that the normal vector of the specular reflection point and the Earth's radial direction are consistent. If there is a deviation between the radial direction of the Earth and the specular point's normal vector, it can be used to correct the position of the reflection point. In reality, there are deviations between the actual Earth's surface and the WGS-84 ellipsoid model, and further correction is needed on the basis of the solution results.

The Earth's surface approximation can choose the rotating ellipsoid surface or the geoid, as shown in Fig. 5.6. The difference between the reference ellipsoid and the instantaneous sea level is on the order of tens of meters. If the geoid model is used, the difference is maximally up to a few meters. Using the global Earth Gravitational Model 1996 (EGM96) geoid, the accuracy can reach the decimeters level at the sea surface; if the more precise local geoid model is used, the accuracy can reach centimeters level.

The reflection point position correction is decomposed into two parts: the incident plane component and perpendicular to the incident plane component. This is shown in Fig. 5.7a, b.

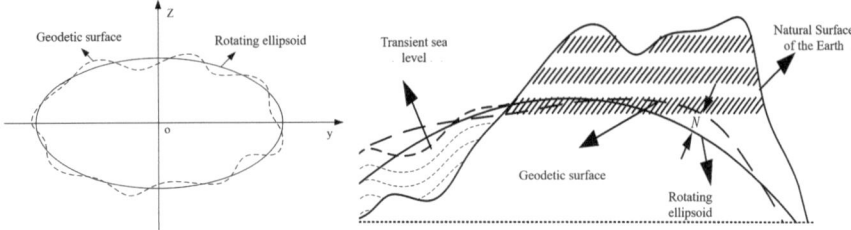

Fig. 5.6 Schematic of geodetic leveling surface

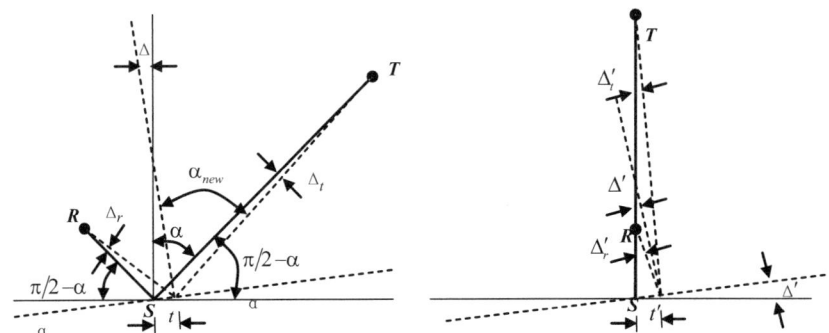

(a) Correction component within the incidence plane

(b) Correction component perpendicular to the incidence plane

Fig. 5.7 Methods of correcting the position of the specular reflection point

In Fig. 5.7a, from the geometric relationship where the angle of incidence equals the angle of reflection and the relationships within and external to the triangle, it is known:

$$\pi/2 - \alpha_{new} = \pi/2 - \alpha + \Delta - \Delta_r = \pi/2 - \alpha - \Delta + \Delta_t$$
$$\Rightarrow \Delta_r + \Delta_t = 2\Delta \quad (5.8)$$
$$\Rightarrow \Delta_r = 2\Delta - \Delta_t$$

Furthermore, according to the common side lengths of right-angled triangles being equal, there are:

$$t \sin(\pi/2 - \alpha_{new}) = TS \sin(\Delta_t) \Rightarrow t = TS \sin(\Delta_t)/\sin(\pi/2 - \alpha_{new}) \quad (5.9)$$

$$t \sin(\pi/2 - \alpha_{new}) = RS \sin(\Delta_r) \Rightarrow t = RS \sin(\Delta_r)/\sin(\pi/2 - \alpha_{new}) \quad (5.10)$$

resulting in $RS \sin(\Delta_r) = TS \sin(\Delta_t)$

Substituting $\Delta_r = 2\Delta - \Delta_t$ into the above equation, and Δ_t is very small $(\cos(\Delta_t) \approx 1)$, then

$$\sin(\Delta_t) = RS \sin 2\Delta / (TS + RS \cos 2\Delta) \tag{5.11}$$

Then the correction is

$$t = TS \sin(\Delta_t) / \sin(\pi/2 - \alpha_{new}) = TS \sin(\Delta_t) / \cos(\alpha + \Delta - \Delta_t) \tag{5.12}$$

Similarly, from the geometric relationship in Fig. 5.7b, we can derive:

$$\Delta'_r - \Delta' = \Delta' - \Delta'_t \Rightarrow \Delta'_r + \Delta'_t = 2\Delta' \Rightarrow \Delta'_r = 2\Delta' - \Delta'_t \tag{5.13}$$

$$\begin{aligned} t' &= RS \cos\alpha \sin\Delta'_r / \cos(\Delta'_r - \Delta') = TS \cos\alpha \sin\Delta'_t / \cos(\Delta' - \Delta'_t) \\ &\Rightarrow RS \sin\Delta'_r = TS \sin\Delta'_t \end{aligned} \tag{5.14}$$

Let $\cos(\Delta'_t) \approx 1$, have

$$\sin(\Delta'_t) = RS \sin 2\Delta' / (TS + RS \cos 2\Delta') \tag{5.15}$$

$$t' = TS \cos\alpha \sin\Delta'_t / \cos(\Delta' - \Delta'_t) \tag{5.16}$$

The corrections t' and t are re-solved for Δ and Δ' and iterated until t' and t are less than a certain threshold, thus obtaining the exact position of the reflection point.

5.2.4 Methods for Improving the Signal-To-Noise Ratio

Reflected GNSS Signals have a longer propagation path than direct signal, resulting in greater attenuation. A common method to improve the signal-to-noise ratio (SNR) when computing the two-dimensional correlation function of reflected signals is to combine coherent and incoherent accumulation.

As shown in Fig. 5.8, the combination process of coherent and incoherent accumulation is illustrated. The correlator outputs an I and Q value every 1 ms (note: 1 ms corresponds to one pseudocode period), with data from n ms being coherently accumulated, followed by m instances of incoherent accumulation.

The relationship between the items in Fig. 5.8 can be expressed by Eqs. (5.17) and (5.18) as follows:

$$(I_2)_j = \sum_{i=1}^{n} (I_1)_i \quad (Q_2)_j = \sum_{i=1}^{n} (Q_1)_i \quad j = 1, 2, 3 \ldots \tag{5.17}$$

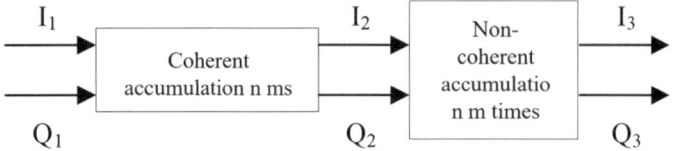

Fig. 5.8 Combination of coherent and incoherent accumulation

$$I_3 = \sqrt{\sum_{j=1}^{m}(I_2)_j^2} \quad Q_3 = \sqrt{\sum_{j=1}^{m}(Q_2)_j^2} \qquad (5.18)$$

That is, coherent accumulation is the direct accumulation of the outputs of the I- and Q-channel integrators, while incoherent accumulation involves adding the squares of their outputs.

(1) Coherent accumulation

Taking a 1 ms pseudocode period as the baseline for coherent accumulation, when increasing the coherent accumulation duration, it is equivalent to narrowing the noise bandwidth. The process gain, also known as the coherent gain, is the amount of signal-to-noise ratio enhancement after the coherent accumulation, denoted as:

$$G_c = 10 \lg n \qquad (5.19)$$

Although coherent accumulation can improve SNR, it requires signal coherence over the duration of accumulation, with phase continuity ensured. For a 50bps navigation message transmission rate, with data bit length being 20 ms, coherent accumulation time typically does not exceed 20 ms. The coherence of the reflected signals depends mainly on the motion state of the receiver, and the coherent integration time at the specular reflection point can be defined by Eq. (5.20) [22, 23].

$$\tau_{\text{coh}} = \left(\frac{\lambda}{2v_r}\right)\sqrt{\frac{h_r}{2c\tau_c \sin\theta}} \qquad (5.20)$$

where, λ is the GNSS signal carrier wavelength, h_r is the receiver height, $c\tau_c$ is the length of 1 code chip, and v_r is the receiver's velocity. It can be seen that the higher the receiver's velocity, the shorter the coherence time; at the same speed, the higher the receiver's altitude, the longer the coherence time. Table 5.1 lists the coherence time for several typical scenarios.

Compared to direct signals, reflected signals have a shorter coherent time, and the coherent time for reflected signals also varies at different delay moments. In engineering applications, the accumulation time is generally chosen to be slightly larger than the coherent time.

(2) Incoherent accumulation

5.2 Reflected Signal Processing Methods

Table 5.1 Coherence times for several typical scenarios

Platform height km	Altitude angle	Speed/km/s	Coherence time/ms
1	90°	0.1	1.3
10	90°	0.1	4.0
500	90°	7.0	0.4

After coherent accumulation, further processing gain improvements can be achieved through incoherent accumulation of the coherent accumulation results. Unlike coherent accumulation, incoherent accumulation does not maintain carrier phase continuity. When Ti is equal to 1 ms, the processing effects of different non-coherent accumulation times m are shown in Fig. 5.9.

The SNR in the figure refers to the signal-to-noise ratio after considering the coherent gain. The SNR improvement brought by M instances of incoherent accumulation is known as the incoherent accumulation gain, as follow:

$$G_{nc} = 10 \lg m \tag{5.21}$$

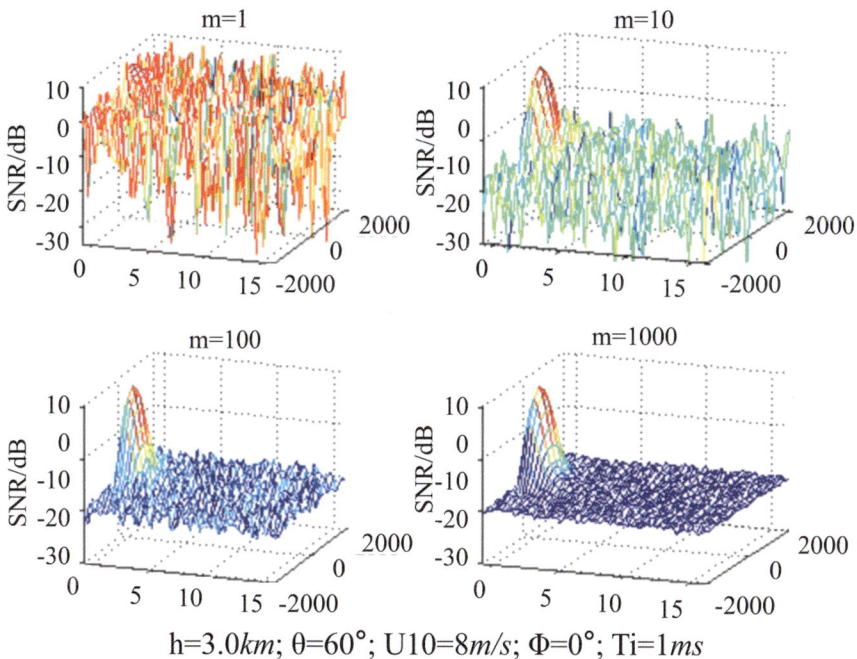

Fig. 5.9 Effect of different incoherent accumulation times

In general, the maximum value of incoherent accumulation time should be maintained in the range where the transmitter–receiver geometry does not change significantly. The incoherent accumulation is obtained modulo the complex result of the I and Q coherent integrals, where the noise mean value is no longer zero, and the modulo process is accompanied by the generation of new random variables. Therefore, in addition to providing a certain degree of signal-to-noise gain, the incoherent accumulation also introduces SNR loss, called the squaring loss $L(m)$. Analysis of squaring loss is more complicated and readers can refer to the references for more details [24–26]. In general, the squaring loss is closely related to the coherent integral SNR. It is worth noting that the squaring loss in incoherent accumulation may "enhance" rather than "reduce" the incoherent accumulation gain if the signal strength is sufficiently substantial.

(3) Total Gain

After combining coherent and incoherent accumulation, the signal processing gain can be expressed as

$$G = 10 \lg n + 10 \lg m - L(m) \tag{5.22}$$

(4) Eliminating the effects of data bit transitions

Navigation satellite signals contain navigation message data bits, and transitions in the data bits will reverse signal polarity, affecting the results of coherent accumulation. For example, the navigation message modulated on the L1 carrier has a rate of 50bps, and the length of the data bits is 20 ms; if the neighboring data are not the same, there will be a phase change in the 20 ms interval, and the length of the coherent accumulation time should consider this impact. To address this issue, a collaborative processing approach for direct and reflected signals can be used. Firstly, the carrier phase of the direct signal is tracked, and after carrier phase synchronization is achieved. Then, the sign of the in-phase component I_D output from the correlation channel of the direct signal is used to compensate for the two-dimensional correlation value matrix of the reflected signal, to eliminate the influence of the data bit transitions. Figure 5.10 illustrates the principle of implementing the direct signal data bit compensation algorithm.

5.3 Hardware Receiver

5.3.1 Overall Architecture of the Receiver

The overall hardware system architecture of a receiver for Reflected GNSS Signals, as shown in Fig. 5.11, includes components such as Right Hand Circular Polarization (RHCP) antenna, Left Hand Circular Polarization (LHCP) antenna, and dual RF front

5.3 Hardware Receiver

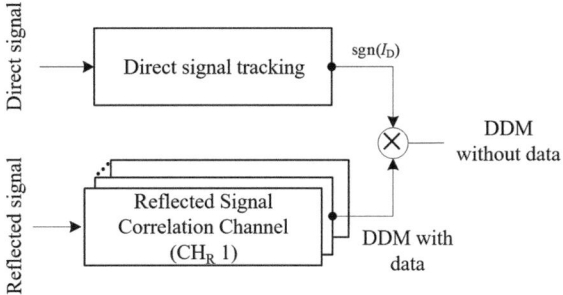

Fig. 5.10 Schematic diagram of the implementation principle of the data bit compensation algorithm for direct-radiation signals

end, High-speed A/D converter, Field Programmable Gate Array (FPGA) dedicated correlator, DSP (Digital Signal Processor), high-speed data transmission interface (multiple can be configured as needed) and data storage devices.

The RHCP antenna can be a general GNSS antenna used to receive direct signals from GNSS satellites; the LHCP antenna could be a multi-element array antenna or a conventional microstrip antenna, depending on the requirements for gain and beamwidth in the application domain, used for receiving GNSS satellite signals reflected off various surfaces. Task monitoring and data acquisition storage devices are usually equipped with serial and USB (Universal Serial Bus) interfaces for uploading data processed by the receiver. The specific processing flow of the GNSS reflection signal receiver is shown in Fig. 5.12.

Direct and reflected signals are received through the RHCP and LHCP antennas, respectively, and then filtered and down-converted to intermediate frequency analog signals by the dual RF front-end. They are sampled by dual-channel high-speed

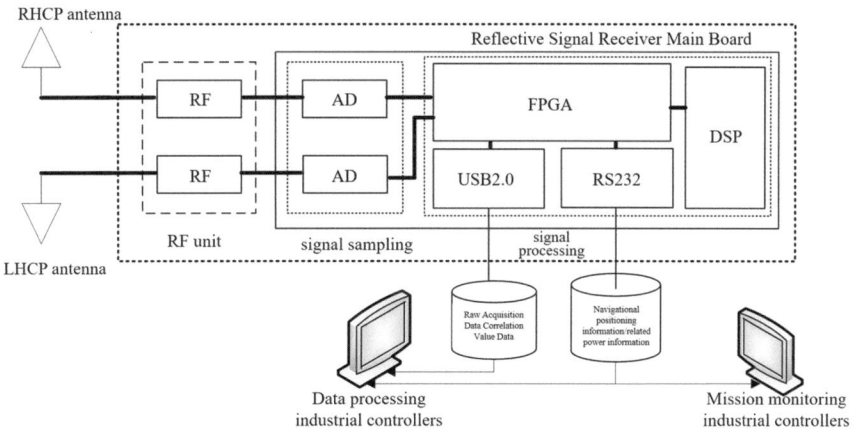

Fig. 5.11 Schematic of the receiver's overall architecture

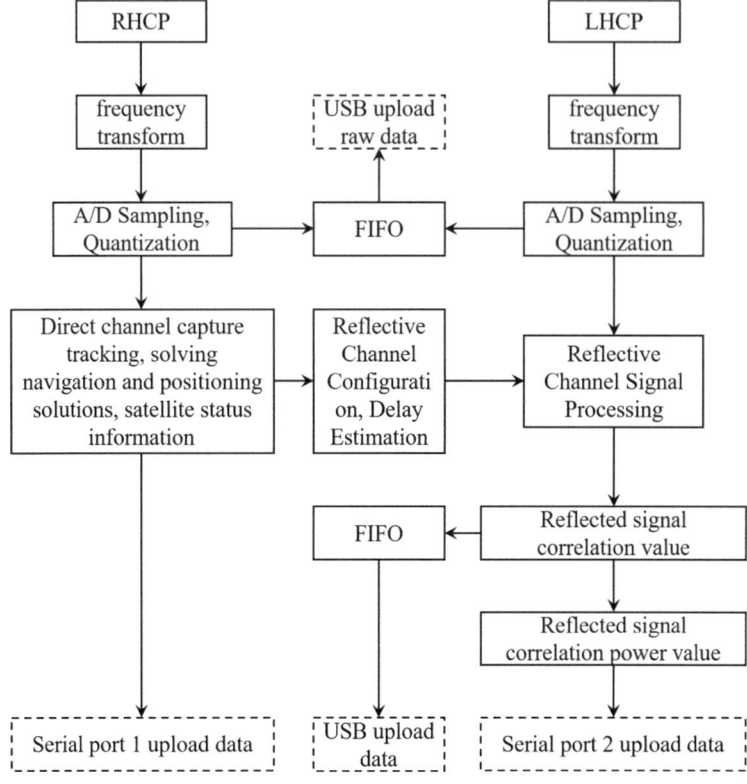

Fig. 5.12 Specific processing flow of Reflected GNSS Signal receiver

A/D converters and input into the FPGA's digital quantization module for 2-bit quantization coding. For sampling of raw intermediate frequency data, the quantized information from both channels is combined into frames, processed, and stored in First Input First Output (FIFO) buffer, then uploaded to the host computer via USB interface for storage.

In the direct channel, combined with the DSP, satellite signal capture, tracking, and solving of position information and satellite status are conducted. The information solved is used to configure the reflected channel, enabling control of the time delay and thereby obtaining the correlation values and/or correlation power values of satellite reflected signals at different time delays. The navigation positioning solution and satellite status information from the direct channel are uploaded through serial port 1 (RS232 interface in Fig. 5.11), the correlation values of reflected signals are uploaded through the USB interface in Fig. 5.11, and the correlation power values obtained through incoherent accumulation are uploaded through another RS232 serial port 2 in Fig. 5.11. All obtained data are stored in the data processing industrial PC.

5.3.2 Main Components of the Receiver

The components of the hardware of the GNSS reflection signal receiver mainly include the LHCP, the RHCP, the RF unit, and the baseband processing circuit mainboard composed of sampling and signal processing units, as shown in Fig. 5.13a–d respectively.

5.3.2.1 Signal Receiving Antennas

(1) LHCP antenna

In Fig. 5.13a the left antenna for receiving reflected signals is a quad-array high-gain antenna, which is designed for receiving reflected signals. Its design specifications include a center frequency 1575.42 MHz, gain 12 dB, beamwidth of 30°, which can be used to receive GPSL1/Beidou B1/Galileo E1 signals. The antenna has the following characteristics:

– Utilizing a single feed point structure to realize the array of antenna elements;
– Using a rotating feed structure to reduce the coupling coefficient between antenna elements;
– Employing serial feeding technology to increase the antenna impedance bandwidth, reduce sidelobes in the E-plane and H-plane, and use its parasitic radiation to improve the antenna's circular polarization characteristics.

The structure and top view of the LHCP antenna are shown in Fig. 5.14, with four antenna elements in the upper layer, a metal baseplate in the middle layer, and an antenna synthesis network in the bottom layer, equipped with a signal output Threaded Neill -Concelman (TNC) standard interface.

The Voltage Standing Wave Ratio (VSWR) of the antenna voltage and antenna array orientation diagram are shown in Fig. 5.15, revealing a maximum antenna gain about 13 dB and a beamwidth of approximate 38° (3 dB), which meets the requirements of the under airborne flight test.

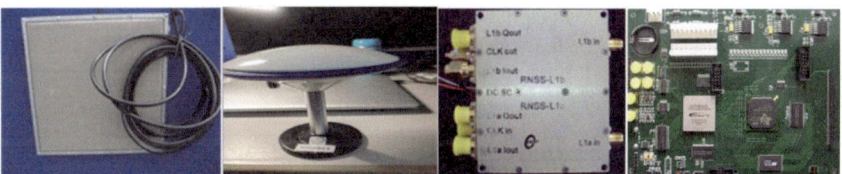

(a) LHCP Antenna (b) RHCP Antenna (c) Dual RF Front-End (d) Circuit Mainboard

Fig. 5.13 Hardware photos of the Reflected GNSS Signal reception and processing

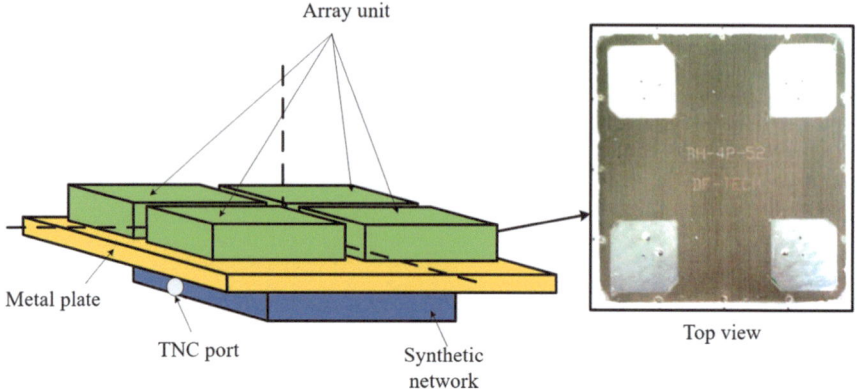

Fig. 5.14 Structure and top view of a LHCP antenna

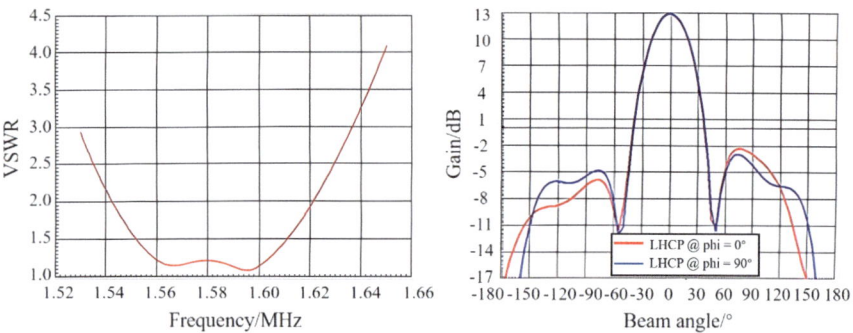

Fig. 5.15 VSWR and pattern of the LHCP antenna

The technical specifications of the left-hand array antenna are shown in Table 5.2.

(2) Right-hand antenna

The right antenna is selected according to the working frequency and bandwidth requirements of the navigation satellite system. To meet the signal reception under the airborne flight conditions, an aviation-grade GNSS antenna with model number S67-1575-39, can be used, with its technical specifications are shown in Table 5.3.

5.3 Hardware Receiver

Table 5.2 Technical specifications of the left-hand array antenna

Indicator name	Indicator
Working frequency	1575.42 MHz
Antenna form	4-Array
Antenna gain	12 dB
Polarization characteristics	LHCP
VSWR	≤1.8: 1
Power	3–5 V
Operating current	22 mA
Output impedance	50Ω
Operating temperature	−40 to + 80 °C
Connector type	TNC
Dimension	200 × 200 × 40 mm
Weight	≈1.2 kg
Height	<15000 m

Table 5.3 Technical specifications of Right-hand antenna

Indicator name	Indicator
Working frequency	1575.42 MHz
Antenna form	Microstrip
Antenna gain	3 dB
Polarization characteristics	RHCP
VSWR	2:1
Power supply	+4 to +24 V
Supply current	25 mA
Output impedance	50Ω
Operating temperature	−55 to + 85 °C
Output interface	TNC
Dimension	90 × 90 × 30 mm
Weight	≈102 g
Height	≤20000 m

5.3.2.2 Dual RF Front-End

Navigation satellite signals are deeply buried below the thermal noise level, requiring the front end of the receiver to have precision frequency conversion, amplification, filtering and gain control circuitry . The RF unit circuit structure design is shown in Fig. 5.16, where an on-chip phase-locked loop generates a 2456 MHz local oscillator signal, which is mixed with the received 1575.42 MHz signal to produce a signal of 880.58 MHz; this signal is mixed with a 927 MHz local oscillator signal to produce an analog IF signal of 46.42 MHz. The output level meets the requirement of 0dBm ±1 dB/50Ω.

The RF module also integrates a 10 MHz temperature-compensated crystal oscillator to provide a reference clock for the backend digital circuit, with a reference frequency stability of $\pm 5 \times 10^{-7}$. This module connects to the signal processing backend using an SMA (Small A Type) interface and implements physical shielding isolation, effectively reducing interference and noise between high-frequency analog and digital circuits, further optimizing signal quality. Automatic Gain Control (AGC) is an important component of the RF unit, maintaining a fixed output voltage level of the RF unit when the input signal voltage varies within a certain range. As a feedback control loop, its basic components include detection, low-pass filtering, and DC amplification.

The main technical parameters and indicators of the RF front-end include input frequency, input level, output frequency, noise figure, IF output amplitude and 3 dB bandwidth, etc., as shown in Table 5.4.

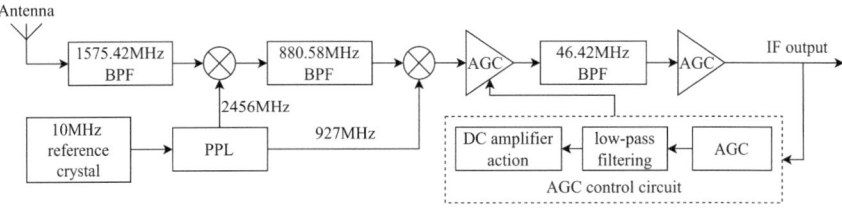

Fig. 5.16 RF front-end circuit structure

Table 5.4 Main technical parameters of the RF front-end

Technical parameters	Technical indicators	Technical parameters	Technical indicators
Input frequency	1575.42 ± 2MHz	Input level	−65 to −115dBm
Output frequency	46.42 MHz	Output level	1.5dBm
Phase noise	−65dBc@1kHz	Phase Noise	−75dBc@10kHz
Out-of-band rejection	@30 MHz:64 dB	Out-of-Band Rejection	@62 MHz:63 dB
Amplitude imbalance	0.5 dB	Phase Error	1.1°
Automatic gain control range	≥ 55dB	Noise Figure	≤10dB
3 dB bandwidth	5.6 MHz	Clock Amplitude	1.5 V
Operating voltage	5V ± 10%	Power Consumption	1W
Temperature range	−20 to +70°C	Control Port Definition	Asynchronous/LVTTL

Note LVTTL stands for Low Voltage Transistor-Transistor Logic

5.3 Hardware Receiver

Table 5.7 Efficiency analysis of direct signal processing software

number	file	call methods / count	Time Used / %
1	correlatorprocess.cpp	5	77.4
2	acqCA.cpp	2	10.6
3	fft.cpp	3	8.3
4	share_mem.cpp	2	0.1

5.3.2.3 Baseband Processing Circuit Mainboard

The circuit design of baseband processing in a reflected signal receiver is constantly evolving with the emergence of new devices and technologies. Although different research organizations have different design styles, but the basic principles remain consistent. Below, we highlight two typical types of baseband signal processing circuits developed by the author's research team.

(1) Signal processing circuit based on dedicated chip GP2010/GP2021

In this circuit, in this circuit, the processing of RF signals is accomplished using two GP2010 chips to form a dual RF front-end, while the correlator is made up of GP2021 chips. Figure 5.17 shows the Printed Circuit Board (PCB)diagram of its circuit mainboard [22].

The GP2010 is a complete RF front-end down-converter mixer, with an integrated phase-locked loop synthesizer, low-noise amplifier, mixer and A/D converter. The GP2010 converts the received 1575.42 MHz L1 carrier signal into a 4.309 MHz analog IF signal through three-stage of down-conversion. The chip generates a 5.714 MHz sampling clock, which, through a fourth-stage conversion, transforms

Fig. 5.17 Signal processing circuit based on GP2021

the intermediate frequency signal into a 1.405 MHz 2-bit digital signal, processed by the GP2021 correlator for correlation operations.

The GP2021 has 12 independent channels. In this receiver, each channel can select the input either as direct or reflected signal, and the demodulated digital signal by the carrier and local PRN code are correlated. The local PRN code is 1023 bits long, with controllable generation rate and phase, where different code phases correspond to different time delays. The coherent integration time used in the GP2021 is set to 1 ms.

(2) Signal processing circuit based on FPGA chip

As the GP2021 sets a fixed coherent integration time of 1 ms, the circuit in Fig. 5.17 uses two parallel GP2021 chips to provide 24 correlator channels, with a fixed signal delay unit of 0.5 code chips. To meet the application requirements in different scenarios and improve the output rate of digital intermediate frequency signals, the second version of the design replaced the dedicated RF chip GP2010 with a dual RF front-end. In the correlator structure, it enables more precise signal sampling and more correlator resources, and its PCB physical photo is shown in Fig. 5.18.

The baseband processing circuit based on FPGA mainly uses FPGA chips to achieve multi-channel dedicated correlator, the correlation operation between the received signal and the local signal, obtaining original observation of direct signals and interface transmission, among other tasks.

The FPGA chip is EP2S60F672C5 selected from Altera, designed and developed in Verilog hardware description language under Quartus II software environment. The

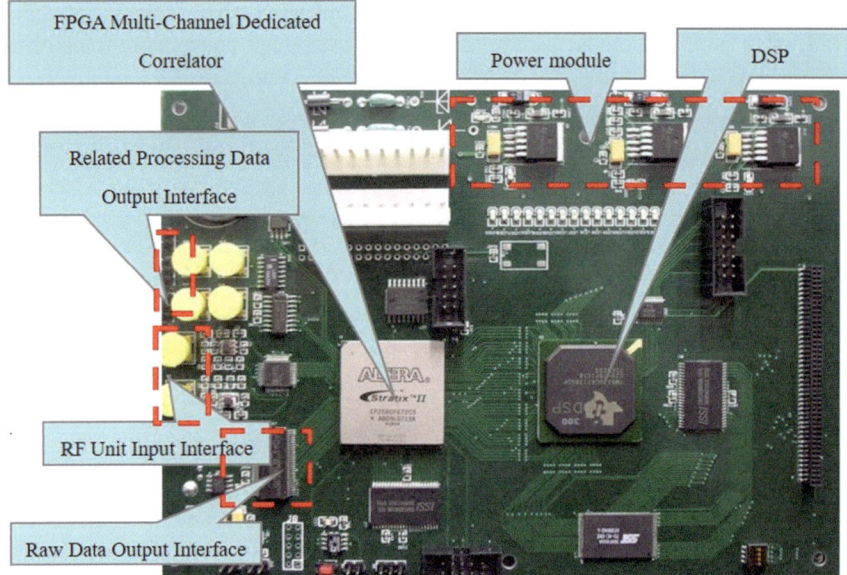

Fig. 5.18 Baseband processing circuit based on FPGA

5.3 Hardware Receiver

DSP chip used is TMSC320C6713 from Texas Instrument (TI), which configures parameters for the dedicated correlator of the FPGA, reads I/Q accumulation data, and performs tasks such as capturing, tracking, positioning calculations for direct signals, and delay control calculations for reflected signals.

5.3.3 Multi-Channel Correlation Unit

The reflected signals from the navigation satellite received by the left hand antenna are down-converted, quantized via A/D encoding and then enter into the reflected signal processing channel of the receiver. Here, the signal undergoes carrier stripping, is correlated with the corresponding delayed C/A code, and the coherent cumulative and incoherent cumulative operations are executed, resulting in outputs of complex correlation values and correlation power.

The reflected signal processing channel consists of a carrier generation module, a delayed C/A code generation module and a power calculation module. The structure of the reflected signal processing channel is shown in Fig. 5.19. Under the joint action of the Doppler control word and carrier control word, the reflected signal carrier generation module 1, 2, …, N generates local carriers with different Dopplers, which are multiplied by the digital IF reflected signal to implement carrier stripping. Correlation calculations are performed using local reflected signal C/A codes generated after multiple delay estimations to create complex time-delay-Doppler two-dimensional correlation values. These are then uploaded to the host computer via the corresponding interface after coherent and incoherent accumulation.

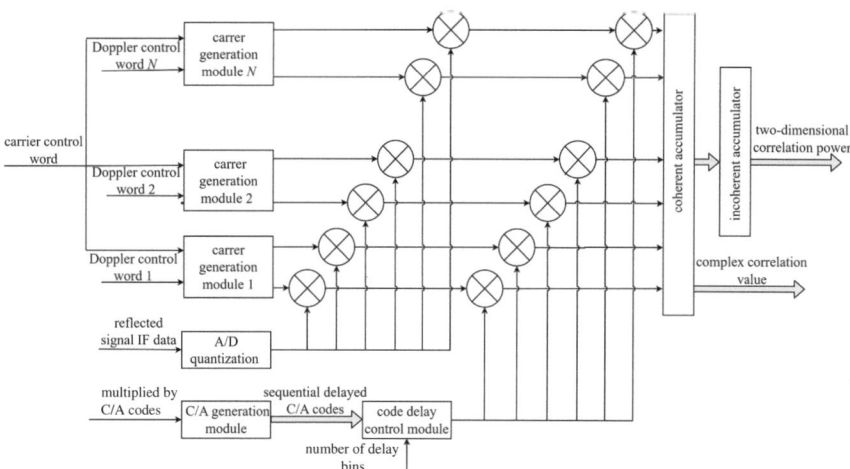

Fig. 5.19 Reflected signal processing channel structure. *Note* I—in-phase (cos), Q—quadrature (sin)

5.3.3.1 Carrier Generation Module

The carrier generation module in the reflected channel differs from the process in the direct channel, mainly because the dynamically outputted carrier control word needs to add or subtract a specified interval Doppler frequency shift control word, input together into an accumulator, to achieve the satellite carrier frequency tracked plus the specified resolution of Doppler shift. The specific implementation is shown in Fig. 5.20.

The carrier generation module is generated by a Direct Digital Synthesizer (DDS), described as follows: An ideal sine wave signal $S(t)$ can be represented as

$$S(t) = A\cos(2\pi ft + \phi) \tag{5.23}$$

After the amplitude A and the initial phase ϕ are determined, the frequency can be uniquely determined by the phase shift. Noting $\theta(t) = 2\pi ft$, after differentiation between the two ends we have $\frac{d\theta}{dt} = 2\pi f$, which is

$$f = \frac{\omega}{2\pi} = \frac{\Delta\theta}{2\pi\Delta t} \tag{5.24}$$

where, $\Delta\theta$ is the phase increment within a sampling interval Δt, the sampling period $\Delta t = \frac{1}{F_{CLK}}$, F_{CLK} is the input sampling clock frequency, so the above equation can be rewritten as

$$f = \frac{\Delta\theta F_{CLK}}{2\pi} \tag{5.25}$$

It can be seen that different frequency outputs can be achieved by controlling $\Delta\theta$. Assuming that $\Delta\theta = \frac{F_{CW} 2\pi}{2^L}$ is controlled by a carrier control word of a length L, a change in F_{CW} will result in a different frequency output f. In other words, the principle equation of DDS can be described as

$$f = f_R \cdot F_{CW} \tag{5.26}$$

where, f_R is the frequency resolution of DDS, defined as the output frequency when the control word $F_{CW} = 1$ is used.

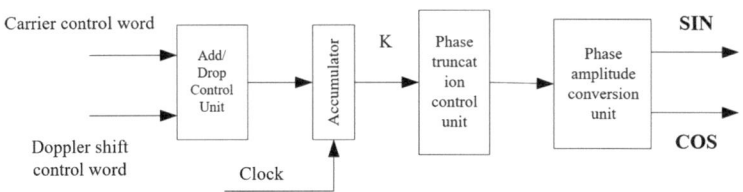

Fig. 5.20 Reflected signal carrier generation module

5.3 Hardware Receiver

$$f_R = \frac{F_{CLK}}{2^L} \quad (5.27)$$

$$f_0 = \frac{(F_{CW} \pm F_{dp})}{2^L} F_{CLK} \quad (5.28)$$

where f_0 is the frequency to be obtained and F_{dp} is the divided Doppler shift control word.

For example, if the carrier NCO uses a 27-bit word length and the input sampling clock frequency is $f_{CLK} = 20.456 \text{MHz}$, the frequency resolution is $f_R = \frac{F_{CLK}}{2^L} = \frac{20.456 \times 10^6}{2^{27}} = 0.15240908 \text{Hz}$.

The Reflected Signal Carrier Generation Module generates two orthogonal carriers with an additional Doppler shift, i.e., a sine wave and a cosine wave with a phase difference of $\frac{\pi}{2}$.

5.3.3.2 Delayed C/A Code Generation Module

Delayed C/A code in the reflection channel can be generated in two different ways; one is to generate the C/A code by directly multiplexing the corresponding direct-reflection channel, and then obtain the delay of the C/A code of the reflected channel through a shift register; another method is to design the delayed C/A code generator in the reflection channel, generating the delayed C/A code required by the reflection channel under DSP control. Each method has its advantages and disadvantages, and the choice depends on the applications.

(1) Shift register shift to generate delayed C/A code

According to certain reflected signal decision criteria, the direct channel corresponding to the reflected channel is selected, the C/A code generated by the direct channel is multiplexed and input into the reflected channel, and different delays C/A code are generated through shift registers.

The rough calculation formula for the delay distance of the reflected signal with respect to the direct signal is given in Eq. (5.29) as

$$\rho_{r-d} = (2h + h_0) \cdot \sin \theta = N\tau \quad (5.29)$$

Here, ρ_{r-d} is the delay distance; h is the height of the receiver relative to the reflecting surface; h_0 is the vertical distance between the left-handed antenna and the right-handed antenna; θ is the altitude angle of the satellite; τ is the distance corresponding to the delay of one code-piece, and the length of the code chip $1\mu s$ corresponds to 300 m; N is the number of code chips delayed by the reflected signal relative to the direct signal.

The number of shift registers depends on the number of delay chips N N and the code chip interval Δ, determined by the maximum altitude of the receiver. For example, if the satellite elevation angle is 80°, with the reveiver altitude of 6000 m

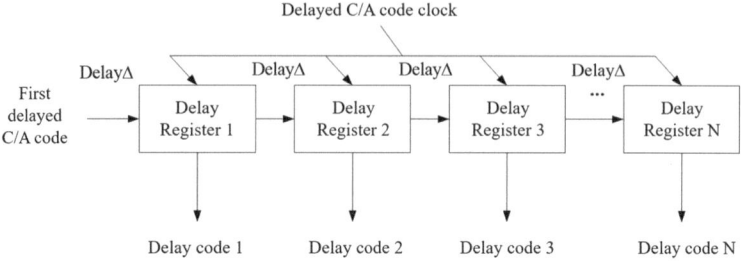

Fig. 5.21 Principle diagram for generating delayed C/A code in the reflection channel

and the shift clock of twice the C/A code rate (1.023 MHz ×2 in this case), the delay distance of the reflected signal relative to the direct signal is about 11818 m, and the number of delayed code chips N is about 40. If the code chip interval is 0.5 code chips, then 80 shift registers are needed to cover the usage altitude of the receiver platform.

The DSP calculates the number of code chip delays N estimated for the receiver from the specular reflection point, thereby the Nth delay code in the shift register set is selected to generate the first delayed C/A code required for the reflection channel. The generation of the delayed C/A code in the reflection channel is achieved through the delay operation of the shift register, generating a series of C/A code delayed successively, under driving by the delayed C/A code clock, as shown in Fig. 5.21.

(2) Delayed C/A code generator

The design of the delayed C/A code generator does not require the reuse of C/A codes from the direct channel but directly generates delayed C/A codes under the control of the DSP in the reflection channel, as shown in the block diagram in Fig. 5.22.[35].

The navigation positioning solution obtained by the reflected signal receiver contains the altitude information of the receiver, the elevation angle of the satellite tracked by the direct channel, the C/A code phase SV_{phase} and the satellite number. Based on this information, the delay distance of the satellite's reflected signal relative to the direct signal can be solved, thereby obtaining the delay time. The processing

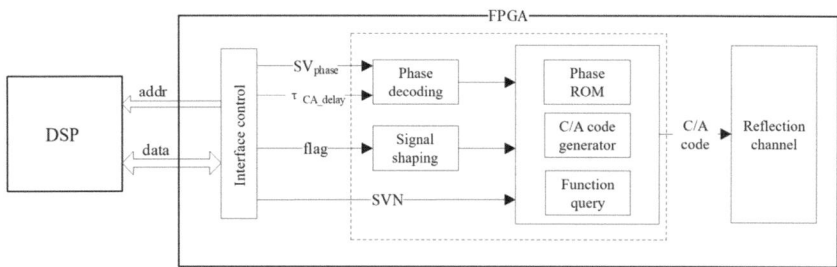

Fig. 5.22 Delayed C/A code generator

5.3 Hardware Receiver

results of the DSP, through the interface, are passed to the delayed C/A code generator (the FPGA part in Fig. 5.22), and its phase decoder translates τ_{CA_delay} into the number of phasesNum$_{phase}$, delayed by the code chip, as follows:

$$M = SV_{phase} - Num_{phase} \tag{5.30}$$

$$P = \begin{cases} M & M \geq 0 \\ Z - (-M - 1) & M < 0 \end{cases} \tag{5.31}$$

where P is the C/A code phase of the reflected signal, and Z is the maximum C/A code phase value (the starting phase value starts at 0 and for a 1023-chip code is 1022).

The delayed C/A code generator contains two 10-bit shift registers, G1 and G2, each of which can generate a sequence of length 1023. Both The G1 and G2 registers have 1023 states, stored in a Read-Only Memory (ROM) table with a width of 20 bits and a depth of 1023 according to their address. Based on the phase address value passed by the DSP, the corresponding initial register values (reg1, reg2) are read from the ROM and assigned to the shift registers G1 and G2 to generate the delayed C/A code.

When the flag bit for generating a C/A code (e.g., set to *flag*) is set high level, the G1 and G2 registers are loaded with the new reg1 and reg2 values, and the C/A code generator immediately stops the current code generation operation and re-generates a new C/A code based on the newly set G1 and G2 values. At the 1023rd cycle, when the *flag* value is 0, the G1 and G2 values are all set to 1, also regenerating the C/A code. The principle diagram for generating the delayed C/A code is shown in Fig. 5.23.

(3) Comparison of the two design methods

Delayed C/A code is generated by the delay operation of shift registers, which is simple to program and easy to implement. However, the number of registers must be adjusted according to changes in the receiver's design altitude, which makes it inflexible. When the design altitude is too high, using a large number of shift registers may even become unfeasible during FPGA synthesis and routing. Therefore, this method is suitable for shore-based reflected signal receiving and processing scenarios where the receiver height does not change significantly.

The design of Delayed C/A code generator does not need to consider the application altitude of the receiver, only requires the DSP processor to calculate the delay path of the C/A code and assign it to the delayed C/A code generator, and subsequently generates the delayed C/A code required by the reflection channel. This method is suitable for airborne and satellite applications where the height of the receiver varies greatly, reducing the consumption of hardware logic resources.

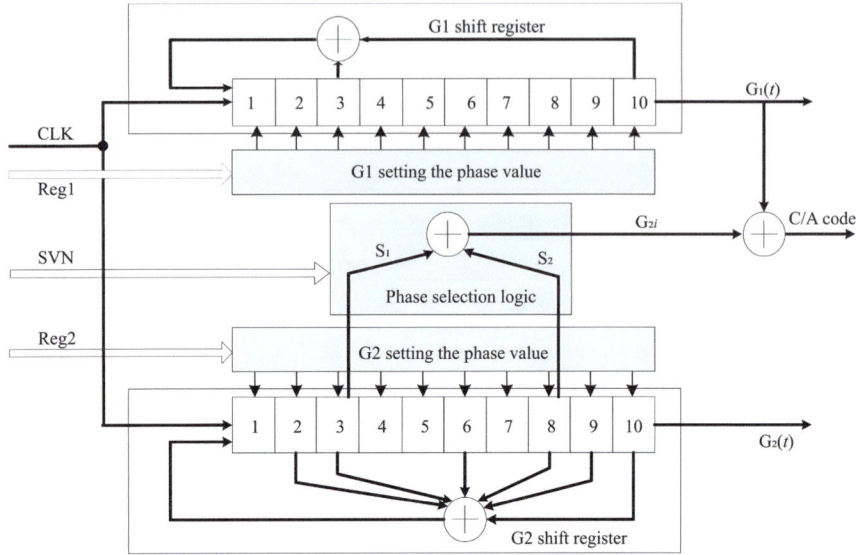

Fig. 5.23 Principle diagram for generating the delayed C/A code

5.3.3.3 Complex Correlation Value Output Module

After correlation operation of GNSS reflection signals, like carrier demodulation and delayed C/A code by the reflection channel, a 1 ms coherent accumulation is performed to get $N \times M$ two-dimensional complex correlation value (of which N is the number of units of Doppler delay, and M is the number of units of code delay). Under the control signal, the data is stored in the corresponding RAM by the serial-to-parallel conversion module and written into FIFO in frames of 8 bits width per frame for buffering by a fast clock, while the next millisecond correlation values arrive. The data are then uploaded to the host computer via the USB interface.

5.3.3.4 Correlation Power Calculations Module

The implementation method of the reflection channel correlation power calculation module is shown in Fig. 5.24. The 1 ms I and Q complex correlation values output by each correlation channel are squared, summed ($I^2 + Q^2$) and then accumulated. The two-dimensional correlation power values of the reflection signals are obtained under the control of the integral-clearing signal (e.g., with a 1 s period).

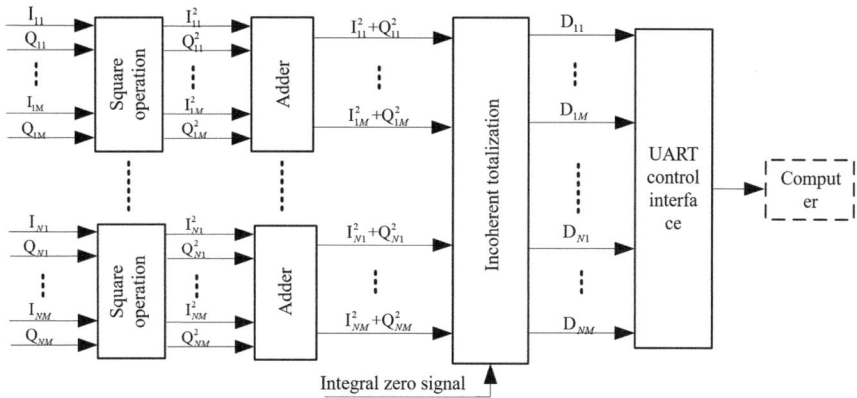

Fig. 5.24 Correlation power calculation block diagram

5.3.4 Multi-channel Control Unit

Data interchange between DSP and FPGA is achieved through the External Memory Interface (EMIF), which offers the advantages of large addressing space and convenient asynchronous timing configuration. It facilitates data exchange and command communication between the two processors. The EMIF interface also provides timing for state machine programming, thereby completing DSP read and write operations to FPGA under different read and write timing control signals. The read operation involves reading observations from the 12 direct channel correlators within the FPGA, such as carrier phase and code phase, as well as the I and Q values of the lead, prompt, and lag branches, and the status of the carrier and code loops. There are three types of write operations: direct channel write operations, reflection channel write operations, and serial port write operations. For the direct channel, the results from the phase detectors are used to dynamically configure. the carrier NCO and code NCO of the channel through the carrier control word and code control word via the interface. After positioning, according to certain satellite selection criteria (such as the highest elevation criterion, etc.), the direct channel number of a tracked satellite is assigned to the channel multiplexer, and the estimated delay distance is converted into code chip delay numbers passed to the reflection channel to select the appropriate code delay. The serial port write operation transmits navigation positioning information (such as longitude, latitude, and altitude), channel information (satellite number, elevation angle, and azimuth), and status information (search, capture, and tracking) from the direct channel to the serial port for uploading, displaying, and storage.

The state machine for DSP read and write operations consists of 7 states, with specific state transitions shown in Fig. 5.25. Write operations are completed in the WRITE_strobe, and read operations in the READ_strobe. Specific read and write contents require a set of address decoders for addressing, with the address in the DSP matching the corresponding address in the FPGA.The WRITE_hold and READ_ hold states ensure correct timing for write-after-read and read-after-write operations, transitioning to write wait and read wait states, respectively. Moreover, due to write latency, writing is done in a single operation, while reading can be repeated.

CE in Fig. 5.25 is the chip select signal, AOE is the output enabling signal, AWE is the write select signal, and ARE is the read select signal, all of which are active low; The symbol "&&" denotes the conjunction operation of the signal; and "!" indicates the inverse operation of the signal.

The USB control interface handles the collection and storage of digital IF data for both direct and reflected signals, which is completed by the FPGA chip and the USB interface chip. The main functions include A/D quantized data decoding, packing, data buffering and timing control. A/D quantized data for both direct and reflected paths, each 2 bits, are combined into a byte and written into FIFO buffers. To prevent data loss when the USB interface chip is busy, a large capacity FIFO memory can be expanded in the FPGA for data buffering before writing to the USB chip.

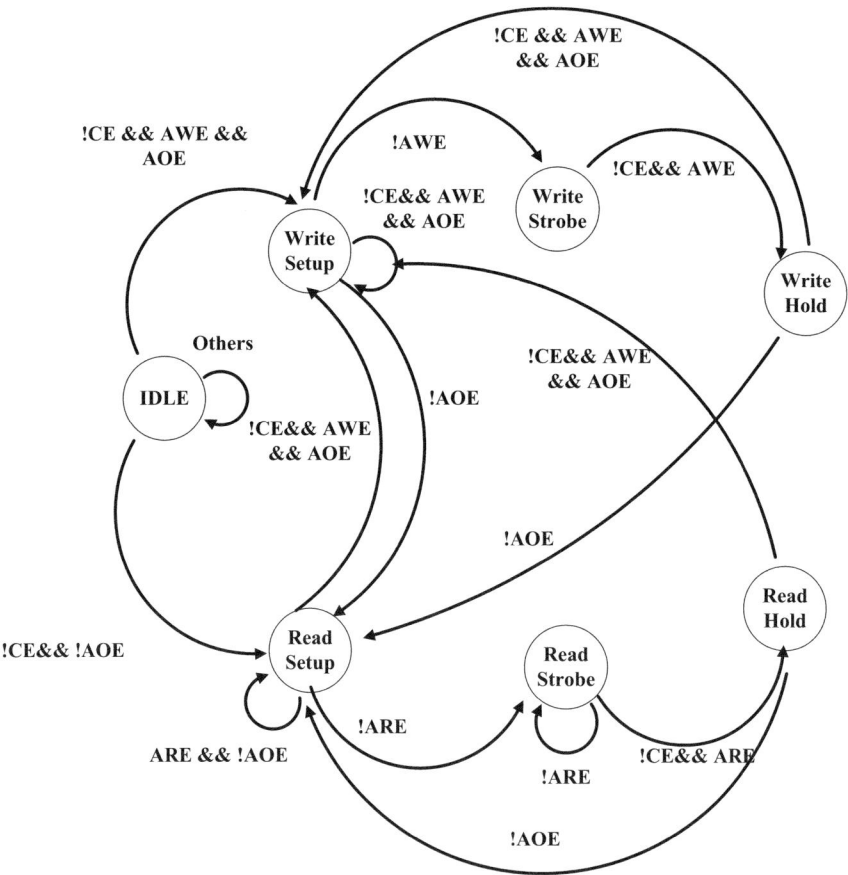

Fig. 5.25 DSP data read and write operation status transition diagram

5.3 Hardware Receiver

For example, the acquired direct signal is $A'_2 A'_1 A_2 A_1 = 1011$, the reflected signal is $B'_2 B'_1 B_2 B_1 = 0100$, and the packed data is $B'_2 B'_1 A'_2 A'_1 B_2 B_1 A_2 A_1 = 01100011$.

The timing control module is used to control data read from the FIFO and written to the USB chip, monitoring the status of the USB chip as well as the status of the internal FIFO of the FPGA, and following the FIFO interface's read/write timing to write into the USB FIFO, thus achieving USB uplink data transmission. Its state transition diagram is shown in Fig. 5.26.

- IDLE: The receiver starts, and the state machine transitions to State1.
- State1: When the USB FIFO is not full and the FPGA on-chip FIFO has unread data, it transitions to State2; otherwise, it remains in State1.
- State2: Drives the bus to write data into the USB FIFO. After writing the data, it transitions back to State1.

Direct channel navigation positioning solutions, channel status, and satellite status information are uploaded via two independent serial ports (i.e., serial port 1 and serial port 2) due to their minimal temporal correlation with the reflection channel's output correlation power values. Data is uploaded and stored every 1 s for easy playback and analysis later. Since the direct and reflection channels are processed differently, the data structures output by the serial ports also differ.

The direct channel primarily perform capture and tracking operation of for satellite direct signals, and transmitting observation results to the DSP for loop control, calculation of positioning results, and extraction satellite status. Serial port 1 uploads the DSP processing results and the direct-channel status using a specific data structure. The specific processing is as follows: after satellite tracking, the channel status can display information such as satellite elevation angle, azimuth, and Doppler shift; after obtaining positioning solutions, the DSP writes the positioning information, processing status of the 12 channels, and satellite status into registers. After data writing is completed, a flag bit sends the data in parallel to the caching module for parallel-to-serial conversion, converting 32-bit data into 8-bit serial data for transmission to the serial transmission module. After data transmission is completed, the serial transmission module sends a clear flag bit, clearing the DSP's previous flag bit, and waits for the DSP's next flag bit arrival. Serial port 1's baud rate is set to 115200bps, meeting the data transmission rate requirement, with the data transmission format shown in Fig. 5.27.

The processing results of the reflection channel, i.e. correlation power value, are uploaded via serial port 2. The specific processing is as follows: The data send flag is triggered by the reset signal of the correlator, indicating that processing for the current second is complete and the data is ready for transmission. The flag bit is

Fig. 5.26 USB Interface state transition diagram

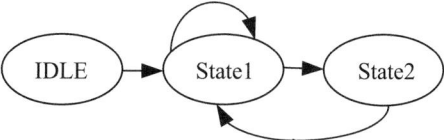

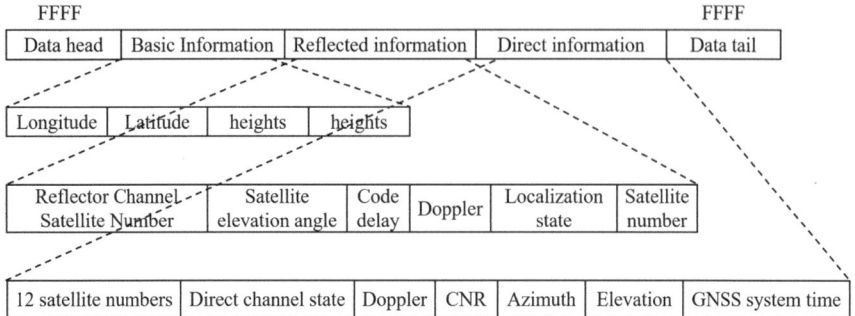

Fig. 5.27 Data transmission format

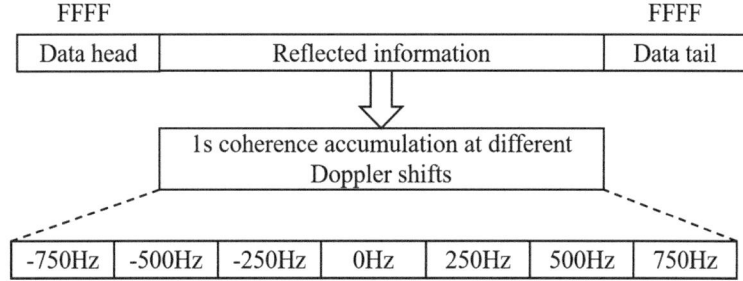

Fig. 5.28 Data format for serial port 2

active on the rising edge. When the rising edge arrives, the transmission module begins transmitting the buffered data until completed. In addition to containing the reflected signal's correlation power values, the data format also includes the direct signal's correlation values for energy comparison, as shown in Fig. 5.28.

5.4 Software Receivers

5.4.1 Basic Structure

With the development of integrated circuit technology and the generational upgrades of processors in the computers, A software receiver technically processes IF data from the RF front-end using software to achieve signal acquisition, tracking, navigation solution, and other functions. Its architecture is shown in Fig. 5.29. Instead of using the downconverter, the RF signal can be sampled directly. The implementation of the software not only has the advantage of cost saving, but also has great flexibility in the development process, which reflects the irreplaceable superiority in the algorithm

5.4 Software Receivers

function performance test, the realization of different configurations of the system, and the research of new signal processing algorithms.

Correspondingly, the architecture of a reflected signal software receiver is shown in Fig. 5.30. Compared to the direct signal processing, the reflected signal software receiver exhibits the following two distinct differences:

(1) For tracked satellites, different channels need to be allocate for their corresponding reflected signals, and the processing of the reflected channels needs to be done in conjunction with the corresponding direct channels.

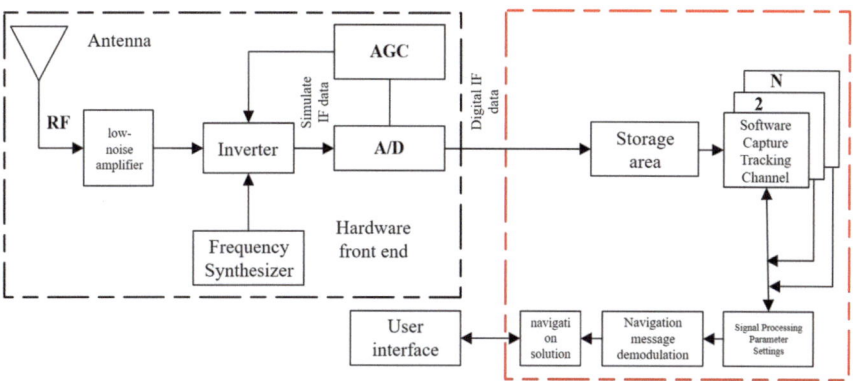

Fig. 5.29 Software receiver architecture

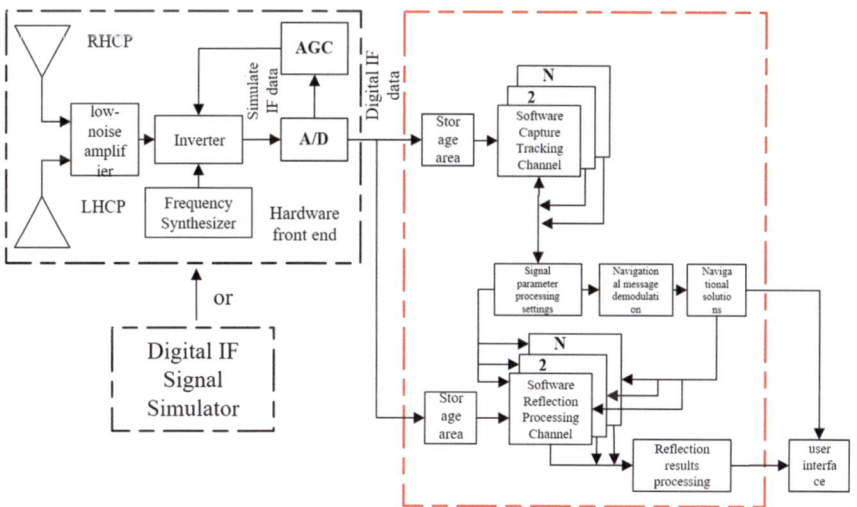

Fig. 5.30 Reflected signal software receiver architecture

The processing of reflected signals is also a correlation process, hence adding reflected signal correlation branches in the signal processing.

5.4.2 Processing Flow

Direct signal processing includes basic modules such as signal capture, tracking, and position solution algorithms, which are relatively rich and mature. Interested readers can refer to the contents of Chapter 5 in literature [27]. the reflected signal processing algorithm is similar to that implemented in the hardware receiver, and its flowchart is shown in Fig. 5.31.

The initialization includes reflection channel initialization, correlator initialization, and correlation data table initialization, among others. Based on the direct signal's positioning, the tracking results of the tracked direct channel are read, including satellite number, code phase, and carrier Doppler information. The corresponding reflection channel for the tracked satellite is then opened, and reflected signal data is read. Local code and local carrier phases for the reflection channel are estimated. By sliding the local Doppler and local spreading code, the correlation power of the reflected signal is calculated for each Doppler and code phase bin. Depending on the application requirements of the reflected signal, the I and Q branch correlation values (or complex correlation values) of the reflected signal can also be directly output.

5.4.3 Software Functionality

Figure 5.32 shows the main interface of 12-channel reflected signal software receiver, which can process GPS L1/BD B1 collection data using different setup parameters. The software runs on an Intel Pentium dual-core processor at 1.6 Ghz, with 2Gbyte of memory, and the operating system is Windows XP.

The main interface is divided into six different functional areas:

(1) Receiver platform information: Located at the upper left, displaying the receiver platform's location information (longitude, latitude, and altitude) and speed information.
(2) Accuracy factor information: Located at the upper right, showing the accuracy factors of the satellite distribution, including horizontal factor, vertical factor, geometric factor, and time factor.
(3) Direct channel information: Located in the middle left, listing the satellite number, azimuth, elevation angle, Doppler, signal-to-noise ratio, and channel status for the 12 direct channels.
(4) Sky view: Located in the middle right, visually displaying the distribution of tracked satellites in the sky.

5.4 Software Receivers

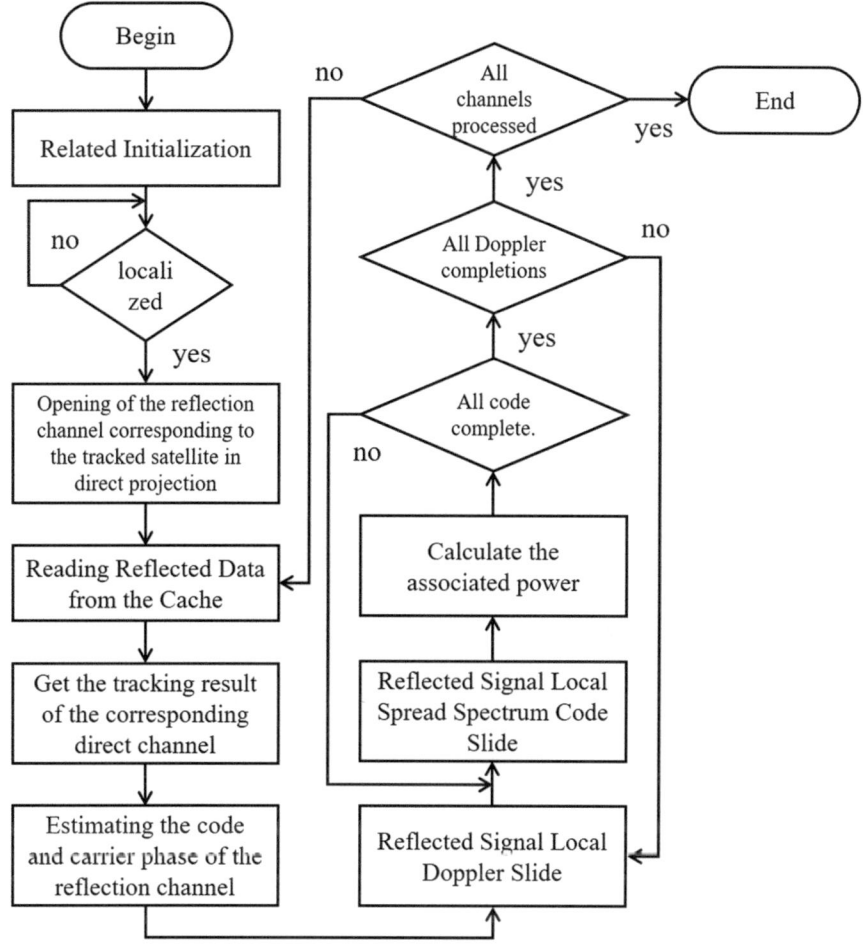

Fig. 5.31 Reflected signal processing flow

(5) Reflection channel information: Located at the lower left, listing the satellite number, elevation angle, path delay, Doppler, and correlation power values at certain code delays for the 12 reflection channels.

(6) Correlation power graphical information: Located at the lower right, providing a graphical relationship between code delay, Doppler, and correlation power for the selected reflected satellite, dynamically changing with calculation results.

The menu bar contains baseband settings, which are used to set the sampling rate, digital IF frequency, number of quantization bits and channels allowing the software receiver to be compatible with different types of hardware front-ends; Environmental parameter settings are mainly used to set the height of the reflection surface (to solve the path difference between the reflected signal and the direct signal),

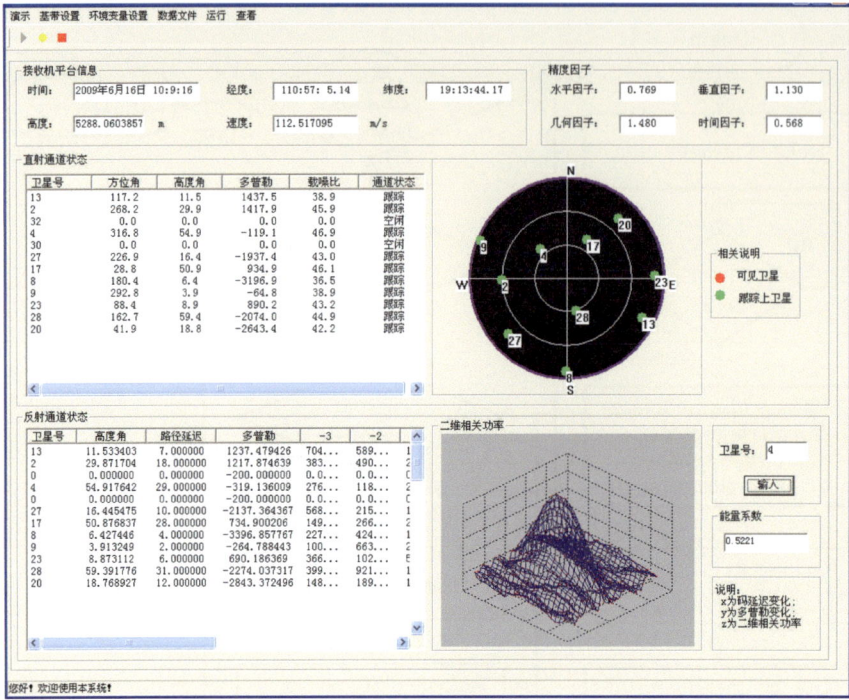

Fig. 5.32 Reflected signal software receiver main interface

the code phase (or code delay) resolution, and the Doppler resolution required for processing the reflected signal. The data file option is used to read stored raw data for post-processing, computing the reflected signal's correlation power. under different parameter conditions, facilitating the inversion requirements of different application fields (such as measuring sea surface height, sea surface wind, and soil moisture, etc.).

5.4.4 Efficiency Analysis of Implementation

In the signal processing modules of GNSS receivers, the required time and space resources are not the same, which is similar in both hardware-implemented and software-implemented receivers. For software implementations utilizing personal computers, processing time is the most important factor affecting real-time performance, especially for time-delay Doppler mapping receivers, which must capture and track direct signals as well as solve for the correlation power of reflected signals. For direct signal processing, software computations and storage are concentrated in the capture and correlator modules. Table 5.5 lists the main functional modules that were run and the time they occupied, showing that the correlator-related computing process (correlatorProcess.cpp) consumes the majority of the processing time and is the main factor limiting the real-time performance of the entire receiver. Capture (acqCA.cpp), fast Fourier transform calculations (fft.cpp), and memory sharing (share_mem.cpp) respectively occupy 10.6%, 8.3%, and 0.1% of the time.

The processing of reflected signals primarily involves correlation calculations under code chip sliding and Doppler sliding. When the Doppler sliding resolution is 100 Hz, with a range of -1 kHz to $+1$ kHz, there are 21 Doppler sliding operations. If the code chip sliding resolution is set to 1/16, with 10 sliding operations, the total number of two-dimensional bins is 210. Therefore, each sampling point requires 210 table lookups and 420 accumulations. For correlators in the software receiver, with 2 ms data as a computational unit, and approximately 32,735 sampling points, each correlation operation involves 6,874,350 table lookups and 13,748,400 accumulations, which is a significant computational load. This is an important consideration in the reflected signal software receiver.

In terms of storage, by reusing the lookup tables from direct signal processing , only space allocation for 420 correlation values and necessary intermediate variables is needed, resulting in relatively modest additional storage requirements.

5.4.5 *Real-Time Solution Research*

The flexibility of software receivers has attracted increasing attention, and many researchers have proposed effective methods to address their lack of real-time performance. Here we briefly introduce three aspects for the reader's consideration:

(1) Reducing sampling rate

In each accumulation loop, only one sampling point can be operated on, then the higher the sampling frequency, the more computationally intensive it is, and the more time is consumed by the table look-up and accumulation statements. According to the bandpass signal sampling theorem

$$f_s = \frac{4f_0}{2n+1} \quad (5.32)$$

where $f_0 = (f_H + f_L)/2$, n are taken as the largest integers satisfying ($f_s \geq 2BB = f_H - f_L$). The sampling rate can be reduced to as low as twice the bandwidth, and the corresponding amount of operations can be reduced to half of the original.

(2) Optimizing algorithm

B. M. Ledvina used a bit-by-bit parallel algorithm in a 12-channel real-time GPS L1 software receiver [28] to store multiple samples in 32-bit words in symbol and magnitude bits. For a 2-bit quantized signal, 32 sampling points are stored in 2 32-bit words: one for sign bits and one for amplitude bits. The corresponding local carrier signal and spreading code use the same storage method, then binary XOR operations are performed between stored 32-bit words to obtain the final I and Q values of each branch. For GPS L1 signals, the general algorithm requires 6 multiplications and 4 additions for 2 sampling points, whereas this algorithm needs only 6 XORs and 52 logical additions for 32 sampling points, with accumulation performed within the 32-bit words. Its calculation speed can be increased nearly 4 times [29].

(3) Optimizing procedures

Compared to hardware, software struggles with parallel processing, which is a key reason for the low efficiency of software correlators. In direct signal processing, since each satellite signal has different Doppler and code offsets, each channel must be equipped with a corresponding correlator, each containing multiple correlation branches. For reflected signal processing, different code chip and Doppler two-dimensional sliding can only be implemented serially, consuming a significant amount of time. Another limitation to the running speed of software correlators is the use of high-level programming languages, which makes direct calls to hardware-related memory and I/O ports difficult, and efficiency is somewhat limited by the compilation system.

MMX's Single Instruction Multiple Data (SIMD) stream operations can pack 8 8-bit, 4 16-bit, or 2 32-bit integers into one MMX register, then perform parallel operations on all data within the register. SIMD instructions are highly suited for parallel processing, able to execute multiple branch carrier stripping, code stripping, and accumulation operations simultaneously, significantly increasing execution speed [30] The bit-wise parallel algorithm described in the previous section can be implemented using SIMD operations.

In addition to spatial parallelism achieved through SIMD operations, the development of multi-core processors provides conditions for temporal parallel computing. For software correlator calculations, multi-core and multi-threaded programming will also be an effective way to increase speed.

5.5 Summary

This chapter introduces the general model of a reflected signal receiver, detailing the methods for processing reflected signals from their discrete form, through multi-channel correlation processing algorithms, to calculating the specular reflection point and methods for improving the signal-to-noise ratio. Based on these methods, it discusses two common approaches for processing reflected signals: hardware receivers and software receivers, elaborating on their structures and implementation methods from their respective perspectives.

References

1. Treuhaft RN, Lowe ST, Zuffada C, et al. 2-cm GPS altimetry over Crater Lake [J]. Geophys Res Lett. 2001;28(23):4343–6.
2. Zuffada C, Fung A, Okolicanyi M, et al. The collection of GPS signal scattered off a wind-driven ocean with a down-looking GPS receiver: polarization properties versus wind speed and direction [C]. In: IGARSS 2001. Scanning the present and resolving the future. Proceedings. IEEE 2001 international geoscience and remote sensing symposium (Cat. No. 01CH37217). IEEE;2001(7). pp. 3335–3337.
3. Martín-Neira M, Caparrini M, Font-Rossello J, et al. The PARIS concept: an experimental demonstration of sea surface altimetry using GPS reflected signals [J]. IEEE Trans Geosci Remote Sens. 2001;39(1):142–50.
4. Rius A, Aparicio J M, Cardellach E, et al. Sea surface state measured using GPS reflected signals [J]. Geophys Res Lett. 2002;29(23):37-1–37-4.
5. Ruffini G, Soulat F, Caparrini M, et al. The eddy experiment: accurate GNSS-R ocean altimetry from low altitude aircraft [J]. Geophys Res Lett. 2004;31(12).
6. Rivas MB, Martin-Neira M. GNSS reflections: first altimetry products from bridge-2 field campaign [C]. In: Proceedings of NAVITEc, 1st ESA workshop on satellite navigation user equipment technology. 2007:465–479.
7. Lowe ST, Kroger P, Franklin G, et al. A delay/doppler-mapping receiver system for GPS-reflection remote sensing [J]. IEEE Trans Geosci Remote Sens. 2002;40(5):1150–63.
8. Lowe S T, Zuffada C, Chao Y, et al. 5-cm-Precision aircraft ocean altimetry using GPS reflections [J]. Geophys Res Lett. 2002;29(10):13-1–13-4.
9. Heckler GW, Garrison JL. Architecture of a reconfigurable software receiver [C]. In: Proceedings of the 17th international technical meeting of the satellite division of the institute of navigation (ION GNSS 2004). 2004;947–955.
10. Elfouhaily TS, Thompson DR, Linstrom L. Delay-doppler analysis of bistatically reflected signals from the ocean surface: theory and application [J]. IEEE Trans Geosci Remote Sens. 2002;40(3):560–73.
11. You H, Garrison JL, Heckler G, et al. The autocorrelation of waveforms generated from ocean-scattered GPS signals [J]. IEEE Geosci Remote Sens Lett. 2006;3(1):78–82.
12. Nogués O, Sumpsi A, Camps A, et al. A 3 GPS-channels Doppler-delay receiver for remote sensing applications[C]. In: IGARSS 2003. 2003 IEEE international geoscience and remote sensing symposium. Proceedings (IEEE Cat. No. 03CH37477). IEEE;2003(7):4483–4485.
13. Akos D. Software radio architectures for GNSS [C]. In: Proceeding of 2nd ESA workshop satellite navigation user equipment technologies (NAVITEC);2004. pp. 8–10.
14. Dunne S, Soulat F. A GNSS-R coastal instrument to monitor tide and sea state [C]. In: GNSS reflection workshop. Surrey University;2005. pp. 9–10.
15. Gleason S, Hodgart S, Sun Y, et al. Detection and processing of bistatically reflected GPS signals from low earth orbit for the purpose of ocean remote sensing [J]. IEEE Trans Geosci Remote Sens. 2005;43(6):1229–41.
16. Brown A K. Remote sensing using bistatic GPS and a digital beam-steering receiver [C]. In: Proceedings. 2005 IEEE international geoscience and remote sensing symposium, 2005. IGARSS'05. IEEE;2005(1). p. 4.
17. Cardellach E, Ruffini G, Pino D, et al. Mediterranean balloon experiment: ocean wind speed sensing from the stratosphere, using GPS reflections [J]. Remote Sens Environ. 2003;88(3):351–62.
18. Kelley C. OpenSource GPS open source software for learning about GPS [C]. In: Proceedings of the 18th international technical meeting of the satellite division of the institute of navigation (ION GNSS 2005);2005. pp. 2800–2810.
19. Helm A, Beyerle G, Reigber C, et al. The OpenGPS receiver: remote monitoring of ocean heights by ground-based observations of reflected GPS signals [C]. In: GNSS reflection workshop. Surrey University;2005. pp. 9–10.

20. Nogués-Correig O, Galí EC, Campderrós JS, et al. A GPS-reflections receiver that computes Doppler/delay maps in real time [J]. IEEE Trans Geosci Remote Sens. 2006;45(1):156–74.
21. Liu W, Beckheinrich J, Semmling M, et al. Coastal sea-level measurements based on gnss-r phase altimetry: a case study at the onsala space observatory, sweden [J]. IEEE Trans Geosci Remote Sens. 2017;55(10):5625–36.
22. Zuffada C, Zavorotny V. Coherence time and statistical properties of the GPS signal scattered off the ocean surface and their impact on the accuracy of remote sensing of sea surface topography and winds[C]. In: IGARSS 2001. Scanning the present and resolving the future. Proceedings. IEEE 2001 international geoscience and remote sensing symposium (Cat. No. 01CH37217). IEEE;2001(7). pp. 3332–3334.
23. You H, Garrison J L, Heckler G, et al. Correlation time analysis of delay-Doppler waveforms generated from ocean-scattered GPS signals [C]. In: IGARSS 2004. 2004 IEEE international geoscience and remote sensing symposium. IEEE;2004(1). pp. 428–431.
24. Strassle C, Megnet D, Mathis H, et al. The squaring-loss paradox [C]. In: Proceedings of the 20th international technical meeting of the satellite division of the institute of navigation (ION GNSS 2007);2007. pp. 2715–2722.
25. Tsui JBY. Fundamentals of global positioning system receivers: a software approach [M]. John Wiley & Sons;2005.
26. Van Diggelen FST. A-gps: Assisted gps, gnss, and sbas [M]. Artech house, 2009.
27. Gleason S, Gebre-Egziabher D, Egziabher D G. GNSS applications and methods [J]. 2009.
28. Ledvina BM, Powell SP, Kintner PM, et al. A 12-channel real-time GPS L1 software receiver1 [C]. In: Proceedings of the 2003 national technical meeting of the institute of navigation. 2003:767–782.
29. Tian J, HongLei Q, JunJie Z, et al. Real-time GPS software receiver correlator design [C]. In: 2007 second international conference on communications and networking in China. IEEE. 2007:549–553.
30. Baracchi-Frei M, Waelchli G, Botteron C, et al. Real-Time GNSS software receiver: challenges, status, and perspectives [J]. GPS world. 2009;20(ARTICLE):40–47.

Open Access This chapter is licensed under the terms of the Creative Commons Attribution-NonCommercial-NoDerivatives 4.0 International License (http://creativecommons.org/licenses/by-nc-nd/4.0/), which permits any noncommercial use, sharing, distribution and reproduction in any medium or format, as long as you give appropriate credit to the original author(s) and the source, provide a link to the Creative Commons license and indicate if you modified the licensed material. You do not have permission under this license to share adapted material derived from this chapter or parts of it.

The images or other third party material in this chapter are included in the chapter's Creative Commons license, unless indicated otherwise in a credit line to the material. If material is not included in the chapter's Creative Commons license and your intended use is not permitted by statutory regulation or exceeds the permitted use, you will need to obtain permission directly from the copyright holder.

Chapter 6
Marine Remote Sensing Applications

The ocean covers about 70% of the Earth's surface, and oceanographic physical parameters such as sea surface wind field, mean sea level, wave height, seawater salinity have a great impact on human production and life. Traditionally, oceanographic physical parameters mainly come from buoys, ships and sporadic atmospheric soundings, which provide limited data volume, spatial and temporal resolution, and lack systematic completeness. With the development of microwave remote sensing technology, especially the rapid development of satellite remote sensing technology, methods of obtaining oceanographic physical parameters have become more diverse. Microwave scatterometers, microwave radiometers, radar altimeters and Synthetic Aperture Radar (SAR) are commonly used equipment today among the satellite-based ocean remote sensing.

Since the mid-1970s, the United States and other countries have launched a series of ocean satellites (such as Seasat, Topex and Jason, etc.), which utilize onboard radar altimeters to continuously transmit radar pulses to the Earth and to receive sea surface echoes in order to extract information on physical parameters of the ocean.

However, radar altimetera only detects in the direction perpendicular to the Earth's surface, receiving broadened pulse signals, which can only invert to get the wind speed but not the wind direction, with a measurement range of 2–15 m/s. Altimeters can only obtain the height of the subsatellite point and not the height information of a certain range at the same time, leading to low spatial coverage and long repeat cycles. Furthermore, because both transmission and reception equipment are on the same satellite, the cost of a single satellite is high. Microwave radiometers can measure wind speeds in the range of 4–50 m/s, but its spatial resolution is low, suitable for large and medium-scale detection of sea surface wind fields, and require very high calibration precision and polarization measurement. Microwave scatterometer is currently more mature remote sensing devices that can provide wind speed and direction simultaneously, but the inversion of wind direction suffers from directional ambiguities. Despite continuous improvements in antenna and polarization methods, it remains challenging to solve this issue fundamentally. Synthetic Aperture Radar

(SAR) can obtain high spatial resolution (generally 12.5–40 m) sea surface wind field information, but they are usually costly, have narrow swath widths, and are rarely used for high-resolution sea surface wind field acquisition.

Marine remote sensing based on GNSS-R is a branch of the GNSS-R remote sensing technology field. In 1993, Martin-Neira first pointed out the feasibility of using GPS reflection signals for altimetry [1], and applied for a patent [2]. The first results of the PARIS experiment published in 2001 [3]. In 1994, J. C. Auber first reported the accidental discovery during a flight test in July 1991 that GNSS-R signals could be detected by conventional navigation receivers [4]. In 2000, V. U. Zavorotny and A. G. Voronvich analyzed more systematically the use of GPS reflected signals for sea surface wind field detection, presented the theoretical model of sea surface reflection signals (Z-V model) [5] and indicated the potential to improve wind field inversion accuracy by using delay/Doppler mode to enhance sampling spatial resolution [6]. In 2002, J. L. Garrison described the structure of a delay mapping receiver and gave a method for wind speed inversion [7]. In 2003, Hajj et al. Conducted a systematic analysis of using GPS reflected signals for altimetry [8]. Starlab in Spain conducted a series of experiments called Eddy [9–12] for sea surface roughness and elevation measurements. Key experiments in satellite observation of GNSS-R include: the SuRGE (student reflected GPS experiment) satellite observation experiment plan formulated jointly by the University of Colorado and several other universities, NASA, and NOAA, [13] and the UK-DMC disaster monitoring satellite launched by the United Kingdom in September 2003, which carries GPS reflective signal reception equipment provided by Surrey Satellite Technology Ltd, aiming to study the feasibility of remote sensing of sea state parameters, snow and ice, and land using satellite-borne GNSS-R equipment [14]. To date, numerous bridge, airborne, and satellite experiments have been carried out internationally on the application of GNSS-R signals in marine remote sensing, achieving significant research outcomes.

Compared with the traditional remote sensing technologies, GNSS-R marine remote sensing technology significantly compensates for the deficiencies of existing marine remote sensing technologies, with its abundant signal sources, low cost, low power consumption, all-weather capability and high real-time performance. Table 6.1 provides a comparison of sea surface wind field measurement capabilities, showing that GNSS-R technology has certain advantages over other satellite remote sensing technologies.

Table 6.2 Comparison of GNSS-R altimetry technology with other remote sensing methods.

Marine remote sensing involves a large number of physical quantities, including sea surface physical parameters, seawater composition parameters, and sometimes oceanographic physical state, such as the presence of typhoons, oil spills. In this section, three typical marine remote sensing applications are selected: sea surface wind field inversion, sea surface height measurement and storm surge simulation, with corresponding results already validated in engineering practice. The following introduction includes principles, experimental scenarios and data analysis results.

6.1 Wind Field Inversion

Table 6.1 Comparison of GNSS-R wind measurement technologies with other remote sensing tools [1, 15]

Methods of observation	Wind speed accuracy (m/s)	Wind direction accuracy	Measurement range (m/s)	Time resolution (h)	Spatial resolution (km)
Satellite photo	±2.1	±40° or ±50	0–44	12	1.1
Microwave radiometer	±1.9	N/A	2–50	24	25
Radar altimeter	±1.6	N/A	2–18	240	6.7 (along the trail)
Microwave scatterometer	±2.0	±20°	4–26	24–28	10–50
Synthetic aperture radar (SAR)	±2.0	±20°	2–15	>72	12.5–40
GNSS-R technology	±2.0	±20°	2–50	12	1

Table 6.2 Comparison of GNSS-R altimetry with other remote sensing tools [1]

Observation mode	Sea level error (cm)	Flight level	Surface type	Time resolution (h)	Spatial resolution (km)
Altimeter	2.5 (Jason)	1300 km	Marine/land	240	15 (along the trail)
GNSS-R technology	2	480 m	Ocean or lake	12	1
	5	3000 m			
	50	30 km			
	<10	700 km			

6.1 Wind Field Inversion

6.1.1 Sea Surface Wind and Waves

The oceans and the atmosphere are two fluids of different densities on Earth, constituting a complex coupled system that exchanges physical quantities, such as heat, momentum, salinity and moisture, over a vast interface. The mutual influence, mutual constraints and mutual adaptation between the two is called the sea-air relationship, including different time scales and spatial scales of the sea-air relationship. Because of its heat capacity, the oceans absorb 70% of the total solar radiation entering Earth, with the surface layer of the oceans (mixing layer) storing about 85% of the absorbed solar radiation. This energy is transferred it to the atmosphere in the form of long-wave radiation, latent heat, and sensible heat. The uneven distribution of ocean heat in time and space creates temperature gradients and pressure gradients in the atmosphere, driving atmospheric motion and forming sea surface winds. The sea surface wind field generally refers to the motion of the atmosphere at a height of 10 m from the sea surface. The sea surface wind field is a vector defined by its speed and direction. Wind speed refers to the rate of atmospheric motion relative to a fixed point on Earth. Wind direction is the direction of atmospheric motion relative to a fixed point, measured meteorologically by the azimuth angle between the direction from which the wind blows and the north direction of the northeast celestial coordinate system, ranging from 0° to 360°.

Seawater, a liquid with a free surface, undergoes periodic motion and propagates outward when local water particles are disturbed by factors such as sea surface wind fields, causing regular periodic undulations on the sea surface, known as waves. When the wind force is very small (<0.2 m/s), the sea surface remains calm; as the sea surface wind force gradually increases (0.3–1.5 m/s), capillary waves are generated; with further increase in wind force, reaching a critical value (1.6 m/s), wind waves begin to form on the sea surface [16]. The essence of wave motion is the forward propagation of wave shapes and energy, while water particles do not advance with the waves. Sea winds change the sea surface roughness through waves, thereby altering the sea surface scattering coefficient.

6.1.1.1 Statistical Characteristics of the Sea Surface

As shown in Fig. 6.1, important parameters describing the statistical characteristics of the sea surface include the standard deviation of surface height variable δ_ζ, the surface correlation length l, and the sea surface mean-square slope σ_s^2.

Let $\zeta = \zeta(\mathbf{r})$ be a random variable of sea surface height at horizontal position vector $\mathbf{r}$ with mean height of $\overline{\zeta}$, using $\langle \cdot \rangle$ to denote the sign of the averaging, then the standard deviation of the surface height is

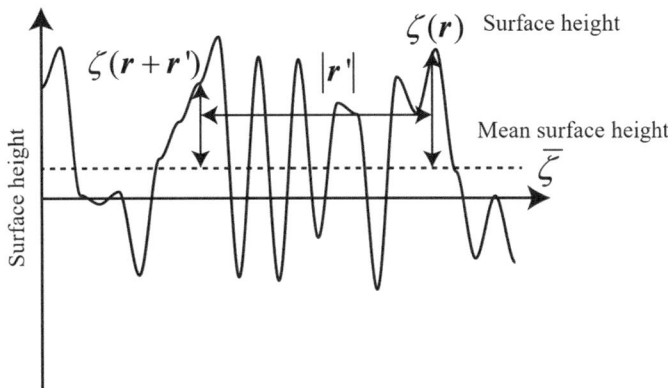

Fig. 6.1 Schematic diagram of random sea surface

6.1 Wind Field Inversion

$$\delta_\zeta = \sqrt{\left\langle (\zeta - \overline{\zeta})^2 \right\rangle} \tag{6.1}$$

The normalized surface autocorrelation function along a given direction is

$$A(\mathbf{r}) \equiv \frac{\langle \zeta(\mathbf{r}')\zeta(\mathbf{r}+\mathbf{r}') \rangle_{\mathbf{r}'}}{\langle \zeta(\mathbf{r}')\zeta(\mathbf{r}') \rangle_{\mathbf{r}'}} \tag{6.2}$$

The surface correlation length l is given by the following equation.

$$A(l, \mathbf{u}_l) = 1/e \tag{6.3}$$

where $\mathbf{u}_l$ is the unit vector in the direction of the surface correlation length. The value of the surface correlation length, l, provides a benchmark for estimating independence of two points on the sea surface, i.e., if two points are separated by a distance greater than l at a horizontal distance, then the height statistics of these two points are relatively independent from a statistical standpoint.

The mean-square slope of the sea surface is obtained from the wave spectrum

$$\begin{aligned} \sigma_{su}^2 &= \int_0^\infty \int_{-\pi}^\pi (k \cos \psi)^2 S(k, \psi) d\psi dk \\ \sigma_{sc}^2 &= \int_0^\infty \int_{-\pi}^\pi (k \sin \psi)^2 S(k, \psi) d\psi dk \end{aligned} \tag{6.4}$$

$$\sigma_s^2 = \int_0^\infty \int_{-\pi}^\pi k^2 S(k, \psi) dk d\psi \tag{6.5}$$

where k is the wave number; ψ is the wave direction; $S(k, \psi)$ is the wave spectrum; σ_{su}^2 and σ_{sc}^2 represent the sea surface mean-square slope in the downwind (upwind) and sidewind directions, respectively, which are two components of the total sea surface mean-square slope, i.e.

$$\sigma_s^2 = \sigma_{su}^2 + \sigma_{sc}^2 \tag{6.6}$$

6.1.1.2 Sea Surface Roughness Criterion

The sea surface roughness is typically defined using the Rayleigh criterion. According to the Rayleigh criterion, if the phase difference between the reflected distance traveled at two points is less than $\pi/2$ radians, then the surface can be considered smooth, and vice versa for roughness.

Peake and Oliver modified the Rayleigh criterion to categorize sea surface roughness into rough, moderately rough, and smooth cases, with the following criteria conditions [17]:

$$\begin{cases} \delta_\zeta < \lambda/25 \sin\theta & \text{smooth} \\ \lambda/25 \sin\theta \leq \delta_\zeta \leq \lambda/8 \sin\theta & \text{Medium rough} \\ \delta_\zeta > \lambda/8 \sin\theta & \text{rough} \end{cases} \quad (6.7)$$

where λ is the signal wavelength and θ is the altitude angle.

6.1.2 Ocean Wave Spectrum Model

6.1.2.1 Definition of the Wave Spectrum

Ocean waves can be viewed as composed of infinitely many waves of different amplitudes, frequencies, directions, and disordered phases. The distribution of ocean wave energy relative to each constituent wave is known as the ocean wave spectrum, or "energy spectrum," which is the Fourier transform of the sea surface's autocorrelation function, i.e.,

$$S(\boldsymbol{k}) = \text{FT}\{\langle \zeta(\boldsymbol{r}_0)\zeta(\boldsymbol{r}_0 + \boldsymbol{r})\rangle\} \quad (6.8)$$

where FT{·} denotes the Fourier transform and $\boldsymbol{k}$ denotes the wave number vector.

The wave spectrum describes the distribution of internal energy relative to frequency and wave direction, and is a two-dimensional function of frequency and wave direction, also denoted as $S(\omega, \psi)$, where ω is the angular frequency. The angular frequencies of the constituent waves ω are converted to the wave numbers k of the constituent waves, i.e. the wave number spectrum $S(k, \psi)$.

Integrating the two-dimensional wave spectrum along the wave direction yields a one-dimensional wave spectrum, $S(k)$, $S(\omega)$ are functions of wave number and angular frequency, respectively, defined as

$$\begin{aligned} S(k) &= \int_{-\pi}^{\pi} S(k, \psi) d\psi \\ S(\omega) &= \int_{-\pi}^{\pi} S(\omega, \psi) d\psi \end{aligned} \quad (6.9)$$

For simplicity, the two-dimensional ocean wave spectrum $S(k, \psi)$ is often expressed in the following form [18]:

6.1 Wind Field Inversion

$$S(k, \psi) = M(k)f(k, \psi) \tag{6.10}$$

where $M(k)$ denotes the isotropic part of the wave spectrum, and $f(k, \psi)$ is a function corresponding to the directional part, which is generally denoted as

$$f(k, \psi) = \frac{1}{2\pi}[1 + \Delta(k) \times \cos(2\psi)] \tag{6.11}$$

where, $\Delta(k)$ is the ratio of downwind (upwind) direction to the wind speed in the side direction.

6.1.2.2 Elfouhaily Ocean Wave Spectrum

One-dimensional wave spectra include the Neumann spectrum, Pierson-Moskowitz spectrum, and JONSWAP (Joint North Sea Wave Project) spectrum, all of which are gravity spectra. The Pierson spectrum is among the earliest wave spectra to consider both gravity and capillary wave components, while other wave spectra continue to retain their gravity components, with only a modification in the capillary wave component, such as Apel spectrum. However, under some specific natural conditions, the Apel spectrum also fails to accurately describe the capillary wave component [15]. The Elfouhaily spectrum, established in 1997, is a comprehensive result based on previous research, considering factors not covered by the Apel and Pierson spectra and incorporating wind zone effects. It is currently a widely used two-dimensional energy spectrum in sea surface wind field inversion applications [16].

$$S_E(k, \psi) = M_E(k)f_E(k, \psi) \tag{6.12}$$

The function corresponding to the directional part is

$$\Delta_E(k) = \tanh\left(0.173 + 4\left(\frac{v_{ph}}{v_g}\right)^{2.5} + 0.13\frac{u_f}{v_{phm}}\left(\frac{v_{phm}}{v_{ph}}\right)^{2.5}\right) \tag{6.13}$$

$$f_E(k, \psi) = \frac{1}{2\pi}[1 + \Delta_E(k)\cos(2\psi)] \tag{6.14}$$

The isotropic part of the Elfouhaily spectrum is

$$M_E(k) = \frac{k^{-3}}{2v_{ph}}(\alpha_g v_g F_g + \alpha_c v_{phm} F_c) \kappa^{\exp\left[-\left(\sqrt{k/k_p}-1\right)^2/2\delta^2\right]} \exp\left(-5k_p^2/4k^2\right) \tag{6.15}$$

$$\alpha g = 6 \times 10^{-3}\sqrt{\Omega}, \quad v_g = u_{10}/\Omega, \quad F_g = \exp\left[-\Omega\left(\sqrt{k/kp} - 1\right)/\sqrt{10}\right] \tag{6.16}$$

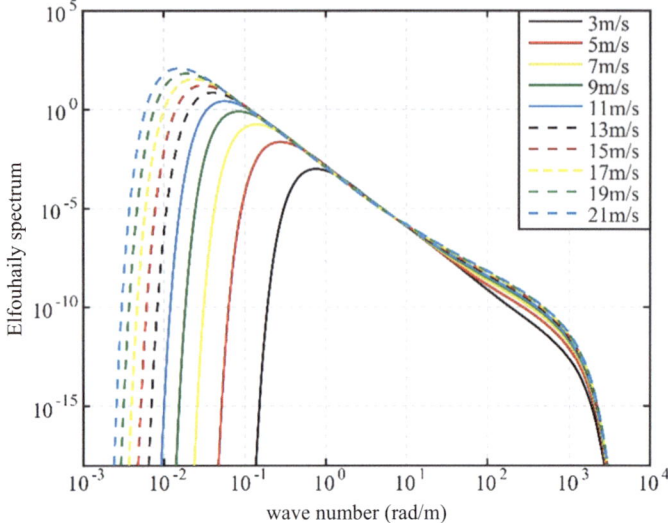

Fig. 6.2 One-dimensional Elfouhaily ocean wave spectrum

$$\kappa = \begin{cases} 1.7 & 0.84 \leq \Omega \leq 1 \\ 1.7 + 6\lg\Omega & 1 < \Omega \leq 5 \end{cases} \quad (6.17)$$

$$\delta = 0.08(1 + 4/\Omega^3), \ kp = \Omega^2 g/u_{10}^2 \quad (6.18)$$

$$\Omega = 0.84 \tanh\left[(X/2.2 \times 10^4)^{0.4}\right]^{-0.75} \quad (6.19)$$

$$\alpha c = 10^{-2} \begin{cases} 1 + \ln(uf/vphm) & uf \leq vphm \\ 1 + 3\ln(uf/vphm) & uf > vphm \end{cases} \quad (6.20)$$

$$Fc = \exp\left[-\frac{1}{4}\left(\frac{k}{km} - 1\right)^2\right] \quad (6.21)$$

$$km = 363 \text{rad/m}, \ vphm = 0.23 \text{m/s}, \ vph = \sqrt{g(1 + k^2/km^2)/k} \quad (6.22)$$

where X is the length of the wind zone (in m); for open, fully mature seas, X is infinity and the Elfouhaily wave spectral function value is close to the Apel spectral function. u_f is the friction velocity . In $ME(k)$, the term with subscript g in each of the terms denotes the gravity wave component. The shape of one-dimensional Elfouhaily ocean wave spectrum in mature sea area, within a sea surface wind speed range of 2–21 m/s, is shown in Fig. 6.2.

6.1.3 Electromagnetic Scattering Models

The ocean wave spectrum represents the distribution of wave energy on the ocean surface. Electromagnetic waves incident to the ocean surface will undergo reflection/scattering and refraction phenomena, and the models describing the changes of electromagnetic waves at this time are collectively known as electromagnetic scattering models, including the Small Slope Approximation (SSA), the Kirchhoff Approximation-Geometric Optics (KA-GO), the Integral Equation Method model (IEM), the Two Scale Model (TSM), and the Small Perturbation Method (SPM), among others. The forward scattering coefficient expressions for all models are integral equations about the carrier frequency, incidence angle, azimuth angle, polarization, and the sea wave directional spectrum, constituting implicit functions of the sea surface wind field. The Kirchhoff approximation (KA) model is suitable for large-scale rough sea surfaces, the SPM model for micro-rough surfaces of all scales, and the IEM and SSA models are suitable for rough surfaces at all scales. The SSA model, based on the assumption of a small-slope surface, can be applied to t any frequency spectrum component and any wavelength of sea waves when the tangent of the incidence and scattering angles exceeds the mean square slope of the rough surface, accurately describing L-band signal scattering. TSM is established based on SSA, but it contains a parameter dividing the wave scale, hence TSM is not a scattering model suitable for any scale. The KA-GO model is suitable for near specular directions, but away from specular directions, the KA-GO model struggles to provide accurate results.

6.1.3.1 SSA Model

The SSA model provides a method to analyze the scattering situation at the sea surface for any frequency component and any range of wavelengths. If a rough sea surface is illuminated by a monochromatic plane wave from the upper half of the sky, $\mathbf{E}^s$ represents the wave spectrum of the reflected field, and $\mathbf{E}^i$ represents the wave spectrum of the incident field, then there are

$$\mathbf{E}^s = \mathbf{S} \cdot \mathbf{E}^i \tag{6.23}$$

where the scattering matrix $\mathbf{S}$ is a random variable associated with a random rough surface. Assuming that the location of the received scattered field is at a distance of R_r from the scattering point, the scattering cross section is [19].

$$\sigma_0 = 4\pi R_r^2 \langle |\mathbf{S}|^2 \rangle \tag{6.24}$$

Numerical simulation results show that as the wind speed increases, the peak of the normalized bistatic scattering cross-section tends to decrease, and this peak occurs where the scattering angle is equal to the incidence angle. When the scattering angle increases, higher wind speeds will lead to large scattering cross sections. The peak values of the normalized scattering cross sections are the same for different incidence angles. When the scattering angle is less than 30°, the smaller the scattering angle, the larger the normalized scattering cross section; this trend reverses for scattering angles greater than 30°. When using the SSA model to invert the wind field in GNSS-R, it is necessary to consider the elevation angle of the satellite being used and to account for scattering signals in non-specular reflection directions [20].

6.1.3.2 KA-GO Modeling

When the radius of curvature of the surface is much larger than the wavelength of the emitted radio waves, the field on the surface can be approximated by the field of tangent plane at each point, a process known as Kirchhoff's Approximation (KA) or Geometrical Optics (GO). It requires that the surface correlation length l in the horizontal direction is greater than the wavelength of the electromagnetic wave, and the vertical direction satisfies the height standard deviation δ_ζ is small enough, i.e., $k_1 l > 6$, $l^2 > 2.76 \delta_\zeta \lambda$, and k_1 is the wave number of the electromagnetic wave in the air.

Using statistical methods to calculate the average number of optically reflective points per unit area of a 2D rough surface n_A and the average radius of curvature of the optically reflective points $\langle |r_1 r_2| \rangle$, the average scattering cross section per unit area σ_{0A} can be obtained as [21].

$$\sigma_{0A} = \pi n_A \langle |r_1 r_2| \rangle |\Re|^2 \tag{6.25}$$

where $\Re$ is the reflection coefficient. For a sea surface with a surface height of $\zeta(x, y)$ obeying a Gaussian distribution, the average number of optically reflective points per unit area is

$$n_A = \frac{7.255}{\pi^2 l^2} \exp\left(-\frac{\tan^2 \theta}{\sigma_s^2}\right) \tag{6.26}$$

The KA-GO model is applicable when the incidence angle is less than 30°. Figure 6.3 shows distribution of bistatic scattering cross-section of the sea surface with the KA-GO model, where the receiver height is 5 km and the satellite elevation angle is 90° (i.e., the incidence angle is 0°).

6.1.4 Wind Speed Inversion Based on Waveform Matching

6.1.4.1 Basic Process of Wind Field Inversion

The general process of inverting the sea surface wind field using Reflected GNSS Signals is shown in Fig. 6.4.

6.1 Wind Field Inversion

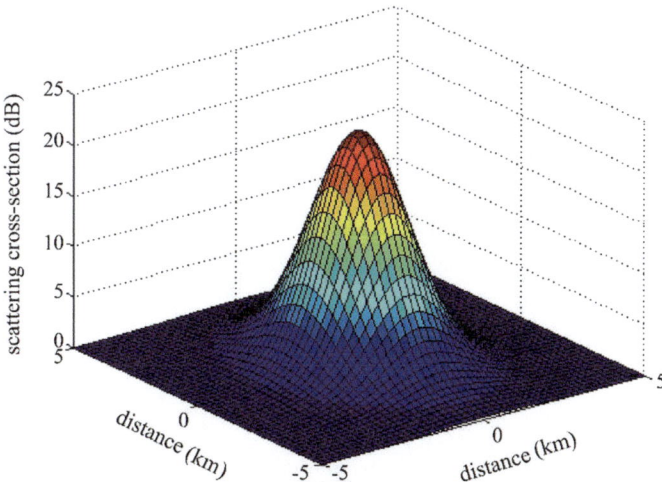

Fig. 6.3 Distribution of the sea surface scattering cross-section with the KA-GO model (Sea water temperature 25 °C, salinity 35‰)

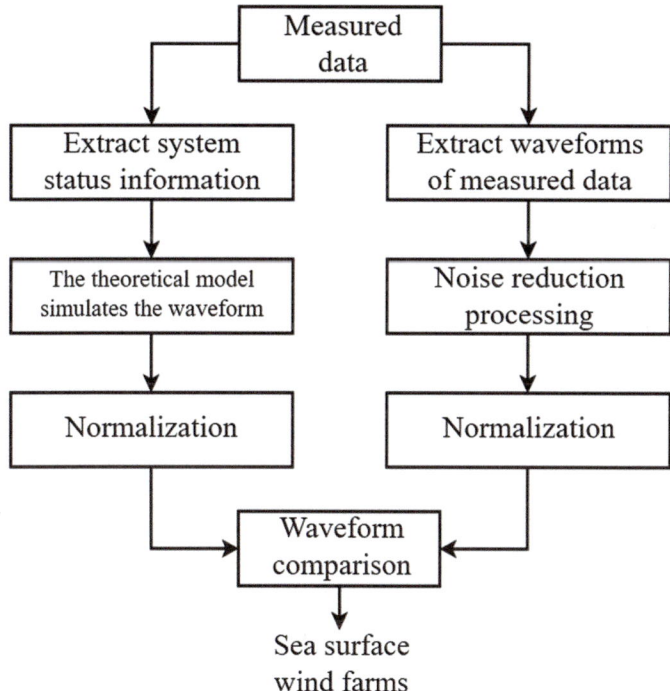

Fig. 6.4 Process diagram of sea surface wind field inversion

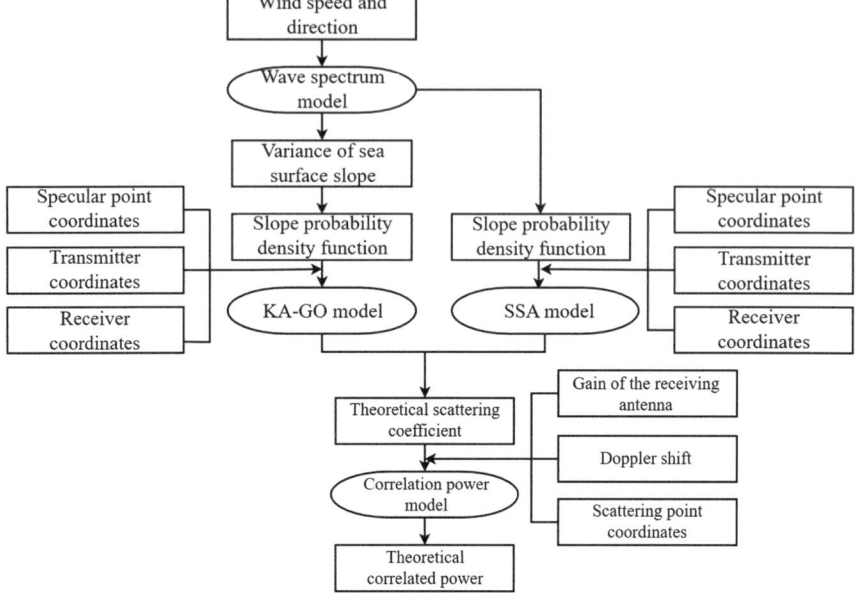

Fig. 6.5 Data processing flow chart for KA-GO and SSA scattering models

In Fig. 6.4, "extracting system state information" means extracting the receiver's height, carrier movement speed, satellite elevation angle, azimuth angle and other information. "Waveform comparison" means the comparison of the trailing slope or the trailing edge of the entire waveform.

Different scattering models may also require different sea surface statistical characteristic parameters, such as the KA-GO model requires the sea surface slope probability density function as an input condition, while the SSA model requires the sea surface correlation function as an input. Therefore, the specific processes for calculating theoretical scattering coefficients and correlation power using the two scattering models are also different, as shown in Fig. 6.5.

6.1.4.2 Wind Field Inversion Algorithm

The sea surface wind field is a vector that includes both magnitude (commonly known as wind speed) and direction (i.e., wind direction). The two can be solved separately or jointly.

1. Wind Speed Inversion Algorithm Based on Direct Matching of Power Curve Trailing Edge

Inversion of wind speed based on the direct matching method at the trailing edge of the theoretical correlation power curve is a relatively simple method, and the methods discussed later, based on the slope of the trailing edge of the correlation power curve

6.1 Wind Field Inversion

and the empirical functions, are developed on this basis. The trailing edge of the correlation power curve (surface) of the scattered signal is relatively sensitive to wind speed, making this method initially commonly used for wind speed inversion and the specific process is described as follows.

(1) Due to differences in gain between the direct and reflected signal receiving antennas and environmental factors, the received reflected signals contain noise that must be eliminated when calculating the reflected signal correlation power. The noise floor can be theoretically calculated or estimated from measured data. The actual reflected signal correlation power obtained is denoised using a moving average method.

(2) Before waveform matching, the code delay of the measured power curve and the theoretical waveform must be aligned, and the number of code delays should be as large as possible to ensure the accuracy requirements for waveform matching.

(3) To make the measured data comparable to the theoretical data, the measured data is normalized. Usually, the maximum power of the direct signal or the total reflected signal power is used as the normalization factor. The direct signal power as a normalization factor is simple, but not very accurate, as the difference in direct and reflected antenna gains also leads to a significant difference in the power received by the direct and reflected signal channels. The total reflected signal power, unrelated to sea surface statistical properties and approximated as a constant, can smooth the waveform as a normalization factor. If compensation is made for the gain of the direct and reflected signal antennas, the direct signal power can also be used as the normalization factor for measured and theoretical waveforms.

(4) The measured power curve is matched to the theoretical power curve using the least squares method. When the mean-square deviation between trailing edges of the two waveforms reaches a minimum, the wind speed corresponding to the theoretical correlation power curve is the initial value of the sea surface wind speed sought. If the error is large, further verification and checking can be performed.

(5) If the wind speed estimate is obtained by matching the time-delay correlation power curve, Doppler correlation power curve can be utilized for calibration and correction; if the wind speed estimate is obtained by using the two-dimensional surface, data from satellites at different elevation angles can be used for calibration and correction, thus improving the accuracy of the wind speed inversion.

2. Wind Speed Inversion Algorithm Based on Matching the Slope of the Power Curve's Trailing Edge

Due to the significant uncertainty in directly matching the trailing edge of the correlation power for actual wind speed inversion, achieving a complete match across all code delays is difficult. However, precise matching within a specific local delay range is relatively easier. Another simple algorithm for wind speed inversion involves comparing compares the slopes of the trailing edges of the measured and theoretical correlation power curves. The steps are as follows :

(1) Noise reduction is performed on the variation curve of the time-delayed power associated with the received real reflected signal;
(2) Normalization of measured and theoretical correlation power;
(3) Calculate the slope of the trailing edge (i.e., the first derivative) of the normalized theoretical correlation power curve;
(4) Calculate the slope of the trailing edge of the normalized measured correlation power curve (the calculation interval is usually taken as $0 - 4\tau_c$);
(5) Match the measured waveform's trailing edge slope to the theoretical waveform's trailing edge slope; then, the wind speed corresponding to the theoretical correlation power curve is the sea surface wind speed value to be inverted;
(6) If the error is large, use Doppler or correlation power information from satellites at different elevation angles to calibrate and correct the estimated values.

3. Wind Direction Inversion Algorithm Based on Known Wind Speed from Theoretical Correlation Power Curves

Utilizing reflected signals from multiple navigation satellites and wind speed inferred by the calculation module as auxiliary information, sea surface wind direction is inverted.

First, perform correlation power denoising, normalization, and code chip alignment processing, with the specific algorithms the same as those used for wind speed inversion. Then, with an initial wind direction set to 0°, calculate the sum of squared residuals. The sum of squared residuals of the theoretical and measured correlation power waveforms from multiple GNSS satellites is derived from Eq. (6.27):

$$SOS = \sum_{i=1}^{m} \sum_{j=1}^{n} \left(P_{ij} - \hat{P}_{ij}\right)^2 \quad (6.27)$$

where i and j denote the index numbers of the visible GNSS satellites and the correlation power sampling point calculation, respectively, and $P_{ij}, \hat{P}_{ij}$ are the theoretical and measured correlation power, respectively. The residual the sum of squared residuals is recalculated for each adjusted wind direction value using Eq. (6.28).

$$\phi_{i+1} = \phi_i + \Delta\phi \quad (6.28)$$

The angle corresponding to the local minimum value of the sum of squared residuals is considered the predicted wind direction solution. Since wind direction is very sensitive to changes in delay correlation power, and the wind direction sensitive area manifests in the latter half of the curve's trailing edge, possible wind direction solutions are clarified through matching the latter half region of the delay correlation power trailing edge, finalizing the wind direction solution.

6.1 Wind Field Inversion

4. Wind Field Inversion Method Based on Multiple Navigation Satellites

Using the sum of squared residuals of the theoretical and measured waveforms from multiple navigation satellite reflections as the objective function, a joint inversion of wind speed and wind direction is performed. Given that the ocean wave spectrum changes with wind speed and direction, with a fixed wind direction, there is only one wind speed corresponding to the ocean wave spectrum. Therefore, possible wind speeds for each wind direction can be inverted by finding the minimum value of the objective function, followed by seeking local minimum values along the wind direction to determine possible wind direction solutions.

This algorithm differs from the wind direction inversion method based on known wind speed from theoretical correlation power curves, as it requires calculating all possible wind speed solutions under different wind directions. Initially assume wind speed and wind direction values, typically choosing moderate wind speeds prevalent at the sea surface and starting with an initial wind direction of 0°. Based on these initial values, determine the corresponding theoretical correlation power waveform and calculate the sum of squared residuals, the objective function, between it and the measured waveform. Multiple objective functions are generated by varying the search step length for wind speed. Continue searching until the relationship between the target function size of the neighboring wind speed and the current wind speed changes direction, thus obtaining the current possible wind speed solution. Then, seek local minimum values along the wind direction to determine possible wind direction solutions.

6.1.4.3 Wind Field Inversion Example

1. Yellow Bohai Sea Test Result Analysis

Airborne data collection tests were carried out in September 2004 in the Yellow Bohai Sea within the range of 34°–38°N and 112°–124°E. The GNSS-R receiver used was the first-generation processing circuit board developed by our research group, and the antenna for receiving the reflected signals is a left-hand circularly polarized microstrip antenna with a gain of 3 dB. Taking the flight on September 11 as an example, its flight altitude and satellite elevation angle are shown in Fig. 6.6, with the flight altitude of above 3,000 m and a satellite altitude angle of between 20° and 90°.

Based on the analysis mentioned above, the following processing was performed on the data collected from the flight experiment to obtain sea surface wind field data.

(1) Data preprocessing

Data preprocessing includes two processes: denoising and normalization. Denoising is achieved by selecting a suitable noise floor, and normalization is done by using the total reflected signal power or direct signal power. In practice, there are three choices for noise floor:

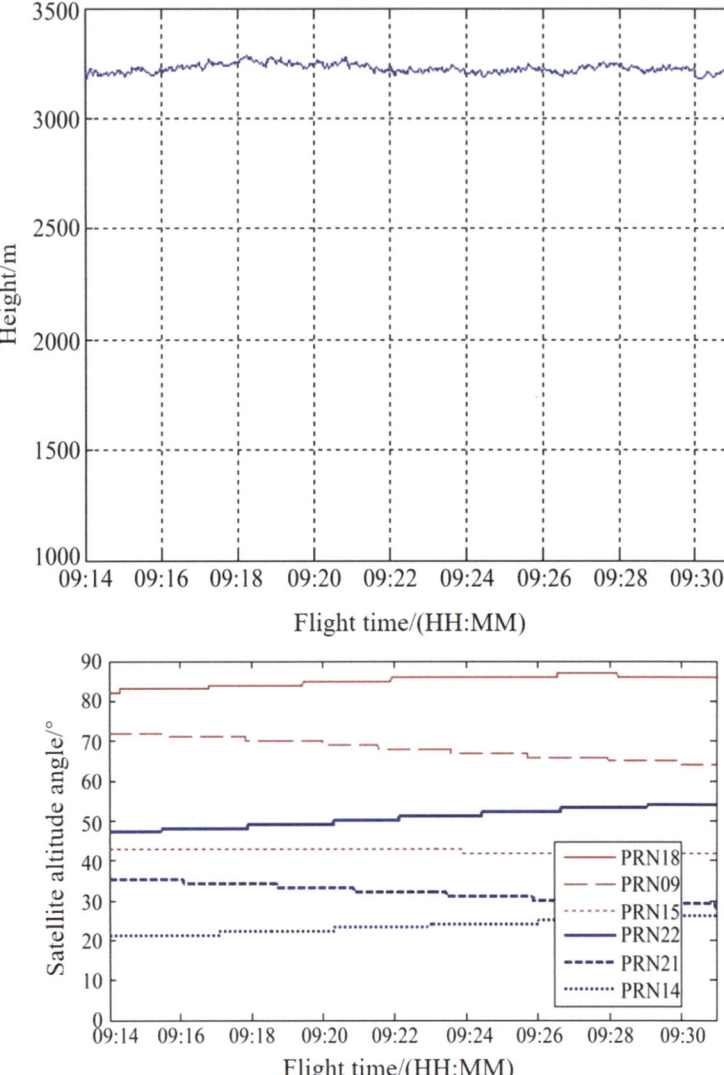

Fig. 6.6 Basic information on the GNSS-R flight experiment (September 11, 2004)

- Calculate the average of the reflected signal output data of 2 code chips before the specular reflection point;
- Choose a position away from the specular reflection point, such as the average the last few delay code chip corresponding data output by the receiver.
- Select the noise floor value corresponding to the processing chip.

6.1 Wind Field Inversion

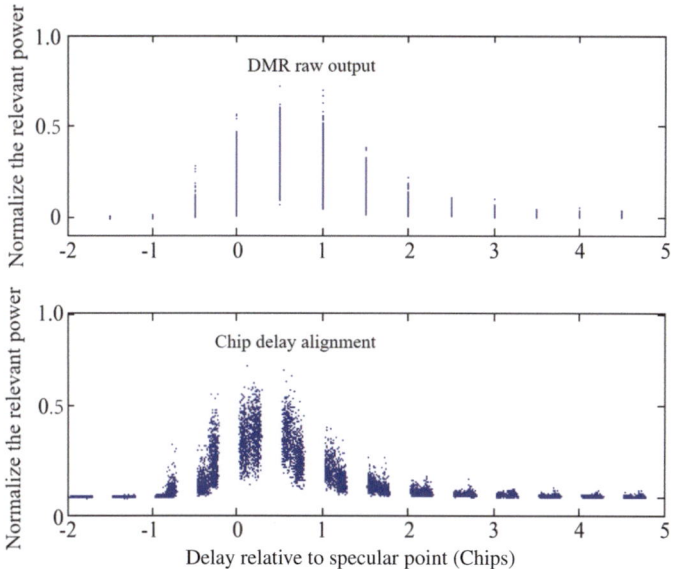

Fig. 6.7 Code chip delay alignment processing (PRN18)

(2) Delayed code-chip alignment processing

Code chip delay alignment processing refers to correlating the reflected signal power output values with the value corresponding to the specular reflection point, matching the reflected signal power to the code chip delay. Figure 6.7 shows the code chip delay alignment results for the reflected signal power from GPS satellite PRN18 (with the highest elevation angle).

(3) Incoherent Accumulation

Incoherent accumulation can improve the signal-to-noise ratio of weak reflected signals. Figure 6.8 shows the cases of accumulation times of 30, 120 and 480, in which the curves represent the theoretical waveforms of one-dimensional time-delayed correlation power for wind speeds of 3, 5, 7, 9, and 11 m/s, respectively, and the discrete dots are the waveforms of the measured data. From this, it can be observed that the incoherent accumulation can make the reflected signal normalized waveforms more concentrated, aiding in the accuracy of wind speed inversion.

(4) Wind speed inversion

The theoretical model is used to generate the theoretical one-dimensional time-delay correlation power values for different wind speeds in the range of 3–21 m/s Matching is done with the trailing edge data (0.5–2 code chip delays) of the reflected signal power waveform from GPS satellite PRN18 using the least squares method, obtaining wind speed results for the entire flight time, as shown in Fig. 6.9. During the data

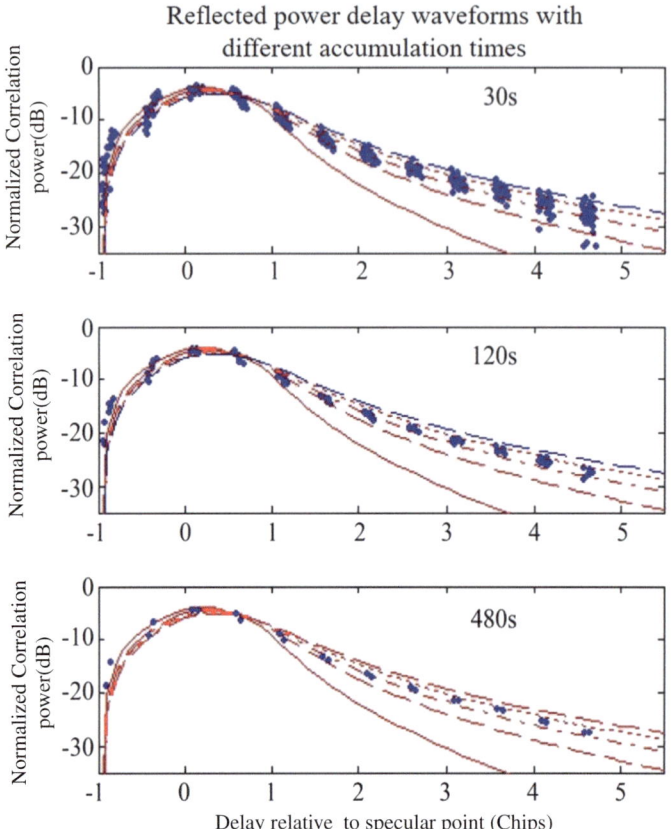

Fig. 6.8 Results of incoherent accumulation (30 s/120 s/480 s)

collection period from 09:14 to 09:30, wind speeds ranged from 7.2 m/s to 10 m/s, with an average of 8.2 m/s and a standard deviation of 0.8 m/s. Using all sampling data over 17 min for averaging, the inverted wind speed is 7.6 m/s.

Figure 6.10 shows the flight path and the wind speed results obtained by the QuickSCAT scatterometer, and the average wind speed within the flight path is about 7.5 m/s (7.3 m/s at the North Godown wind measurement station). The difference between the scatterometer and GNSS-R inverted wind speed averages is 0.7 m/s (8.2 m/s–7.5 m/s = 0.7 m/s).

2. Analysis of Test Results in Certain Sea Area

Several flight tests were conducted in February–March 2009 in the South China Sea. The sea area where the data were collected along the flight route was 19°–19°30′N latitude and 110°50′–111° E longitude, and the collected signals were the sea surface scattering signals of the GPS L1 C/A code. The GNSS-R receiver used was the second-generation baseband processing circuit board developed by the research

6.1 Wind Field Inversion

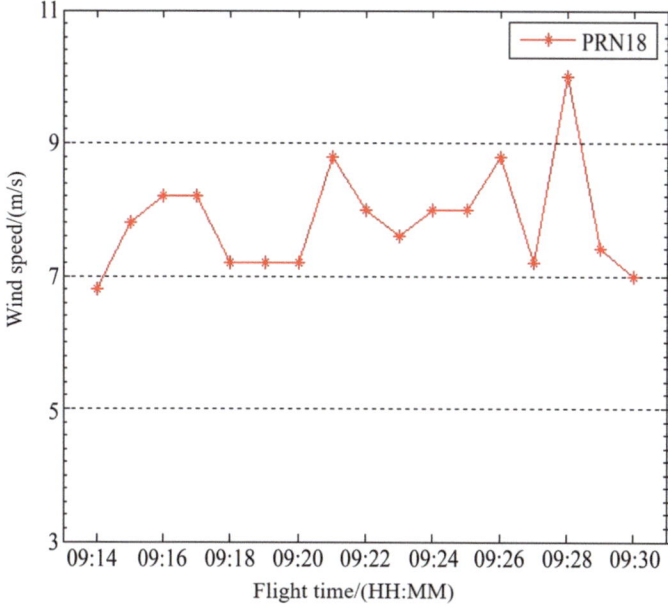

Fig. 6.9 Wind field inversion results (PRN18)

group, with an antenna for receiving the reflected signals being a 12 dB gain left-hand circularly polarized array antenna. Figure 6.11 shows the flight altitude and satellite elevation angles received during the flight experiment.

The delay/Doppler two-dimensional correlation power obtained through related processing of the reflected signals is shown in Fig. 6.12.

The signal-to-noise ratio of the reflected signal was improved using a combination of coherent and incoherent accumulation methods, as shown in Fig. 6.13. Figures (a) and (b) show the processing results of the same 1 s data under different coherent and incoherent accumulation conditions. Figure (a) represents the delay/Doppler two-dimensional correlation power result using 1 ms coherent accumulation and 1000 times incoherent accumulation. Figure (b) represents the result using 10 ms coherent accumulation and 100 times incoherent accumulation. The 10 ms coherent accumulation results show more concentrated energy along the Doppler frequency axis due to the narrowing of the noise bandwidth caused by the increased coherent accumulation time. Along with the Doppler frequency shift, it presents a typical "horseshoe shape," consistent with theoretical analysis.

Wind speed inversion uses a matching algorithm based on the trailing edge of the theoretical correlation power curve. Figure 6.14 shows the matching results of the correlation power after delay for received signals from satellite PRN29, where dotted lines represent measured data, dash-dotted and solid lines represent theoretical curves for wind speeds of 3.37 m/s and 4.37 m/s, respectively, and the dashed line

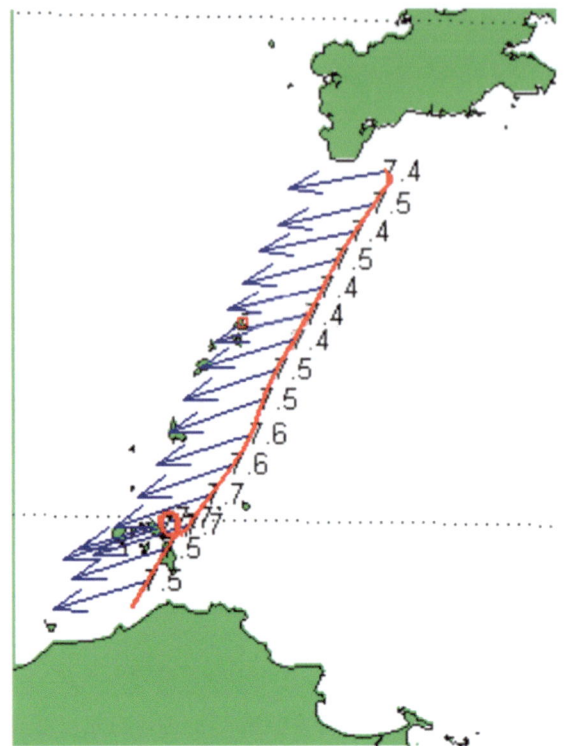

Fig. 6.10 Flight route and QuickSCAT scatterometer wind speed results

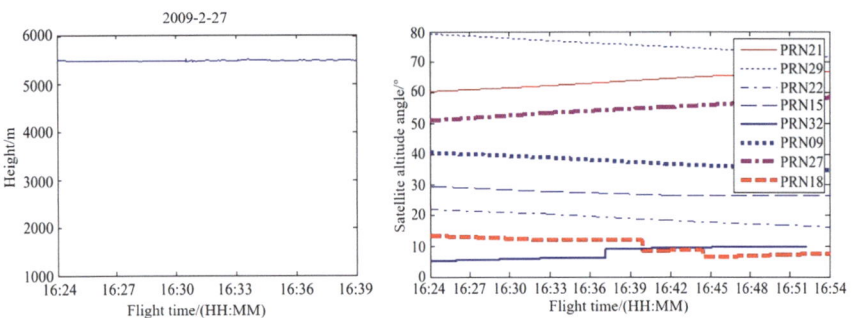

Fig. 6.11 Flight altitude and satellite elevation angles during the flight experiment

represents a theoretical curve for a wind speed of 5.37 m/s. A comparison shows that measured data match well with the 4.37 m/s theoretical curve.

Inversion of the wind direction is performed using a theoretical correlation power curve matching algorithm based on the wind speed being known. Figure 6.15 shows the wind direction matching results for received PRN29 satellite reflection signals

6.1 Wind Field Inversion

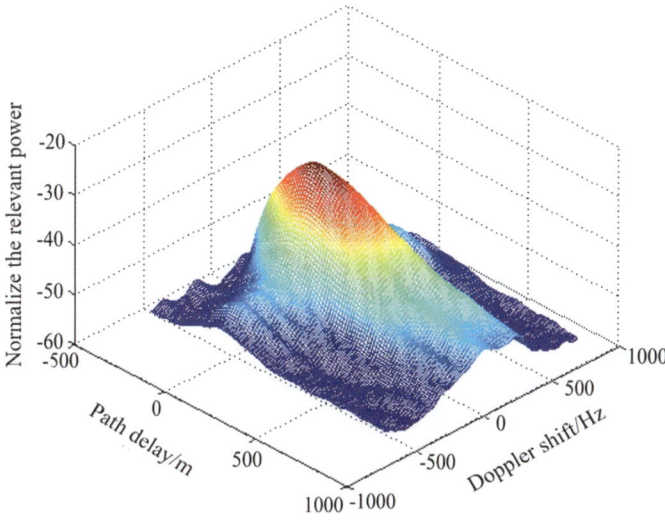

Fig. 6.12 Delay/Doppler two-dimensional correlation power map

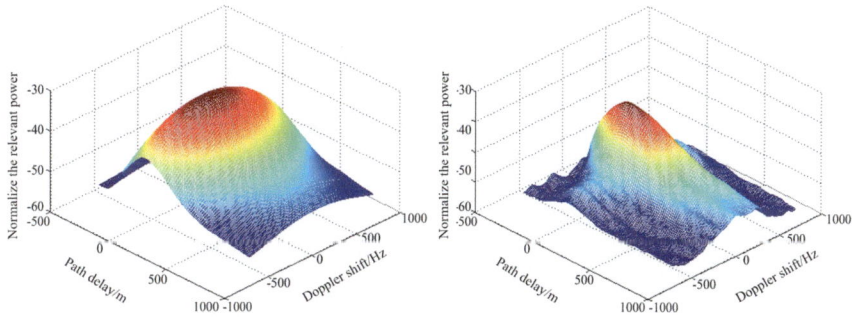

(a) 1ms coherent accumulation, 1000 times incoherent accumulation (b) 10ms coherent accumulation, 100 times incoherent accumulation

Fig. 6.13 Delay/Doppler two-dimensional correlation power maps under different accumulation time conditions

under the condition of known wind speed (4.37 m/s), where dotted lines represent measured data, dash-dotted lines represent a theoretical curve for a wind direction of $-30°$, solid line for $-10°$, and dashed lines for $10°$. Comparison shows that measured data match well with the $-10°$ theoretical curve.

For data of 1000 s on February 27, 2009, the wind field inversion is performed every 100 s. The wind speed inversion method was based on the direct matching algorithm of the theoretical correlation power curve's trailing edge, and the wind direction inversion method was based on the matching algorithm of the theoretical correlation power curve with known wind speed. The inversion results are

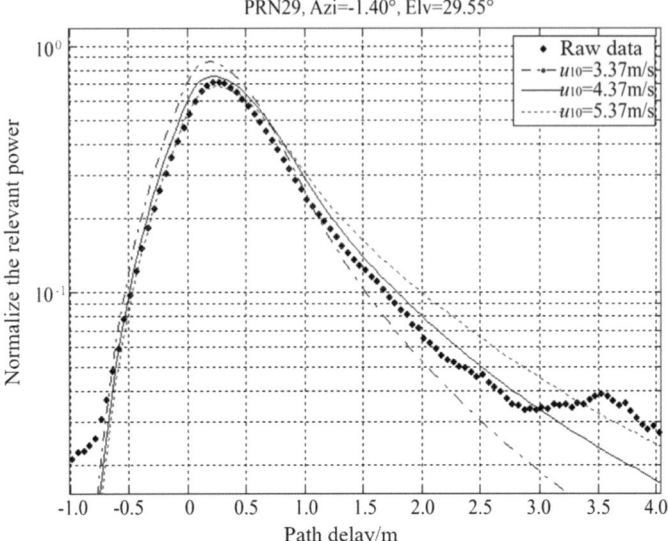

Fig. 6.14 Wind speed matching diagram

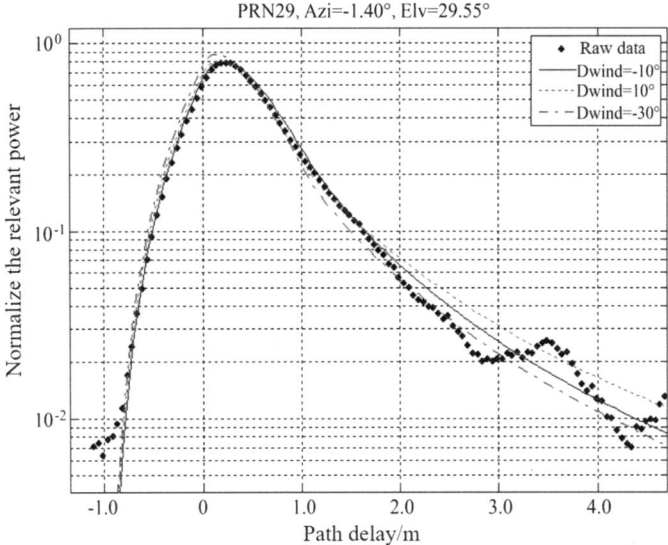

Fig. 6.15 Wind direction matching diagram

6.1 Wind Field Inversion

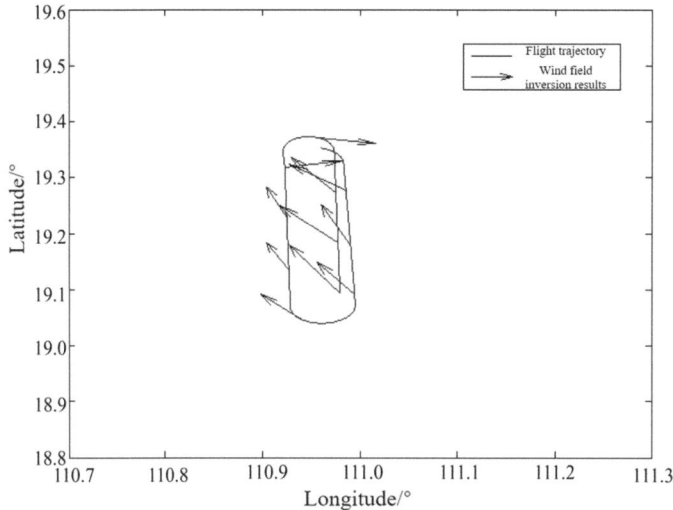

Fig. 6.16 Wind field inversion results

shown in Fig. 6.16. Compared with QuickSCAT satellite data and Boao marine meteorological observation station data, the standard deviation of wind speed inversion is about 1 m/s, and the standard deviation of wind direction inversion is about 20°.

These results provide a validation of the corresponding algorithms with typical wind field inversion outcomes.

6.1.5 Wind Speed Inversion Based on Model Functions

Another algorithm for sea surface wind field inversion is realized based on empirical model functions. In other words, a particular mode function is established by using the quantitative functional relationship between the ocean wind vector and the marine wind-sensitive parameter obtained through theoretical simulation or measurement. By substituting the measured wind-sensitive parameters and corresponding measurement parameters (such as polarization mode, observation azimuth angle, incidence angle, etc.) within the same resolution unit into the modal function, the corresponding marine wind speed and direction can be inverted. Here, the modal function inversion method for marine wind speed is introduced as an example, presenting the research achievements of the author's research group in recent years.

6.1.5.1 Wind Speed Sensitive Parameters

There are a variety of sea surface wind speed inversion models for coastal scenarios, which are broadly categorized into two types from the perspective of observed quantities. One type utilizes the coherent feature parameters of GNSS reflection signals, such as correlation time, [22] effective incoherent counts [23] and coherent to incoherent ratio, [24] etc. The other category defines characteristic parameters sensitive to the sea state in the delay waveform, such as waveform area, [25] trailing edge amplitude, [26] power distribution ratio, [27] etc. Here, we focus on introducing two basic characterization parameters: correlation time and waveform area.

1. Correlation Time

The correlation time τ_{icf} of a time series is defined as the integral width of the autocorrelation function of the series which is the width of the integral of the autocorrelation function of the sequence, i.e.

$$\tau_{icf} = \frac{\int_0^{+\infty} R(\tau) d\tau}{R(0)} \tag{6.29}$$

where $R(t)$ is the autocorrelation function of the time series. The time series of complex correlation value of the GNSS reflection signal is defined as

$$S_{icf}(t) = \frac{I_r(t) + jQ_r(t)}{I_d(t) + jQ_d(t)} \tag{6.30}$$

where I_r and Q_r are in-phase and quadrature correlation values of the Reflected GNSS Signals; and I_d and Q_d are the correlation values of the in-phase and quadrature branches of the GNSS direct signals, which are used to eliminate the effects of navigation data, signal power, etc., on the reflected signal.

When the sea surface mean square height follows a Gaussian distribution, the autocorrelation function of the complex correlation values of GNSS reflection signals is also approximated as the following Gaussian function [28]:

$$R(t) \approx A(h_{swh}, l_z, \theta, G_r) \exp[-(\pi h_{swh} t \sin \theta)^2 / 2(\lambda \tau_z)^2] \tag{6.31}$$

where h_{SWH} is the effective wave height at the sea surface; l_z is the correlation length at the sea surface; τ_z is the correlation time at the sea surface, approximately represented as $a \cdot h_{swh} + b$, and G_r is the gain of the receiving antenna of the reflected signal. Substitute Eq. (6.31) into Eq. (6.29) to yields

$$\tau_{icf} = \frac{\sqrt{2}\lambda}{\pi \sin \theta} \left(a + \frac{b}{h_{swh}} \right) \tag{6.32}$$

Based on the Pierson-Moskowitz sea wave spectrum, the relationship between the effective wave height of a mature sea area and wind speed given in reference [29] is

6.1 Wind Field Inversion

given as

$$h_{\text{swh}} = 0.0235 \cdot U_{10}^2 \tag{6.33}$$

The relationship between correlation time and wind speed in the shore-based scenario can be approximated as

$$\tau_{\text{icf}} = \frac{1}{\sin \theta} \left(a_{U_{10}} + \frac{b_{U_{10}}}{U_{10}^2} \right) \tag{6.34}$$

where, $a_{U_{10}}$ and $b_{U_{10}}$ can be obtained by least squares fitting. It is worth noting that the above model is an empirical and initially $a_{U_{10}}$ and $b_{U_{10}}$ need to be determined from in-situ collocated measurements. Once the two parameters are determined, the inversion of the sea surface wind speed can be performed using the GNSS reflection signals.

2. Normalized Delay Waveform Area

The normalized delay waveform area is defined as the integrated area of the normalized delay waveform exceeding the threshold, i.e.

$$A_N = \int_{W_N(\tau) > T_h} W_N(\tau) d\tau \tag{6.35}$$

where $W_N(\tau)$ is the normalized delay waveform ; T_h is a preset threshold. As shown in Fig. 6.17, for an altitude of 100 m, elevation angle of 30°, and thresholds of 0.2 and 0.5 respectively, the relationship between the normalized waveform area simulated by the Z-V model and wind speed can be approximately represented as

$$A_N = a_{U_{10}} \cdot U_{10}^{c_{U_{10}}} + b_{U_{10}} \tag{6.36}$$

where $a_{U_{10}}$, $b_{U_{10}}$ and $c_{U_{10}}$ can be obtained by least squares fitting.

6.1.5.2 Field Test Validation

To verify the validity of the mode functions, field experiments were carried out at Qingdongwuyan tide station (37°26′51″N; 119°0′36″E) in Dongying City, Shandong Province, in which direct and reflected signals of the Beidou B3I from 16:00 on June 16th, 2021 to 18:00 on June 19th, 2021 are collected. The experimental setup is shown in Fig. 6.18, with both direct and reflected antennas approximately 12 m above the sea surface. The direct antenna was an omnidirectional right-hand circularly polarized antenna, and the reflected antenna was a directional left-hand circularly polarized antenna with a beamwidth of ±60°. The reflected antenna was pointed 45° downward towards the sea surface at an azimuth angle of 250°, stably receiving signals from BeiDou GEO satellites PRN01, PRN02, and PRN03. The receiver's intermediate

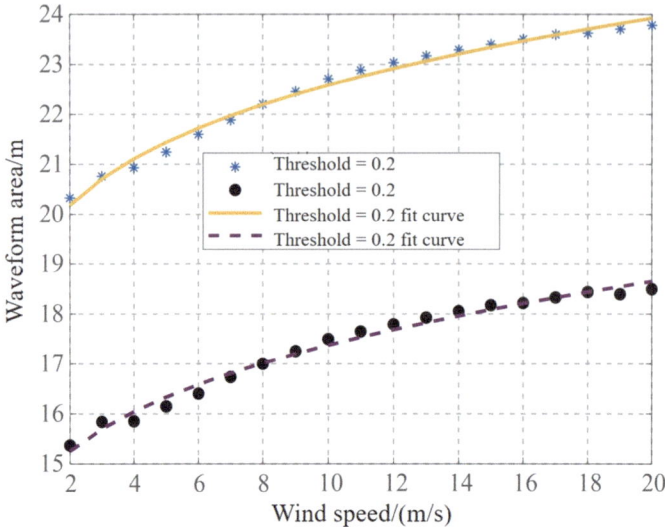

Fig. 6.17 Relationship between normalized waveform area and wind speed

frequency signal was quantized at 4 bits with a sampling rate of 32.738 MHz. To reduce data storage, B3I direct and reflected signals were collected every 10 min, each for a duration of 2 min. A cup anemometer installed at the tide station provided comparative wind speed data.

In shore-based scenarios, the GNSS signals reflected from the sea surface contain both coherent and incoherent components, closely related to the sea state and satellite elevation angle. In order to suppress or eliminate the interference of the direct signal and coherent components in the delay waveform, a combination of coherent and incoherent accumulation is mostly used. The delay waveform of incoherent accumulation with accumulative number of N_{incoh} is expressed as [26].

Fig. 6.18 Experimental setup

6.1 Wind Field Inversion

$$\langle |Y(\tau)|^2 \rangle = \sum_{i=1}^{N_{incoh}} \left| Y_{cohi}(\tau) - \sum_{i=1}^{N_{incoh}} Y_{cohi}(\tau) \right|^2 \tag{6.37}$$

where $Y_{cohi}(\tau)$ denotes the delay waveform of N_{coh} coherent accumulations.

Figure 6.19a shows the delay waveforms of direct and reflected signals from the BeiDou GEO satellite PRN01. The delay equivalent distance of the reflected signal relative to the direct signal is 15.30 m, corresponding to a receiving platform height of 12.05 m, consistent with the actual height. Due to the significant coherent component in the coastal, suppressing the coherent component using Eq. (6.37) will reduce the peak of the reflected signal's delay waveform. The delay of the reflected signal relative to the direct signal is less than one B3I chip, causing interference from the direct signal to the reflected signal. Compared to the unsuppressed coherent component delay waveform, Eq. (6.37) effectively suppresses direct signal interference, making the leading edge of the reflected signal's delay waveform steeper. Figure 6.19b is the autocorrelation function of the complex correlation values of the reflected signal from the BeiDou GEO satellite PRN01, following the Gaussian distribution as shown in Eq. (6.31), with a correlation time of approximately 193.9 ms.

The scatter plot of correlation time and waveform area estimated from the reflected signal of the BeiDou GEO satellite PRN02 against wind speed is shown in Fig. 6.20. From the figure, it can be seen that both correlation time and waveform area are inversely and directly proportional to the wind speed, respectively, and are consistent with the relationships shown in Eqs. (6.34) and (6.36). The fit of correlation time is 0.52, whereas the fit of the normalized delay waveform area against wind speed, both with and without direct signal interference, are 0.19 and 0.35, respectively. The fit of the correlation time is significantly higher than that of the normalized delay waveform area, and it can provide a better inversion result of the wind speed. The fit of the normalized delay waveform area is obviously improved after eliminating the direct signal interference. This indicates that direct signal interference is one of the important sources of error in wind speed inversion, and makes the waveform area not show a stable proportional relationship with the wind speed. After suppressing

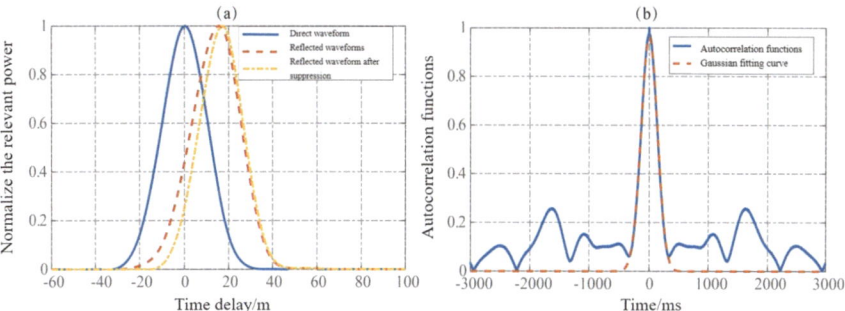

Fig. 6.19 **a** Delay waveform; **b** Autocorrelation function of complex correlation values of the reflected signal

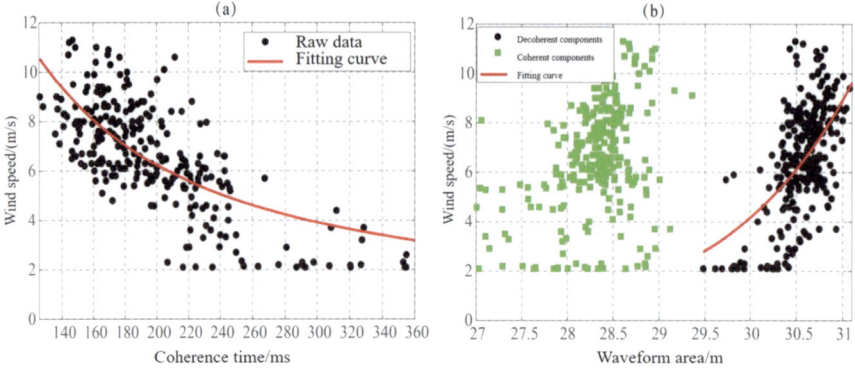

Fig. 6.20 Scatter plot of correlation time and waveform area against wind speed

Table 6.3 Root mean square error of retrieved wind speed

PRN1 RMSE (m/s)		PRN2 "(m/s)"		PRN3 "(m/s)"	
Correlation time	Waveform area	Correlation time	Waveform area	Correlation time	Waveform areaWaveform area
1.26	1.43	1.57	2.07	1.70	1.81

coherent components, waveform area and wind speed show a better proportional relationship.

Table 6.3 gives the root mean square error of the inversion of wind speed by each algorithm. The results show that the inversion results obtained based on the waveform area are poorer compared to the correlation time, as Eq. (6.37) cannot completely suppress coherent components. Figure 6.21 shows the time series of comparative wind speed and the inversion results of correlation time and normalized delay waveform area from GEO PRN01 satellite. It is evident that the wind speed inverted from correlation time is closer to the comparative data than the normalized delay waveform.

6.2 Sea Surface Height Measurement

Sea Surface Height (SSH) refers to the height of the ocean's horizontal surface above (or below) the geoid at a given moment. Global sea surface height is usually provided by altimetry satellites, which measure the round-trip time of radar pulses from the satellite to the sea surface to determine the vertical distance between them. The difference between the satellite's orbit and the geoid distance gives the sea surface height.

6.2 Sea Surface Height Measurement

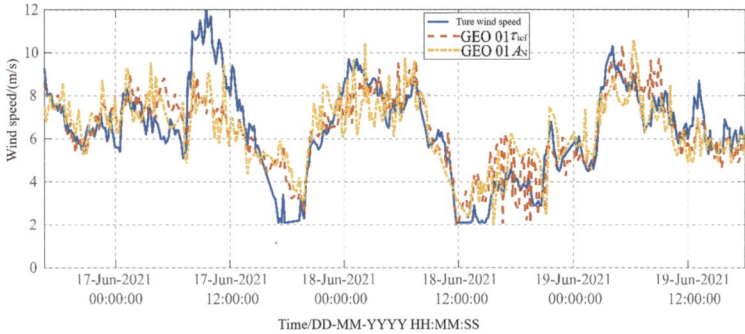

Fig. 6.21 Time series of inverted wind speed and comparative wind speed

6.2.1 Principle of Direct and Reflected Synergy for Measuring Sea Surface Height

There are two definitions of reflecting surface in this chapter. One is to assume the Earth's surface is flat, ignoring the curvature of the Earth, where GNSS signals are reflected off the Earth's surface like a mirror reflection. This method is used for height measurement in ground-based or low-altitude flight scenarios. The other uses a spherical Earth model, incorporating the effect of Earth's curvature on reflection, suitable for high-altitude or satellite-based measurements. For GNSS signals, three methods are usually used for height measurement, including code phase delay of the reflected signal relative to the direct signal, carrier phase delay and carrier frequency change. Among them, the method using code phase delay to calculate the height of the reflecting surface is the least precise, but most widely applied due to its simple model; the method using carrier phase delay offers the highest precision, but requires the reflected signal to maintain phase continuity, which is challenging to achieve on rough surfaces.

6.2.1.1 Height Measurement Geometric Model

Assuming the Earth's surface is flat, the geometric relationship of GNSS-R height measurements is shown in Fig. 6.22. The total delay of the reflected signal relative to the direct signal is

$$\rho_E = \rho_r + \rho_i \tag{6.38}$$

Since $\rho'_r = \rho_r$, there is

$$\rho_E = \rho'_r + \rho_i \tag{6.39}$$

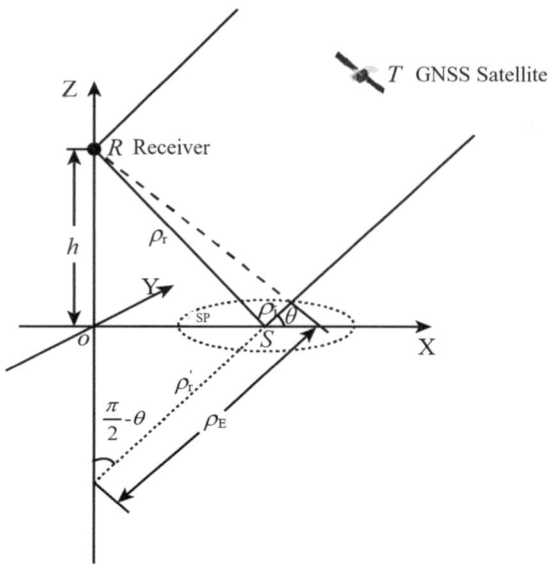

Fig. 6.22 GNSS-R height measurement geometry

$$\rho'_r + \rho_i = 2h\cos(\frac{\pi}{2} - \theta) \tag{6.40}$$

$$\rho_E = 2h\sin\theta \tag{6.41}$$

where h is the height from the receiver's reflecting antenna to the reflecting surface; θ is the satellite elevation angle at the specular reflection point; After accurately measuring ρ_E the height h from the reflection point to the receiving platform can be obtained as

$$h = \frac{\rho_E}{2\sin\theta} \tag{6.42}$$

After obtaining the height h, the sea surface height is

$$h_{ss} = h_{ref} - h \tag{6.43}$$

where h_{ref} is the vertical distance of the receiving platform from the reference plane.

If the GNSS-R receiver is mounted on a low-orbit satellite, due to the influence of the curvature of the earth, the height obtained directly by Eq. (6.42) is the vertical height from the receiving platform to the tangent line at the reflection point, hence sea surface height cannot be solved using Eq. (6.43). Considering the geometric relationship shown in Fig. 6.23, where r_t, r_r, and r_s respectively represent the distance from the center of the earth to the GNSS satellite, altimetry platform, and specular reflection point, γ_t stands for the angle between r_t and r_s, and γ_r stands for the angle between r_r and r_s, respectively. θ_r represents the angle between r_t and r, represents the

6.2 Sea Surface Height Measurement

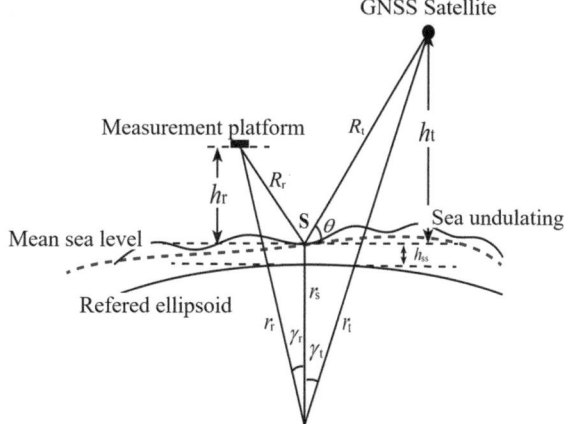

Fig. 6.23 GNSS-R geometry considering earth's curvature

angle between r and r, and is the altitude angle of the GNSS satellite at the specular reflection point. h_r and h_t are the vertical altitudes of the tangent lines from the LEO satellite and the GNSS satellite to the specular reflection point, respectively.

The specular reflection point is on the reference ellipsoidal plane, and the propagation delay of the reflected signal is referred to the theoretical delay, denoted as [28].

$$\delta\rho_{\text{modeled}} = 2(h_r + h_{ss})\sin\theta \tag{6.44}$$

In fact, the specular reflection point is on the sea surface, and the propagation delay of the reflected signal is measured from the delayed waveform or delay-Doppler waveform obtained from the receiver, denoted as

$$\delta\rho_{\text{measured}} = 2h_r \sin\theta \tag{6.45}$$

The theoretical delay minus the measured delay yields

$$\Delta\rho = \delta\rho_{\text{modeled}} - \delta\rho_{\text{measured}} = 2h_{ss}\sin\theta \tag{6.46}$$

From the above equation, the sea surface height can be obtained as

$$h_{ss} = \frac{\Delta\rho}{2\sin\theta} \tag{6.47}$$

It is evident that the sea surface height measurement models differ between coastal/airborne and satellite scenarios. In coastal/airborne, the measurement is the delay of the reflected signal relative to the direct signal, whereas in satellite scenarios, it is the delay of the reflected signal relative to the theoretical value. Regardless of being shore-based/airborne or satellite-based, the sea surface height measurement model is related to the satellite elevation angle at the specular reflection point.

6.2.1.2 Error Analysis

For different sea surface height measurement scenarios, Eqs. (6.43) and (6.47) give sea surface height measurement results related to the arrival time of the reflected signal. Accordingly, the error analysis is also based on the arrival time of the reflected signal [30].

The distance R_t from the GNSS satellite T to the specular reflection point S and the distance R_r from the altimetry platform R to the specular reflection point S can be obtained from the geometrical relations.

$$R_t = \sqrt{r_t^2 + (r_s + h_{ss})^2 - 2r_t(r_s + h_{ss})\cos\gamma_t} \tag{6.48}$$

$$R_r = \sqrt{r_r^2 + (r_s + h_{ss})^2 - 2r_r(r_s + h_{ss})\cos\gamma_r} \tag{6.49}$$

The propagation delay of the reflected signal is

$$\rho_R = R_r + R_t \tag{6.50}$$

Taking partial derivatives of both sides of Eq. (6.50) with respect to h_{ss} yields

$$\begin{aligned}\frac{\sigma_{\rho_R}}{\sigma_{ssh}} &\equiv \frac{\partial \rho_R}{\partial h_{ss}}\Big|_{h_{ss}=0} = \frac{r_s - r_t \cos\gamma_t}{R_t} + \frac{r_s - r_r \cos\gamma_r}{R_r} \\ &= -\cos(\frac{\pi}{2} - \theta) - \cos(\frac{\pi}{2} - \theta) \\ &= -2\sin\theta\end{aligned} \tag{6.51}$$

That is, the measurement error of the mean sea level height is a function of the path delay error σ_{ρ_R} and the altitude angle θ at the specular reflection point S, i.e.

$$\sigma_{ssh} = -\frac{\sigma_{\rho_R}}{2\sin\theta} \tag{6.52}$$

This equation indicates that the sea surface height measurement error is directly proportional to the measurement error of the arrival time of the reflected signal and inversely proportional to the sine of the satellite's elevation angle at the specular reflection point. Therefore, under the same measurement precision for the reflected signal's arrival time, reflected signals at higher elevation angles can achieve higher sea surface height measurement accuracy.

6.2.2 Code Delay Height Measurement Method

6.2.2.1 Path Delays

As known from Sect. 6.2.1.1, the total delay between reflected and direct signals is shown in Eq. (6.38). However, in addition to the geometric path delay ρ_E, the path delay error ρ_{atm} caused by the atmosphere and the peak delay ρ_{sca} caused by the roughness of the sea surface, etc., need to be considered in the direct and reflected signal path delay model is shown in Fig. 6.24, and the total path delay can be expressed as

$$\Delta \rho_E = \rho_E + \rho_{atm} + \rho_{sca} + \rho_n \qquad (6.53)$$

where ρ_n denotes the random measurement error.

6.2.2.2 Atmospheric Propagation Delay

Atmospheric propagation delay can usually be categorized into the dry and wet components. The dry component is mainly related to ground atmospheric pressure and temperature, and constitutes the main component of atmospheric propagation delay, accounting for about 90%. The wet component is mainly related to the atmospheric humidity and altitude of the signal propagation path, which is usually defined as the delay caused by water vapor, and its effect can be ignored in altimetry applications. Assuming a sea level atmospheric pressure of $P_0 = 1.013 \times 10^5$ Pa and the atmospheric pressure at height h is $P(h) = P_0 \cdot \exp(-h/8500)$, the delay value ρ_{atm} due to atmospheric propagation is

$$\rho_{atm}(h, \theta) = \frac{4.6127968(1 - \exp(-h/8500))}{\sin \theta} \qquad (6.54)$$

The relationship between atmospheric propagation delay and the satellite elevation angle θ and height h is given in Fig. 6.25. From the figure, it can be seen that when the height h is less than 100 m, the atmospheric propagation delay is relatively small, about 10 cm or less. For the heights greater than 1000 m, the atmospheric propagation delay decreases gradually with the increase of the satellite elevation angle.

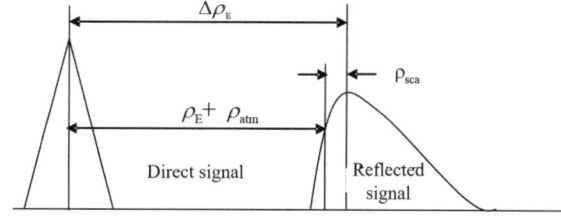

Fig. 6.24 Illustration of reflected signal path delay

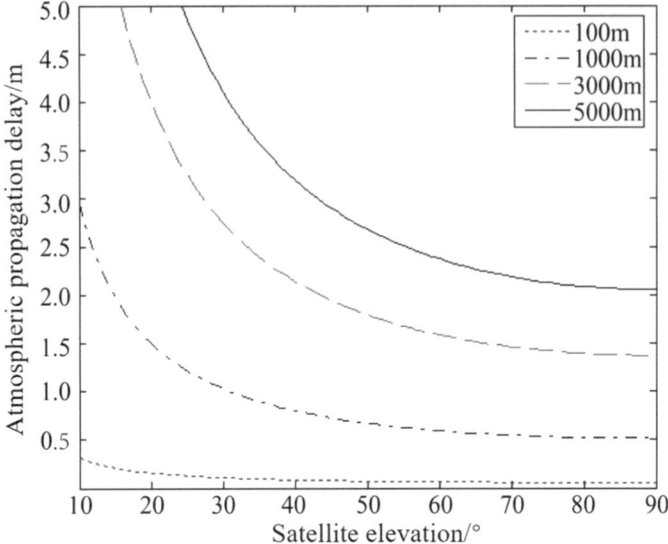

Fig. 6.25 Atmospheric propagation delay versus satellite altitude angle

6.2.2.3 Peak Delay

Due to sea surface roughness, diffuse reflection points around the specular reflection point (note: at this time, the reflected signal is also often referred to as the scattered signal; this book does not distinguish between the two) cause the propagation delay of the reflected signal to increase, shifting the peak position of the delay-Doppler power waveform backward. This peak point difference from the specular reflection point's signal arrival time is denoted as ρ_{sca}, which increases with wind speed, reaching tens to hundreds of meters, as shown in Fig. 6.26.

Random measurement errors ρ_n include effects such as multipath, clock uncertainties, antenna model errors, receiver hardware delays, thermal noise, and other unmodeled error sources.

6.2.2.4 Experimental Analysis

The authors' research team conducted several shore-based and airborne sea surface altimetry experiments using self-developed equipment [31]. The valid results of sea surface height measurements were obtained using code delay, verifying the usability of the above analysis.

The data used in this section were collected in the South China Sea at latitude 19°–19°30′ N and longitude 110°50′–111° E, using the self-developed "GNSS-R Ocean Microwave Remote Sensor" to receive GPS L1 CA code signals. The peak gain of the reflected signal receiving antenna is about 13 dB, with a beam angle of

6.2 Sea Surface Height Measurement

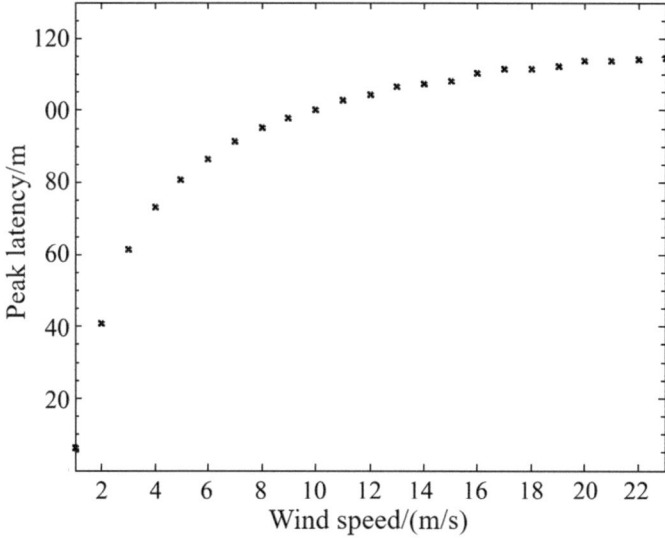

Fig. 6.26 Relationship between peak delay and wind speed

38°. The direct signal receiving antenna has a gain of 3 dB. The test aircraft was a domestically produced Y-7 airplane, flying at a maximum altitude of about 5200 m and a speed of 420 km/h.

After denoising and normalizing the direct and reflected signal power data, one-dimensional delay power curves were obtained, as shown in Fig. 6.27. The first peak in the figure is the direct signal power curve, and the second peak is the reflected signal power curve. The delay difference between the two peaks, inserted into formula, yields the delay distance of the reflected signal relative to the direct signal, from which the height of the aircraft's reflected signal antenna to the reflection surface is obtained using Eq. (6.42), with the satellite elevation angle at 52.13°. When selecting satellite 24 with an elevation angle of 63.40°, the delay distance of the direct and reflected signals is about 9899.73 m, different from the measurement result of 8743.68 m for satellite 26 with an elevation angle of 52.13°. The higher the satellite elevation angle, the greater the delay distance, consistent with Eq. (6.41).

Figure 6.28 shows the altimetry results based on C/A code phase, with an average sea surface height of 5496 m obtained from about 1 min of data, with a measurement standard deviation of 6.435 m. The average altitude during the period was 5509 m.

6.2.3 Carrier Phase Altimetry Method

For the case of a single satellite and receiver, the carrier phase pseudorange observation equations for direct and reflected signals are respectively

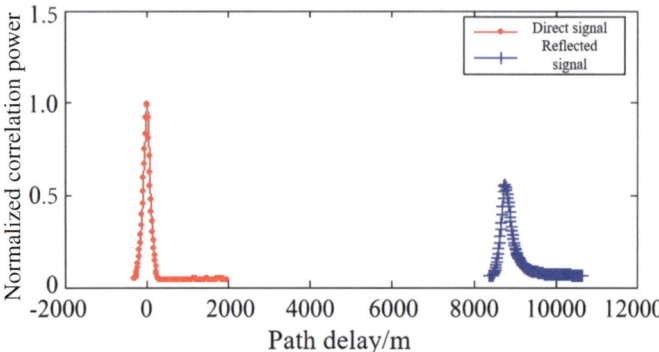

Fig. 6.27 One-dimensional delay power curves of satellite 26 direct and reflected signals

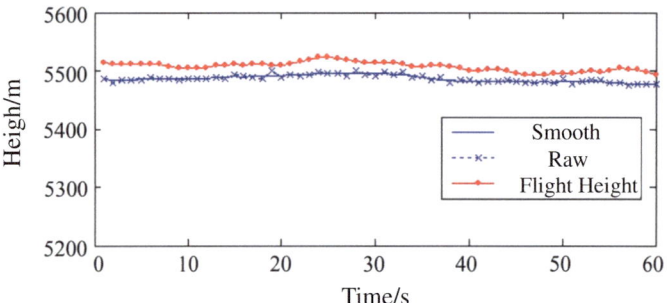

Fig. 6.28 Altimetry results based on C/A code phase

$$\lambda \varphi_{dir}(n) = [D(n) + D_{tro}(n) + D_{ion}(n)] + c\delta t_{uclock}(n) \\ - c\delta t_{dsclock}(n) - \lambda N_{0d} + \lambda \varepsilon_{\varphi d}(n) \quad (6.55)$$

$$\lambda \varphi_{ref}(n) = [R(n) + R_{tro}(n) + R_{ion}(n)] + c\delta t_{uclock}(n) \\ - c\delta t_{rsclock}(n) - \lambda N_{0r} + \lambda \varepsilon_{\varphi r}(n) \quad (6.56)$$

where $x(n)$ represents the measurement of physical quantity x at epoch n; $D(n)$ and $R(n)$ are the propagation distances of direct and reflected signals, respectively; φ_{dir} and φ_{ref} are the carrier phases measurements of direct and reflected signals, respectively; λ is the wavelength of the carriers; D_{tro} and R_{tro} are the propagation errors in the troposphere for the direct and reflected signals; D_{ion} and R_{ion} are the propagation errors in the ionosphere for the direct and reflected signals; δt_{uclock} is the user clock error. $\delta t_{dsclock}$ and $\delta t_{rsclock}$ are the satellite clock differences corresponding to the observation of direct and reflected signals. N_{0d} and N_{0r} are the integer ambiguities of direct and reflected signals; $\varepsilon_{\varphi d}$ and $\varepsilon_{\varphi r}$ are combine the effects of satellite clock

6.2 Sea Surface Height Measurement

and receiver clock phase noise, receiver hardware delay, cycle slip, etc., for direct and reflected signals, in cycles.

Differencing the original pseudorange observation equations for the direct and reflected signals of the same star essentially eliminates the time-synchronized user clocks. The satellite clock differences corresponding to the direct and reflected signals of calendar element n are not perfectly synchronized, so the difference of $\delta t_{rsclock}(n) - \delta t_{dsclock}(n)$ only weakens the effect of the satellite clock differences on the long-term stability, but does not eliminate the effect of the short-term stability. Derivation of the original pseudorange observational equations for the direct and reflected signals of the same star yields

$$\lambda[\varphi_{\text{ref}}(n) - \varphi_{\text{dir}}(n)] = R(n) - D(n) + R_{\text{tro}}(n) - D_{\text{tro}}(n) + \lambda[\varepsilon_{\varphi r}(n) - \varepsilon_{\varphi d}(n)]$$
$$- c[\delta t_{\text{rsclock}}(n) - \delta t_{\text{dsclock}}(n)] - \lambda[\Delta N_{0r} - \Delta N_{0d}] \quad (6.57)$$

(math.) simplification gives

$$\lambda \Delta \varphi(n) = \rho(n) + \Delta tro + \lambda \Delta \varepsilon_\varphi(n) - c\Delta \delta t_{\text{sclock}}(n) - \lambda \Delta N \quad (6.58)$$

where $\Delta \varphi(n) = \varphi_{\text{ref}}(n) - \varphi_{\text{dir}}(n)$ is the phase difference between the direct and reflected signals, $\Delta N = N_{0r} - N_{0d}$ is the difference between the integer ambiguity of the direct and reflected signals, $\rho(n) = R(n) - D(n)$ is the difference between the propagation distances of the direct and reflected signals, and $\Delta tro = R_{\text{tro}}(n) - D_{\text{tro}}(n)$ is the effect of the troposphere on the signal propagation due to the different propagation paths of the direct and reflected signals. $\Delta \delta t_{\text{sclock}}(n) = \delta t_{\text{rsclock}}(n) - \delta t_{\text{dsclock}}(n) \Delta \varepsilon_\varphi(n) = \Delta \varepsilon_{\varphi r}(n) - \Delta \varepsilon_{\varphi d}(n)$ is the difference between the satellite clock difference corresponding to the direct and reflected signals, and is the difference between the random observation errors of the direct and reflected signals.

The above single-difference observations are further difference between epochs, then the satellite clock difference can be basically eliminated, and the whole week ambiguity can be eliminated, obtaining

$$\lambda \Delta \varphi(n_1, n_2) = \rho(n_1, n_2) + \lambda \Delta \varepsilon_\varphi(n_1, n_2) \quad (6.59)$$

where

$$\Delta \varphi(n_1, n_2) = \Delta \varphi(n_2) - \Delta \varphi(n_1) \quad (6.60a)$$

$$\Delta \varepsilon_\varphi(n_1, n_2) = \Delta \varepsilon_\varphi(n_2) - \Delta \varepsilon_\varphi(n_1) \quad (6.60b)$$

$$\rho(n_1, n_2) = \rho(n_2) - \rho(n_1) \quad (6.61)$$

Assuming that the Earth's surface is horizontal, Eq. (6.41) can be substituted into Eq. (6.61) to get

$$\rho(n_1, n_2) = 2h(n_2)\sin\theta(n_2) - 2h(n_1)\sin\theta(n_1) \tag{6.62}$$

Due to the small change in altitude angle over a short period of time (0.1 s), when $n_2T - n_1T < 0.1s$ (T represents the time interval between neighboring calendar elements), $\sin\theta(n_2) \approx \sin\theta(n_1)$ is assumed. Equation (6.62) can be approximated as

$$\rho(n_1, n_2) = 2h(n_1, n_2)\sin\theta(n_1) \tag{6.63}$$

where $h(n_1, n_2) = h(n_2) - h(n_1)$. The value of the estimated reflective surface height change between epochs is obtained from Eqs. (6.60) and (6.63) as

$$\hat{h}(n_1, n_2) = \frac{\lambda\Delta\varphi(n_1, n_2)}{2\sin\theta(n_1)} \tag{6.64}$$

Carrier phase has finer resolution performance compared with code phase, which can provide better measurement accuracy, but there are problems such as integer ambiguity fixing, cycle slips in the solution process of carrier phase, which sometimes makes it difficult to find an effective method to solve it completely.

6.2.4 Carrier Frequency Altimetry Method

The carrier frequency is also used in practice to solve for the height of the reflecting surface based on the derivative relationship between frequency and phase. Let φ and f_E be the carrier phase difference and Doppler shift between the reflected and direct signals, respectively, then we have [37]

$$\frac{d\rho_E}{dt} = \lambda\frac{d\varphi}{dt} = \lambda \cdot 2\pi f_E \tag{6.65}$$

That is, the rate of change of path difference is a function of the frequency difference. Assuming that the height h *from the* receiver to the reflecting surface does not change at a certain moment, i.e., the change in path delay is caused only by the change in elevation angle, the derivative Eq. (6.41) with respect to time gives:

$$\frac{d(\rho_E)}{dt} = \frac{d(2h\sin\theta)}{dt} = 2h\cos\theta\frac{d\theta}{dt} \tag{6.66}$$

Combining Eq. (6.65) and Eq. (6.66) yields

$$\lambda \cdot 2\pi f_E = 2h\cos\theta d\theta/dt \tag{6.67}$$

6.2 Sea Surface Height Measurement

$$h = \frac{\lambda \cdot 2\pi f_E}{2\cos\theta \, d\theta/dt} \tag{6.68}$$

Equation (6.68) provides the relationship between the height of the receiver to the reflecting surface and the carrier frequency difference. After determining the time, the position of the GNSS satellite and the receiver, the satellite altitude angle θ and the rate of change of the satellite altitude angle $d\theta/dt$ are available. Therefore, by solving for the carrier frequency difference of the reflected signal relative to the direct signal, the height from the receiver to the reflection surface can be determined (Fig. 6.29).

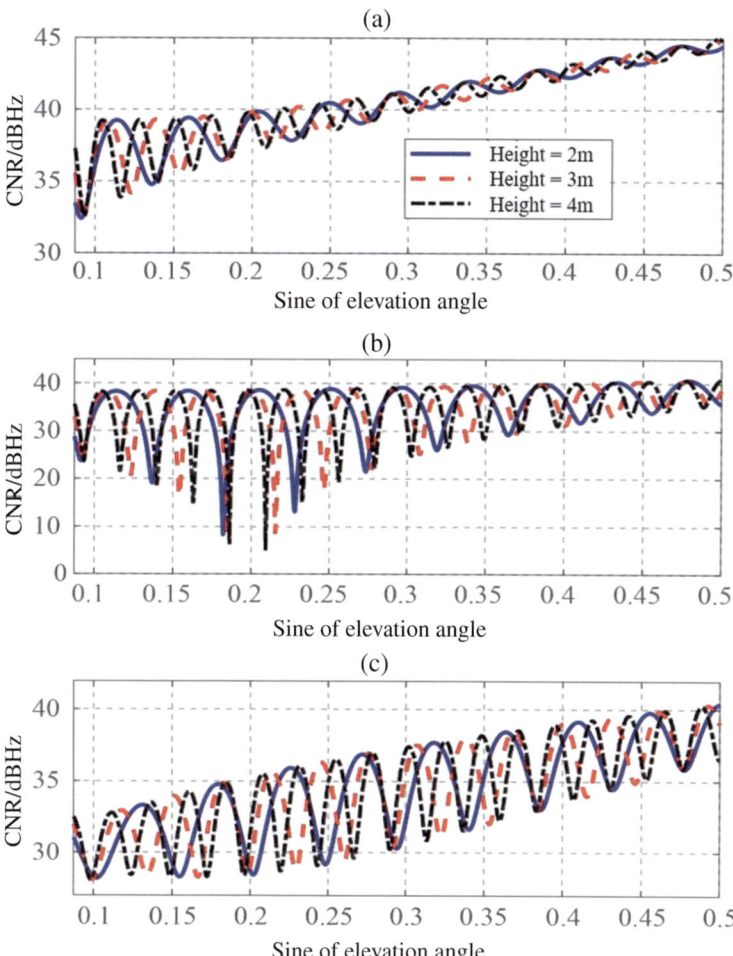

Fig. 6.29 **a** Right-handed circular polarization, **b** vertical polarization, and **c** horizontal polarization signal CNR varying with the sine of the elevation angle

6.2.5 Interferometric Sea Surface Height Measurement

6.2.5.1 Principle of Interferometric Measurement

Methods used for ocean remote sensing applications utilizing the reflected signals from navigation satellites can be categorized into two types: one is the direct-reflective collaborative method, i.e., receiving direct and reflected signals simultaneously for collaborative observation; the other is the interferometric method, i.e., extracting measurements from the direct signal that represent the effect of the reflected signal. A more detailed principle analysis and experimental validation results are provided in the literature [42]. For interferometric applications, in addition to the data realization based on Continuously Operating Reference Stations (CORS), survey-grade receivers are usually used to set up observation stations for remote sensing applications. Smartphones commonly embedded with GNSS receivers are also increasingly being used for such applications, with continuously improving signal processing capabilities. Interferometric ocean remote sensing applications can naturally be implemented on similar smartphones, with sea surface height measurement as an example.

At different heights, i.e., when the distance from the smartphone to the sea surface changes, the Carrier-to-Noise Ratio of navigation satellite signals presents different oscillation characteristics with the change in elevation angle. This forms the basis for using the interferometric method for sea surface height measurement. Figure 6.30 shows the trends of the Carrier Noise Ratio(CNR) with the sine of the elevation angle for right-handed circular polarization, vertical polarization, and horizontal polarization. From the figure, it can be seen that the higher the receiving antenna is from the sea surface, the greater the oscillation frequency of the CNR. Moreover, in both vertical and horizontal linear polarization conditions, the signal oscillation amplitude received by linear polarization antennas is larger than that received by circular polarization antennas. Additionally, as the elevation angle increases, the oscillation amplitude of the signal received by linear polarization antennas decays slower compared to circular polarization antennas.

Fig. 6.30 Test scenario for measuring sea surface height with a smartphone

6.2.5.2 Experimental Validation

Field experiments were carried out from November 02 to November 20, 2021 at Qingdong-5 Tide Gauge Station (latitude 37°26′51″N; longitude 119°0′36″E) in Dongying City, Shandong Province, with a test scenario based on the smartphone measurement of sea surface height as shown in Fig. 6.30. A Xiaomi 6 smartphone was placed approximately 9 m above the sea surface, oriented towards the southwest with an azimuth of about 210°. A float-type tide gauge provided comparative measurement data, and an anemometer provided wind speed data.

Figure 6.31 presents the data processing flow for measuring sea surface height with a smartphone. Firstly, navigation satellites PRN numbers and their signal-to-noise ratio (or carrier-to-noise ratio) sequences are parsed from the Receiver Independent Exchange Format (RINEX) file, with the satellite elevation angle and azimuth angles at the moment of data collection extracted from the ephemeris file. Due to buildings obstructing the signal in the direction the smartphone faced, the azimuth angle was limited to between 100° and 300°, and the elevation angle was not strictly limited. Qualifying carrier-to-noise ratio numerical sequences were processed through empirical mode decomposition to various intrinsic model functions (IMF), which were then iterated over. To obtain the main oscillation frequency of each component, Lomb-Scargle (L-S) spectrum estimation was firstly performed, and then the three characteristic observables of the spectrum, namely, the peak frequency f_{peak}, the spectral width W_{spe}, and the primary-secondary peak ratio R, are extracted. The spectral width threshold T_W was taken as 1.0, the primary-to-secondary peak ratio threshold T_R is taken as 1.5, and the peak frequency limit interval, $[T_{fmin}, T_{fmax}]$, is taken as [0.66, 1.14]. If all the above conditions were met, then the height of the smartphone from the sea surface was obtained using $h_r = f_{peak} \cdot \lambda / 2$. If not, the process was repeated with the next modal component.

The comparison between measured sea surface heights and comparative values from the tide gauge station is shown in Fig. 6.32, where scatter points cluster around the 1:1 line, with a root mean square error of 0.39 m, although errors are larger in the circled area.

Analyzing the characteristics of tides and wind speeds over time at the tide gauge stations, as shown in Fig. 6.33 shows, it is observed that a sudden increase in wind speed occurred at the test site on November 7, leading to an increase in sea surface roughness, weakened coherence of the reflected signals, affected the interference pattern, and thus increased measurement error.

To further analyze the effect of wind speed, wind speed values and the corresponding root mean square error values are plotted in Fig. 6.34. It can be seen that the root-mean-square error (RMSE) of the measured values increases as the wind speed increases, especially when the wind speed exceeds 12 m/s, the error value increases by nearly 60%. It should be pointed out that the vertical height of the smartphone from the sea surface in the test scenario (about 9 m here), obstruction by buildings, and the performance of the GNSS receiver in the smartphone are all possible factors affecting measurement error, and a more detailed analysis can be found in the related literature.

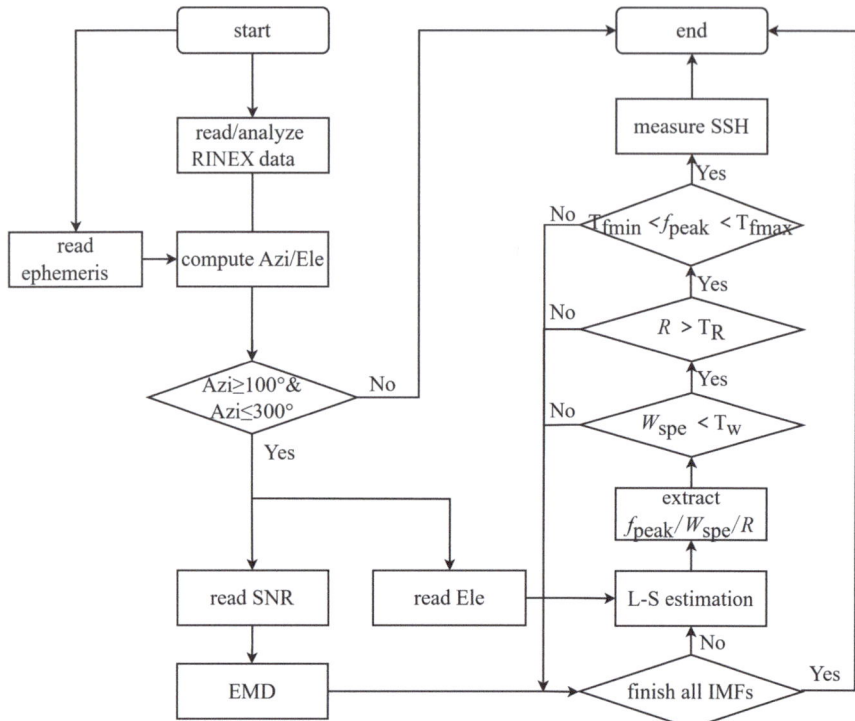

Fig. 6.31 Data processing flow for measuring sea surface height with a smartphone

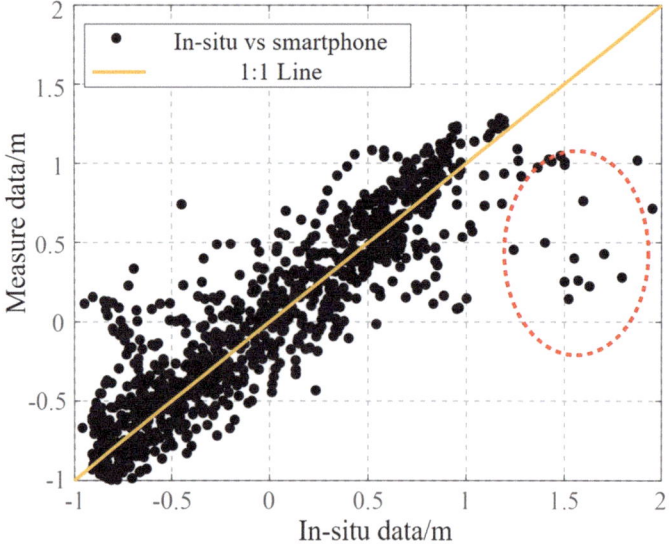

Fig. 6.32 Comparison of measured sea surface heights and comparative values

6.2 Sea Surface Height Measurement

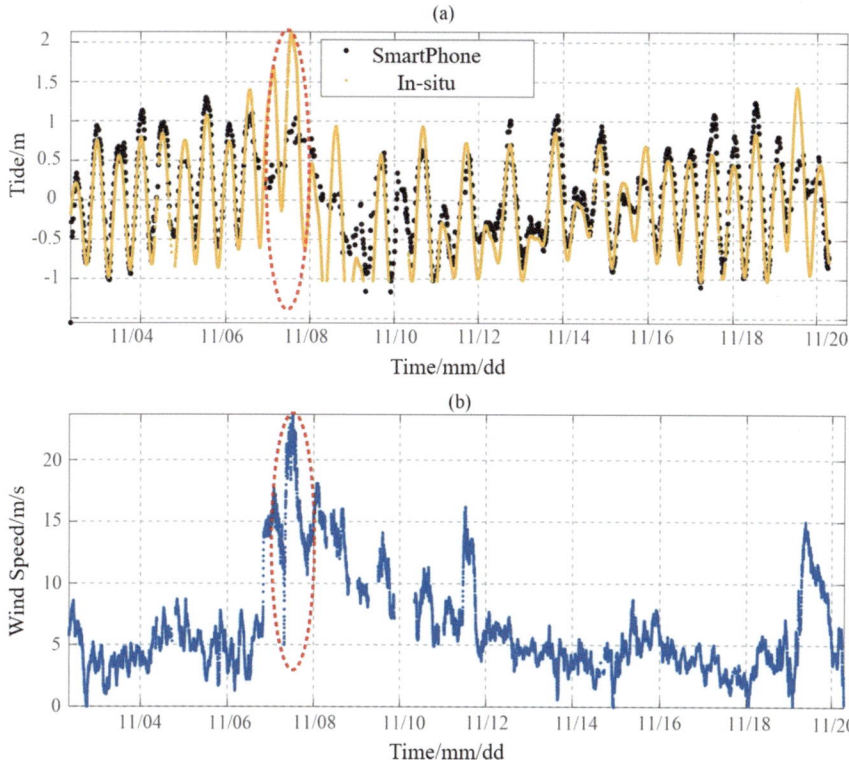

Fig. 6.33 **a** Sequence of measured and comparative sea surface heights **b** Corresponding wind speed at those times

Unlike interference oscillations received by circularly polarized antennas, linear polarization antennas exhibit effective oscillations over a wider range of elevation angles. Figure 6.35 summarizes the statistical results during the 18-day test period. Sequences of carrier-to-noise ratio with elevation angles higher than 30° can still be used to measure the sea surface height, with lower and upper limits of the height angle sequences higher than 30° for 31.9% and 46.8% of the total samples, respectively, reaching up to an 80° the maximum elevation angle. In contrast, survey-grade receivers typically use carrier-to-noise ratio sequences within 30° of elevation angle to infer sea surface height.

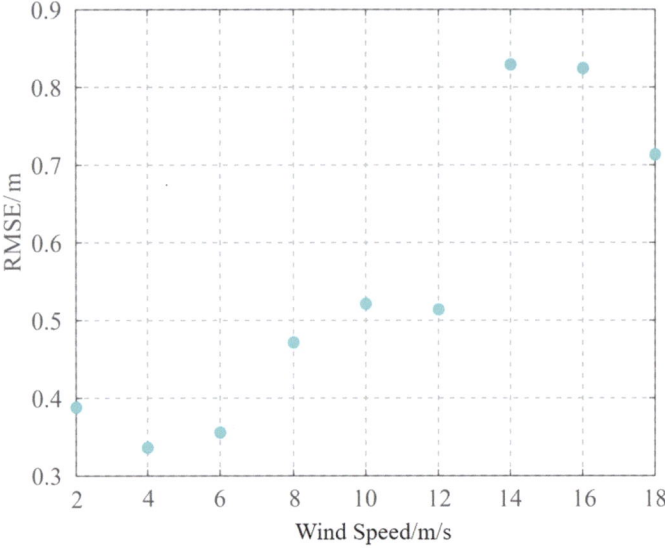

Fig. 6.34 Variation of measurement error with wind speed

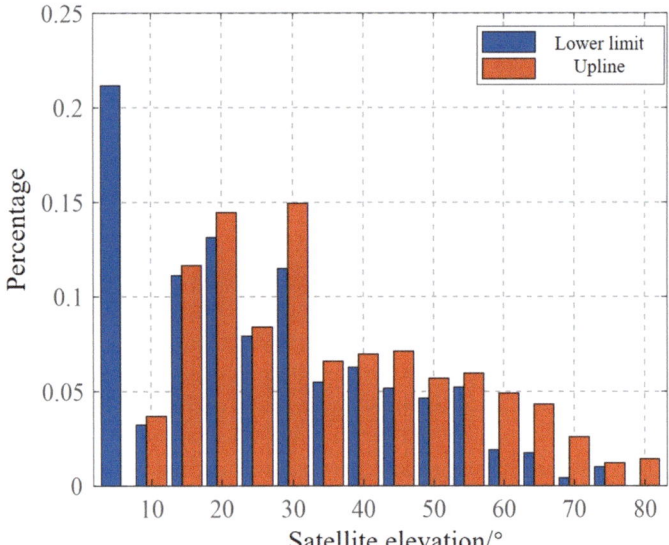

Fig. 6.35 Statistical results of measurement samples at different elevation angles

6.3 Shore-Based GNSS-R Typhoon Wind Field Reconstruction and Storm Surge Simulation

Typhoons are key drivers of storm surge, directly affecting the severity of storm surge disasters. Information on the center location, intensity and structure of the typhoon is crucial for storm surge forecasting. When wind speeds reach typhoon levels, scatterometers and radiometers, for example, cannot provide sufficiently accurate data. As mentioned earlier, shore-based GNSS-R has good performance in sea surface wind field inversion, including in the typhoon wind field reconstruction. This section focuses on related content and its application in storm surge simulation.

6.3.1 Typhoon Wind Field Reconstruction Method

Typhoon intensity is one of the important factors in determining the severity of storm surge disasters, and maximum wind speed is a key parameter in determining typhoon intensity. To enhance the precision of estimating the wind field inside and outside the typhoon, as well as the maximum wind speed, the reconstructed typhoon wind field can be expressed as follows:

$$V(r) = \begin{cases} \left(\frac{r}{RMW} \times ratio + \frac{RMW-r}{RMW}\right) \times V_{BG} & 0 \leq r \leq RMW \\ \left(\frac{r-RMW}{3RMW} + \frac{4RMW-r}{3RMW} ratio\right) \times V_{BG} & RMW < r < 4RMW \\ V_{BG} & r \geq 4RMW \end{cases} \quad (6.69)$$

where, r is the distance from the center of the typhoon; RMW is the maximum wind radius; V_{BG} is the background wind speed (here we use the reanalyzed wind field data, or we can also use the forecast wind field or the theoretical wind field); and *ratio* is a distance correction coefficient necessary for determining the reconstruction of the wind field, which is defined as the ratio of the measured data WS_{obs} (e.g., the wind speed of the on-shore GNSS-R site and the buoy data) to the background field wind speed WS_{BG}, i.e., $ratio = WS_{obs}/WS_{BG}$. In storm surge forecasting, the distance correction factor can be set based on the typhoon intensity. For example, when the forecasted typhoon intensity is too large, make the *ratio* < 1, which can effectively reduce the wind speed value.

Equation (6.69) reconstructs the wind field both with in $1 \times RMW$ (typhoon's interior) and $4 \times RMW$ (typhoon's exterior), respectively, using background field data for all other areas. From the first equation, it can be seen that when $r = 0$, the reconstructed wind speed is the background field wind speed; when $r = RMW$, the background field wind speed is multiplied by the distance revision factor; when $r < RMW$, the wind speed at each point increases or decreases depending on *the ratio* and r, and ensures that the wind speed values are smoother inside the typhoon, and do not have too large a gradient in the spatial distribution. From the second equation, it can be seen that the between $1 \times RMW$ and $4 \times RMW$ is similar to the process of wind speed reconstruction for the interior of the typhoon ($1 \times RMW$). The magnitude of the wind speed at the boundary at RMW is the same as the value of the first equation; the reconstructed wind speed at the boundary at $4 \times RMW$ is the background field wind speed, which ensures the continuity of the spatial distribution of the wind speed at the boundary at $4xRMW$ and the background field wind speed outside the $4RMW$.

6.3.2 Coastal GNSS-R Typhoon Wind Field Reconstruction

The GNSS-R Typhoon Investigation using GNSS Reflected and Interferometric Signals (TIGRIS) conducted in Yangjiang, Guangdong Province, in July–August 2013 is taken as an example to analyze the reconstruction of coastal typhoon wind field [32].

Before the wind field reconstruction, an assessment of the European Centre for Medium-Range Weather Forecasts (ECMWF) data was conducted. As shown in Fig. 6.36, compared to actual measurements at the GNSS-R coastal station, the ECMWF data showed various degrees of underestimation during 12:00 on August 13th to 00:00 on August 14th.

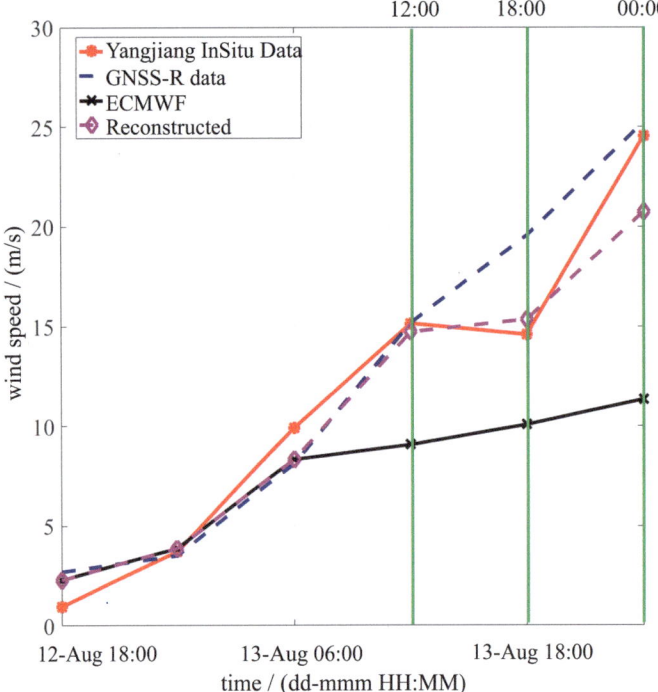

Fig. 6.36 Comparison of GNSS-R inversion results with ECMWF data (Yangjiang station)

6.3 Shore-Based GNSS-R Typhoon Wind Field Reconstruction and Storm ...

The RMW is determined based on the typhoon center location and the position of the maximum wind speed in the background wind field. The distance correction coefficient *ratio* is calculated from the shore-based GNSS-R wind speed $WS_{\text{GNSS-R}}$ and ECMWF wind speed WS_{ECMWF}.

$$ratio = \frac{WS_{\text{GNSS-R}}}{WS_{\text{ECMWF}}} \tag{6.70}$$

The typhoon wind field is reconstructed by substituting the values of *ratio at* each moment into Eq. (6.69). From Fig. 6.36, it can also be seen that at Yangjiang station, the GNSS-R inversion value and the measured data from the meteorological station both reach 25 m/s, while the maximum wind speed before the reconstruction of the ECMWF data is only 11 m/s. After the reconstruction, the wind speed reaches 21 m/s. The reconstructed wind field aligns better with actual measured data, and the accuracy of the wind speed is improved significantly, which can provide a better-quality wind field for storm surge simulation.

6.3.3 Establishment and Verification of the Storm Surge Model

6.3.3.1 Overview of FVCOM

The Finite-Volume Costal Ocean Model (FVCOM) was jointly developed by the Laboratory of Ocean Ecodynamic Modeling led by Prof. Changsheng Chen of the University of Massachusetts (UMass) and Robert C. Beardsley from the Woods Hole Oceanographic Institution, is based on the primitive equations in a Cartesian coordinate system, primarily including the momentum equation, pressure equation, continuity equation, and temperature-salinity equation [33]. Since computers cannot handle continuous problems, FVCOM discretizes these equations using the finite volume method. Unlike finite element and finite difference methods, this approach allows for individual discrete solutions of equations. It computes scalars (such as sea surface height, temperature, and salinity) on triangular nodes, similar to finite element methods, and vectors at the centroid of triangular cells to ensure mass conservation across the entire mesh. Due to the unstructured triangular mesh architecture and finite volume discretization method, FVCOM flexibly depicts complex shorelines and boundaries, finding broad application in inland lakes and rivers as well as estuaries and other geographical areas.

The FVCOM model utilizes irregular, non-overlapping triangular meshes as shown in Fig. 6.37. Compared to structured meshes, this configuration more flexibly describes intricate coastlines and allows for convenient mesh refinement in specific study areas. The number of mesh nodes and elements are counted according to the following equation.

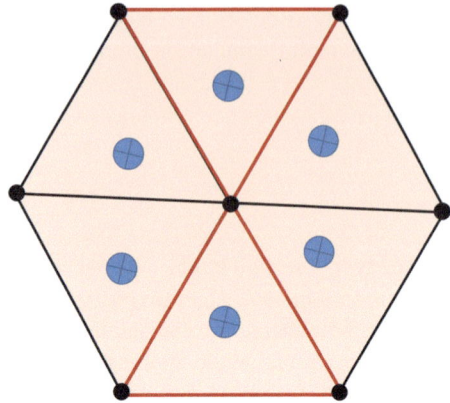

Fig. 6.37 Irregular non-overlapping triangular mesh

$$[X(i), Y(i)]\ i = 1 : N \tag{6.71}$$

$$[X_n(i), Y_n(i)]\ i = 1 : M \tag{6.72}$$

where N is the total number of triangular meshes, $[X(i), Y(i)]$ represents the coordinates of the center of mass of the triangular mesh; each triangle has three nodes, M is the total number of triangular mesh nodes, and represents the coordinates of the mesh nodes of triangular meshes arranged in clockwise order.

In addition, FVCOM features a modular design, which includes not only the basic hydrodynamic modules, but also four-dimensional assimilation, sediment, sea ice, particle tracking, tracers, ecology, and water quality modules. Its core code is continuously being upgraded. Different modules can be integrated based on research focus and objectives to accurately simulate nearshore hydrodynamic phenomena such as tides, storm surges, waves, and sediment transport. The model has been widely applied in the fields of shallow-sea physical oceanography, marine ecosystem dynamics, marine environment, theoretical oceanography, and ocean model development.

6.3.3.2 The Storm Surge Model

To finely reproduce the whole process of storm surge formation, propagation and extinction, the overall composition of the refined storm surge model is depicted in Fig. 6.38. The model couples the storm surge module and astronomical tide module of FVCOM, and realizes the effective simulation of storm surge based on boundary-driven conditions and surface forcing conditions. The core of this model is the construction of refined grid, which requires precise coastline data, based on the global high-resolution geographic data from the National Geophysical Data Center of the United States [32], employing the Surface-water Modeling Solution (SMS) software version 10.1 for generating irregular triangular meshes. Generally, the grid

6.3 Shore-Based GNSS-R Typhoon Wind Field Reconstruction and Storm ...

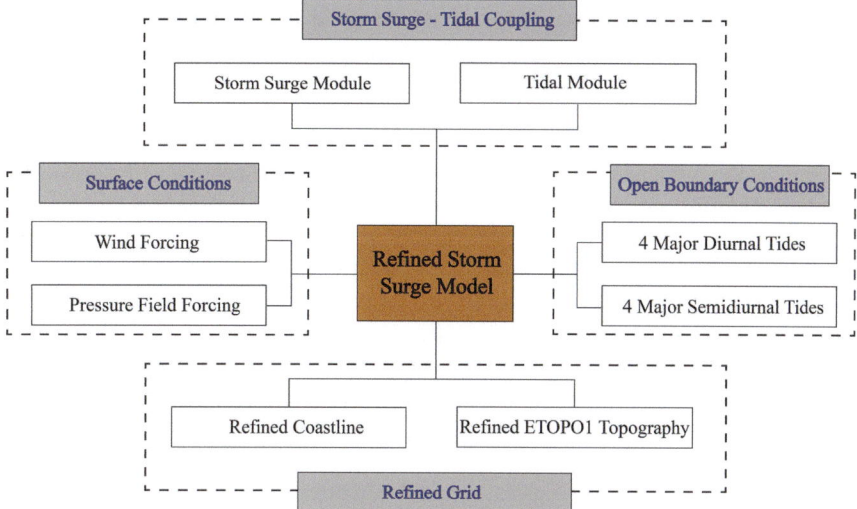

Fig. 6.38 Overall composition of the refined storm surge model

is relatively fine near the coast and sparse near the open boundary, and the transition in the middle region is automatically generated. Meanwhile, considering the range of typhoon landfall in China, the grids are encrypted in areas such as the Pearl River Estuary and the Taiwan Strait. In order to improve the computational efficiency, the horizontal resolution of the grid is set to be 2–5 km in the littoral region as well as the edge of the shelf, and 10–15 km in the open boundary region. In practice, the shape of adjacent triangles may significantly deviate from the ideal due to influences from coastlines or islands.

The ocean bathymetry data (water depth) is sourced from the 1 arc-minute resolution Earth Topography and Bathymetry Database (ETOPO1), with water depth represented by negative values. Interpolating these onto each grid yields the terrain for the entire computational area.

(1) Boundary conditions

The surface condition of the sea includes wind and atmospheric pressure fields, utilizing ECMWF reanalysis data. Open boundary conditions are tide-driven, comprising four diurnal tides (K1, O1, P1, and Q1) and four semi-diurnal tides (M2, S2, N2, and K2), obtained using the Ohio State University Tidal Prediction Software (OTPS).

(2) Initial conditions

The initial velocity and water level fields of the model are both set to zero; the vertical direction is divided into 11 layers; and temperature and salinity are set constants at 15 °C and 35 Practical Salinity Units (PSU), respectively; the model is set to positive pressure, and the effect of baroclinic pressure is not considered for the time being; the nearshore intertidal zone adapts wet-dry treatment; and the coastline boundary is treated as a closed boundary. The operational time step during simulation is 2 s, with a ratio of inner and outer steps of 10. To avoid initial oscillations caused by cold start, data output begins 15 days into the simulation and is used for storm surge analysis.

(3) Simulation program.

To evaluate the influence of reconstructed wind fields on the accuracy of storm surge simulation before and after reconstruction, three numerical simulation schemes are designed. Scheme one is used to test the performance of astronomical tide simulation within the study area; scheme two is used to test the performance of storm surge simulation before wind field reconstruction; scheme three is used to test the performance of storm surge simulation after wind field reconstruction.

Scheme one: Numerical simulation of astronomical tide alone (eight constituents);

Scheme two: Coupled numerical simulation of astronomical tide and storm surge (eight constituents + ECMWF reanalysis wind and pressure fields);

Scheme three: Coupled numerical simulation of astronomical tide and storm surge (eight constituents + ECMWF pressure field + reconstructed typhoon wind field).

6.3.3.3 Analysis and Validation of Storm Surge Model

For numerical models of ocean dynamic parameters, accurately simulating astronomical tides is a fundamental requirement. To test the astronomical tide simulation performance of the storm surge model, tidal data from tide gauge stations is used as reference data.

Based on the tidal data output by the detailed storm surge model, the nearest grid points to the tide gauge station as listed in Table 6.4 are identified. Harmonic analysis is performed separately on the tidal data from these grid points to obtain the amplitude and phase of the M2 constituent. Compared with the harmonic constants obtained from tide gauge stations along the coast of China, the root mean square errors of the amplitude and phase of the M2 constituent are 11.2 cm and 13.2°, respectively. The storm surge model can accurately reproduce the propagation of astronomical tides within the study area. Furthermore, the simulation results have high spatial resolution, providing detailed nearshore inundation information for better analysis of storm surge structure and propagation.

6.3.4 Storm Surge Simulation with Reconstructed Wind Fields

6.3.4.1 Super Typhoon "Utor"

Super Typhoon "Utor" (International Number 1311, Utor) began on August 9, 2013 and experienced two intensifications and two landfalls. Utor first intensified to reach peak intensity at 12:00 on August 11, with 10 min maximum sustained winds of

6.3 Shore-Based GNSS-R Typhoon Wind Field Reconstruction and Storm … 215

Table 6.4 Comparison of Amplitude and Phase of the M2 Constituent (Phase values relative to 120°E)

Longitude/°	Latitude/°	Observed amplitude/cm	Simulated amplitude/cm	Observed phase/°	Simulated phase/°
117.57	23.75	110	120.25	8	44.34
117.87	23.91	150	156.10	1	35.67
120.17	23.70	114	135.53	331	342.15
121.87	24.58	44	41.707	169	203.65
122.23	31.41	125	116.33	309	335.77
122.32	29.85	120	132.41	265	289.86
123.68	25.93	45	47.494	200	214.53
123.00	24.41	42	43.197	173	204.5
123.73	24.33	45	43.974	183	201.37
124.17	24.33	46	44.864	175	199.27

54 m/s and an atmospheric pressure dropping to 925 hPa, and it made landfall in the northern part of Luzon Island at around 19:00 a.m. "Utor" reached the South China Sea on the morning of August 12. "At 7:50 a.m. on August 14, "Utor" made landfall again in the coastal area of Xitou Town, Yangxi County, Yangjiang City, Guangdong Province, China, with a maximum wind force of 14 at the center at the time of landing, accompanied by heavy rainstorms, causing some rivers to exceed the alert levels. This typhoon was characterized by strong wind speeds, high waves, high tides and heavy rainfall, causing serious economic losses and casualties in China and the Philippines. It was delisted by the 46th session of the Typhoon Committee in 2014.

6.3.4.2 Storm Surge Validation

The refined storm surge model in Sect. 6.3.3.3 is used for numerical simulation, the wind fields before and after reconstruction were respectively used to simulate the storm surge caused by Typhoon "Utor" (No. 1311). Comparing the changes in the storm surge between before and after the reconstruction of the wind field reveals anomalies in sea level during the typhoon period. Specific calculations are as follows: (1) Subtracting the simulated water level of scheme two from that of scheme one to obtain the storm surge caused by the wind field before reconstruction. (2) Subtracting the simulated water level of scheme three from that of scheme one to obtain the storm surge caused by the wind field after reconstruction. The simulation results from August 11, 2013, 18:00, to August 14, 2013, 12:00, are used to analyze the storm surge caused by Typhoon "Utor."

Figure 6.39 shows the simulation results of the storm surge increase before and after wind field reconstruction from August 11 to August 14, compared with the measured data at tide gauge stations. The duration of storm surge caused by Typhoon

"Utor" is relatively slow. The storm surge at Xiaozhou Station began to rise water from 23:00 on August 12 and lasted to 17:00 on August 13; at Zhapo station, it started to increase water from 20:00 on August 12 and continued until 01:00 on August 14; Taishan station started to increase from 20:00 on August 12 and continued to 04:00 on August 13; Zhuhai station started to increase water from 21:00 on August 12 and lasted to 01:00 on August 14. At the peak of the storm surge, the simulation results of scheme two (eight constituents + ECMWF wind and pressure fields) severely underestimated the storm surge observed at Xiaozhou Station, Zhapo Station, Taishan Station, and Zhuhai Station. In contrast, the storm surge simulated by scheme three (eight constituents + ECMWF pressure field + reconstructed typhoon wind field) relatively matched the tide gauge data. Except for Xiaozhou Station, the simulation accuracy at the peak surge for the other three stations improved significantly using the reconstructed wind field. At Zhapo Station near the typhoon landfall, the simulation accuracy of the storm surge and the maximum surge height and its time of occurrence were significantly improved compared to the tide gauge data.

To quantitatively evaluate the improvement of simulation accuracy before and after reconstructing the wind field, the root mean square errors (RMSEs) between the simulation results and the measured data ares calculated separately for each tide gauge station. As shown in Table 6.5, the RMSE of the reconstructed wind field is significantly smaller than that of the pre-reconstruction results, and the overall performance of the four stations has improved by nearly 30%.

For storm surge simulation, the magnitude of surge and the time of maximum surge are two important indicators. Table 6.6 compares the simulation results before and after wind field reconstruction with the measured data. From the table, it can be seen that except for the time of maximum surge at Taishan Station, the simulation results with the reconstructed wind field can better reproduce the storm surge magnitude and the time of maximum surge caused by Typhoon "Utor." The RMSE of the storm surge simulation at Zhapo Station compared with the measured data is 14.4 cm, the maximum surge differs from the measured data by 5 cm, and the time of maximum surge is completely consistent. The simulation accuracy at the other three tide gauge stations also improved to varying degrees. For example, the maximum surge observed at Taishan Station is 1.2 m, and after wind field reconstruction, the simulated maximum surge increased from 0.39 to 1.185 m. The maximum surge times at Xiaozhou Station and Zhuhai Station also improved after wind field reconstruction.

6.4 Summary

Sea surface height and sea surface wind field are commonly used oceanic parameters or key parameters for characterizing the ocean. This chapter introduces the basic principles and methods to realize the parameter measurements by using GNSS-R, and verifies the theoretical model through the actual collected data processing. Firstly, a detailed analysis of GNSS-R for sea surface wind field retrieval is carried

6.4 Summary

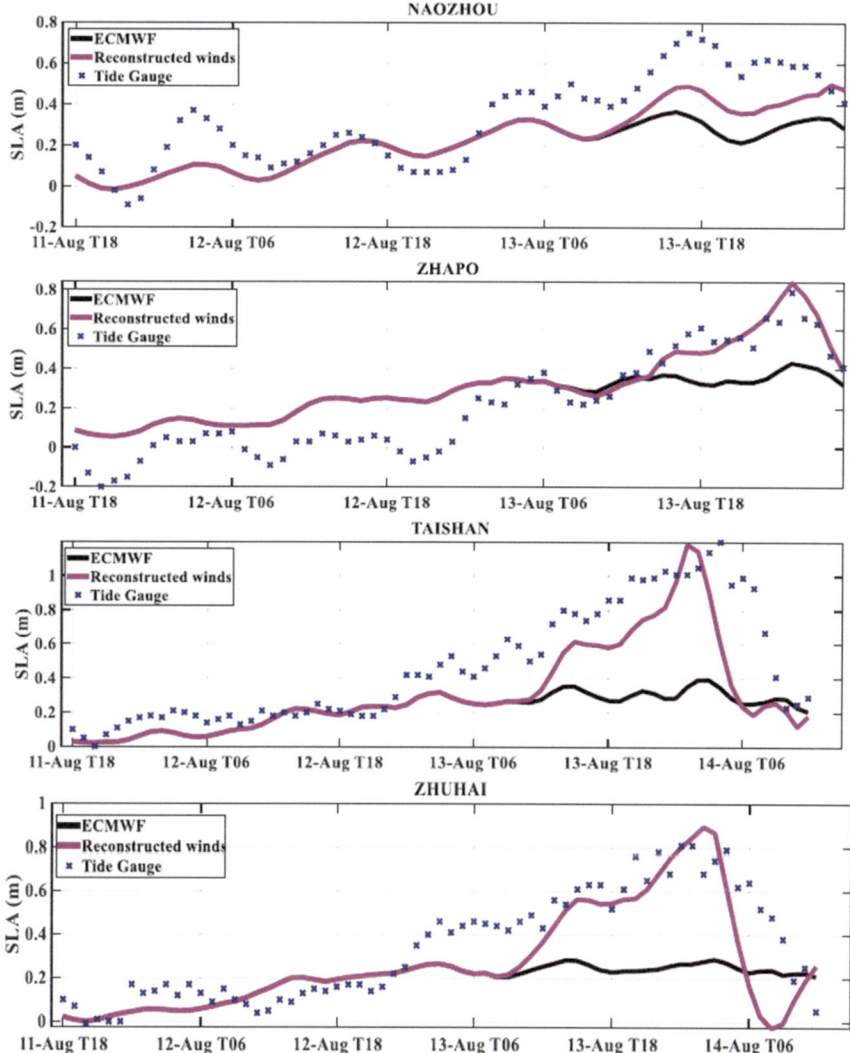

Fig. 6.39 Comparison of storm surge simulation with tide gauge station measured data

out, providing the relationship between the sea surface wind field and the statistical features of the reflected signals, and detailing the algorithm and process for sea surface wind field retrieval, and the actual data processing results. Secondly, the geometric model and error analysis of sea surface height measurement are discussed, and different methods for solving sea surface height using code phase delay, carrier phase delay, and carrier frequency are elaborated with consideration of influencing

Table 6.5 Comparison of simulation results and measured data for at Xiaozhou station, Zhapo station, Taishan station, and Zhuhai station

Tide gauge station	Comparison time range	RMSE/cm before reconstruction	RMSE/cm after reconstruction
Naozhou station	11th 18:00–14th 05:00	19.5	14.7
Zhapo station	11th 18:00–14th 05:00	17.9	14.4
Taishan station	11th 18:00–14th 12:00	35.1	22.4
Zhuhai station	11th 18:00–14th 12:00	24.6	16.1
	Average error	24.28	16.90

Table 6.6 Comparison of maximum surge and time of maximum surge simulation results with tide gauge station measured data

Tide gauge station	Data sources	Time of maximum surge	Maximum surge height / cm	Time deviation/ Hour	Surge height error /cm
Naozhou Station	Before reconstruction	13th 16:00	37	−1	−38
	After reconstruction	13th 17:00	49	0	−26
	Tide Gauge Data	13th 17:00	75	–	–
Zhapo station	Before reconstruction	14th 01:00	43	0	−36
	After reconstruction	14th 01:00	84	0	5
	Tide Gauge Data	14th 01:00	79	–	–
Taishan station	Before reconstruction	14th 03:00	39	−1	−81
	After reconstruction	14th 01:00	118.5	−3	1.5
	Tide Gauge Data	14th 04:00	120	–	–
Zhuhai station	Pre-reconstruction	14th 03:00	29	2	−52
	After reconstruction	14th 02:00	89	1	8
	On-the-spot survey	14th 00:00	81	–	–

factors and experimental verification. Finally, the application of shore-based GNSS-R data in typhoon observation, typhoon wind field reconstruction and storm surge simulation is explored (Fig. 6.40).

6.4 Summary

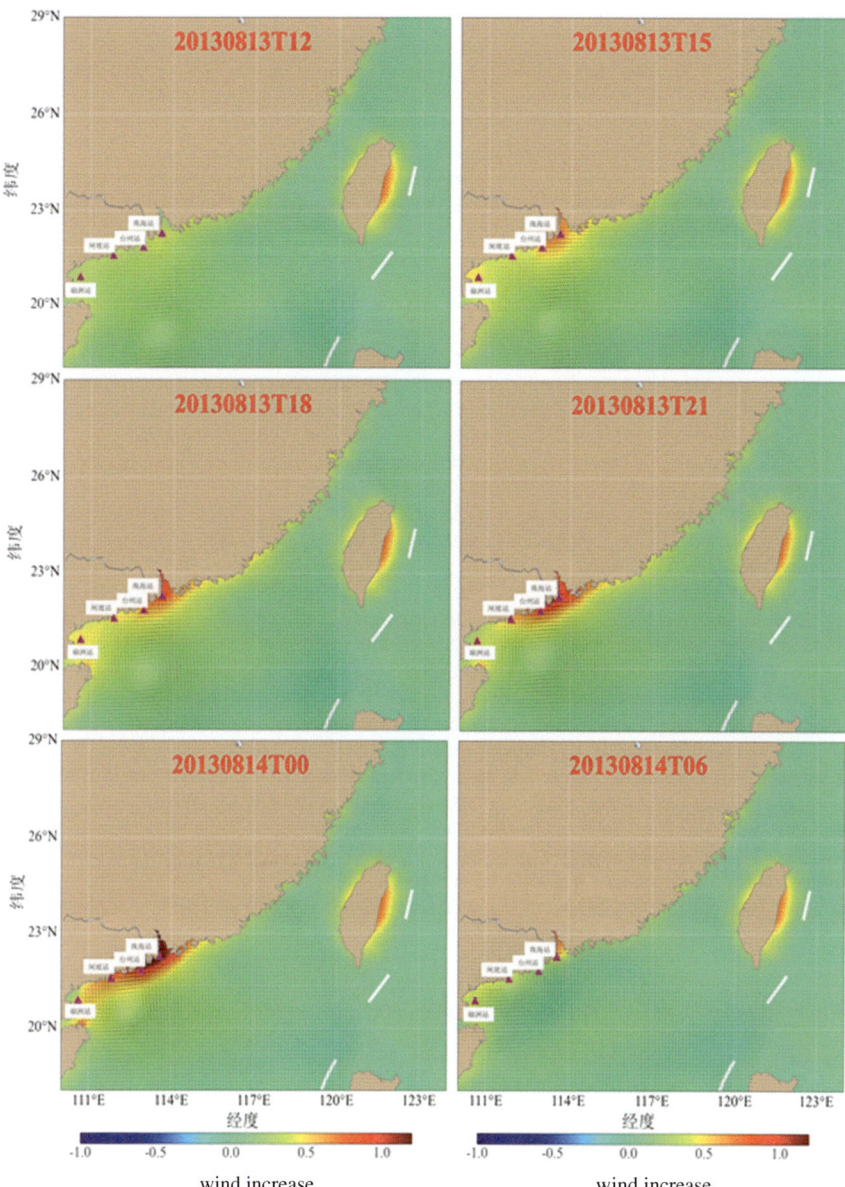

Fig. 6.40 Spatial distribution of storm surge from 12:00 on the 13th to 06:00 on the 14th

References

1. Martin-Neira M. A passive reflectometry and interferometry system (PARIS): application to ocean altimetry [J]. ESA J. 1993;17(4):331–55.
2. Martin-Neira M. Altimetry method [P]. USA: 5546087;1996.
3. Martin-Neira M, Caparrini M, Font-Rossello J, et al. The PARIS concept: an experimental demonstration of sea surface altimetry using GPS reflected signals [J]. IEEE Trans Geosci Remote Sens. 2001;39(1):142–50.
4. Auber J C, Bilbaut A, Rigal J M. Characterization of multipath on land and sea at GPS frequencies[C]. Paris: ION Conference;1994.
5. Zavorotny VU, Voronvich AG. Scattering of GPS signals from the ocean with wind remote sensing application [J]. IEEE Trans Geosci Remote Sens. 2000;38(2):951–64.
6. Armatys M, Komjathy A, Axelrad P, et al. A comparison of GPS and scatterometer sensing of ocean wind speed and direction [C]. Honolulu: IGARSS;2000.
7. Garrison JL, Komjathy A, Zavorotny VU, et al. Wind speed measurement using forward scattered GPS signals [J]. IEEE Trans Geosci Remote Sens. 2002;40(1):50–65.
8. Hajj GA, Zuffada C. Theoretical description of a Bistatic system for ocean altimetry using the GPS signals [J]. Radio Sci. 2003;38(5):1089.
9. Ruffini G, Soulat F, Caparrini M, et al. The GNSS-R eddy experiment I: altimetry form low altitude aircraft [C]. Barcelona: In: Proceedings of the 2003 workshop on oceanography with GNSS-R;2003.
10. Germain O, Rufini G, Soulat F, et al. The eddy experiment: GNSS-R speculometry for directional sea-roughness retrieval from low altitude aircraft [J]. Geophys Res Lett. 2004;31:L21307.
11. Germain O, Ruffini G, Soulat F. The GNSS-R eddy experiment II: L-band and optical speculometry for directional sea-roughness retrieval from low altitude aircraft [C]. Barcelona. In: Proceedings from the 2003 workshop on oceanography with GNSS reflections;2003.
12. Ruffini G, Soulat F, Caparrini M, et al. The eddy experiment: accurate GNSS-R ocean altimetry from low altitude aircraft [J]. Geophys Res Lett. 2004;31:L12306.
13. Emery WJ, Axelrad P, Nerem RS, et al. Student reflected GPS experiment (SuRGE) [C]. Sydney: IEEE Geoscience and Remote Sensing Systems (IGARSS); 2001.
14. Gleason S, Adjrad M. Sensing Ocean, ice and land reflected signals from space: results from the UK-DMC GPS reflectometry experiment [C]. Long beach. In: ION GNSS 18th international technical meeting of the satellite division;2005.
15. Apel JR. An improved model of the ocean surface wave vector spectrum and its effects on radar backscatter [J]. J Geophys Res. 1994;99(C8):16269–91.
16. Elfouhaily T, Chapron B, Kastaros K, et al. A unified directional spectrum for long and short wind-driven waves [J]. J Geophys Res. 1997;102(15):781–96.
17. Peake WH, Oliver TL. The response of terrestrial surfaces at microwave frequencies [R]. Tech Rep;1971. 5.
18. Bourlier C, Saillard J, Bergine G. Instrinsic infrared radiation of the sea surface [J]. Prog Electronmagnetics Res. 2000;27:185–334.
19. Berginc G. Small-slope approximation method: a further study of vector wave scattering comparison with experimental data [J]. Prog Electromagn Res. 2002;37:251–87.
20. Awada A, Khenchaf A, Coatanhay A. Bistatic radar scattering from an ocean surface at L-band [C]. Verona: IEEE Conference on Radar;2006.
21. Barrick DE, Peake WH. A review of scattering from surface with different roughness scales [J]. Radio Sci. 1986;3(8):865–8.
22. Soulat F, Caparrini M, Germain O, et al. Sea state monitoring using coastal GNSS-R [J]. Geophys Res Lett. 2004;31(21):L21303.
23. Martin F, Artin F, Camps A, et al. Significant wave height retrieval based on the effective number of incoherent averages [C]. Milan: IEEE Int Geosci Remote Sens Symp (IGARSS). 2015; 3634–3637.
24. Wang F, Yang D, Zhang B, et al. Wind speed retrieval using coastal ocean-scattered GNSS signals [J]. IEEE J Sel Top Appl Earth Obs Remote Sens. 2016;9(11):5272–83.

References

25. Marchan-Hernandez J, Valencia E, Rodriguez N, et al. Sea-state determination using GNSS-R data [J]. IEEE Geosci Remote Sens Lett. 2010;7(4):621–5.
26. Li W, Fabra F, Yang D, et al. Initial results of typhoon wind speed observation using coastal GNSS-R of BeiDou GEO satellite [J]. IEEE J Sel Top Appl Earth Obs Remote Sens. 2016;9(10):4720–9.
27. Wang Q, Zhu Y, Kasahtikul K. A novel method for ocean wind speed detection based on energy distribution of BeiDou reflections [J]. Sensors. 2019;19:2779.
28. Mashburn JR. Analysis of GNSS-R observations for altimetry and characterization of earth surfaces [D]. Boulder: University of Colorado at; 2019.
29. Anderson C, Macklin J, Gommenginger C. Study of the impact of sea state on nadir looking and side looking microwave backscatter [R]. Technical note;2000.
30. Wu S C, Meehan T, Young L. The potential use of GPS signals as ocean altimetry observable [C]. National Technical Meeting, Santa Monica, CA;1997.
31. Yang D, Wang Y, Lu Y, et al. GNSS-R data acquisition system design and experiment [J]. Chinese Sci Bull. 2010;55(33):3842–6.
32. Wessel P, Smith W. A Global. Self-consistent, hierarchical, high-resolution shoreline database [J]. J Geophys Res Solid Earth. 1996; 101(4):8741–8743.
33. Chen C, Beardsley RC, Cowles G. An unstructured grid, finite-volume coastal ocean model: FVCOM user manual, 3rd Edition [M]. Massachusetts Dartmouth: SMAST/UMASSD-11–1101;2011.

Open Access This chapter is licensed under the terms of the Creative Commons Attribution-NonCommercial-NoDerivatives 4.0 International License (http://creativecommons.org/licenses/by-nc-nd/4.0/), which permits any noncommercial use, sharing, distribution and reproduction in any medium or format, as long as you give appropriate credit to the original author(s) and the source, provide a link to the Creative Commons license and indicate if you modified the licensed material. You do not have permission under this license to share adapted material derived from this chapter or parts of it.

The images or other third party material in this chapter are included in the chapter's Creative Commons license, unless indicated otherwise in a credit line to the material. If material is not included in the chapter's Creative Commons license and your intended use is not permitted by statutory regulation or exceeds the permitted use, you will need to obtain permission directly from the copyright holder.

Chapter 7
Land Soil Moisture Retrieval

Soil moisture, also known as soil water content, is an indicator used to characterize the status of water in soil. Soil moisture is an important component of soil and plays an important role in the formation, development, and transportation of matter and energy in the soil. It has a significant impact on the growth of plants and their distribution on the land surface. As shown in Fig. 7.1, soil moisture participates in the exchange of water between the ground and the atmosphere through evapotranspiration, serving as a vital link in the global water cycle and playing an important role in the process of water conversion and global climate change. Therefore, understanding soil moisture at different scales can have positive impacts on human production and life.

Soil moisture (water content) is usually expressed in terms of gravimetric water content and volumetric water content.

(1) Gravimetric water content

The ratio of the weight of water in soil to the weight of the corresponding dry soil is known as gravimetric water content, which is dimensionless and usually expressed in percentage or g/kg.

$$m_w(\%) = \frac{M_w}{M_s} \times 100\%$$
$$m_w(g/Kg) = \frac{M_w}{M_s} \times 1000 g/Kg \quad (7.1)$$

where m_w is the soil gravimetric water content, M_w is the weight of the water in the soil, and M_s is the weight of dry soil matter.

(2) Volumetric water content

The ratio of the volume occupied by water in soil to the total volume of soil is called volumetric water content, again dimensionless and usually expressed as a percentage or m^3/m^3 (cm^3/cm^3).

© The Author(s) 2026
D. Yang and F. Wang, *Fundamentals and Applications of GNSS Reflectometry*, Navigation: Science and Technology 16,
https://doi.org/10.1007/978-981-96-4554-1_7

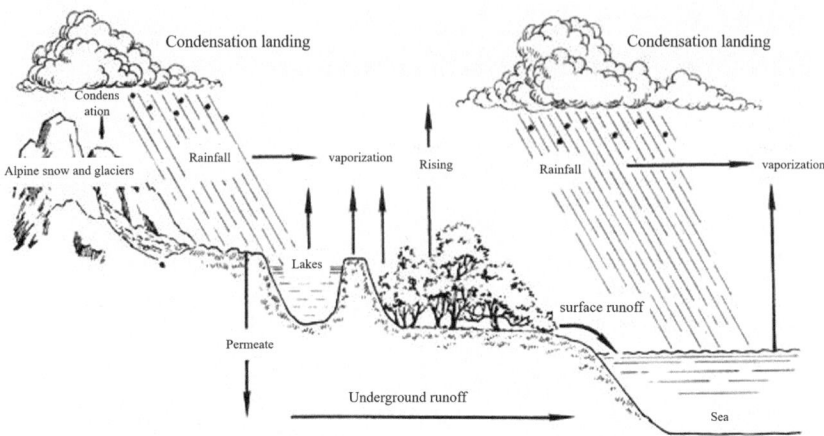

Fig. 7.1 Schematic of the water cycle

$$m_v(\%) = \frac{V_w}{V_s} \times 100\%$$
$$m_v(m^3/m^3) = \frac{V_w}{V_s} \times 1m^3/m^3 \qquad (7.2)$$

where m_v is the volumetric water content, V_w is the volume of water in the soil, and V_s is the total volume of soil.

The gravimetric and volumetric water content of soil can be converted by the following equation:

$$m_v \rho_w = m_w \rho_s \qquad (7.3)$$

where, ρ_w is the water density of the soil (unit: kg/m³), ρ_s is the bulk density of the soil (unit: kg/m³).

Soil moisture can be measured in a variety of ways, which can be generally categorized into in-situ and remote sensing measurements. Typical in-situ measurement methods include the oven-dry method, time-domain reflectometry, and frequency-domain reflectometry, etc. Among them, the oven-drying method is the only standard method that can directly measure soil moisture. The accuracy and reliability of in-situ measurement methods are generally good, but due to the significant spatiotemporal heterogeneity of soil moisture, it is very difficult to carry out long-term and large-scale observation by using in-situ measurement. Typical remote sensing measurement methods mainly include active radar [1] and passive radiometry [2, 3], which are considered to be effective solutions for long-term, large scale observation of soil moisture. However, they are more sensitive to atmospheric conditions, surface undulation and vegetation conditions, and their spatio-temporal performance and measurement accuracy need to be improved. Among the above methods, except for the oven-drying method, which can directly measure the gravimetric water content, the other methods indirectly measure the volumetric water content.

GNSS-R is characterized by not requiring a transmitter, having multiple signal sources, and being relatively less influenced by atmospheric conditions and surface features, etc. [4, 5].In addition to being used for ocean remote sensing, GNSS-R can also complement existing methods for soil moisture measurement and provide a new method for remote sensing of soil moisture.

7.1 Basic Principles of Soil Moisture retrieval

As can be known from Sect. 4.2.2, the polarization mode of GNSS signals changes after reflecting off the soil surface. According to the theory of electromagnetic wave decomposition, this reflected wave can be decomposed into a pair of right-handed and left-handed circularly polarized waves. The reflection coefficients of these polarized waves are given by Eqs. (4.28) and (4.29), both closely related to the soil dielectric constant and the electromagnetic wave incident angle. The incident angle is determined by the geometric relationship between the GNSS satellite, the GNSS-R receiver, and the reflection surface, whereas the soil dielectric constant is related to several factors including its moisture, temperature, salinity, composition, structure, and the frequency of the electromagnetic waves [6, 7]. According to the existing soil dielectric constant models, for a soil medium of known composition and structure, the soil dielectric constant is mainly dependent on soil moisture at the specified electromagnetic wave band. The following equation gives an empirical model of relating the soil dielectric constant to soil moisture at frequency of 1.4 GHz [8]:

$$\begin{aligned}\varepsilon' &= (2.862 - 0.012S + 0.001C) + (3.803 + 0.462S - 0.341C)m_v \\ &\quad + (119.006 - 0.500S + 0.633C)m_v^2 \\ \varepsilon'' &= (0.356 - 0.003S - 0.008C) + (5.507 + 0.044S - 0.002C)m_v \\ &\quad + (17.753 - 0.313S + 0.206C)m_v^2\end{aligned} \quad (7.4)$$

where ε' and ε'' are the real and imaginary parts of the soil dielectric constant, S and C are the sand and clay content of the soil, respectively, and m_v is the volumetric soil moisture.

Figure 7.2 shows the dielectric constant curves of soil with volumetric moisture ranging from 0.05 to 0.5 cm^3/cm^3 for two different soil compositions, illustrating that soil composition does impact the dielectric constant. For a given soil medium, changes in soil moisture will alter the dielectric properties of the soil, ultimately causing changes in the characteristics of the Reflected GNSS Signal.

Figure 7.3 shows the changes of the modal value and phase angle of the reflection coefficients of an ideal sandy loam medium (51.51% sand, 13.43% clay, 35.05% loam) with a flat surface under different satellite altitude angles and soil humidity,

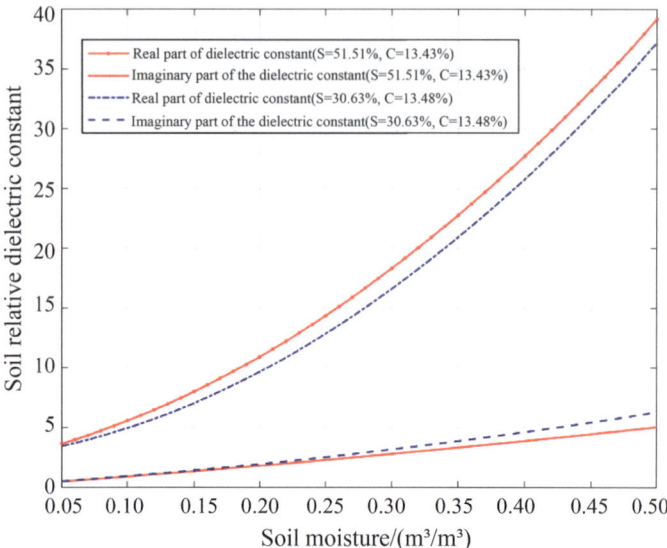

Fig. 7.2 Hallikainen model curve

and it can be seen that for the modal value of the reflection coefficients, the modulus of the reflection coefficient of $|\Re_{RR}|$ decreases with the increase of satellite altitude angle and soil humidity, and the phase angle of $|\Re_{RL}|$ increases with the increase of satellite altitude angle and soil humidity; For the reflection coefficient phase angle, the angles for both reflection coefficients change slightly with increasing satellite elevation angle and soil moisture, basically remaining within 10 degrees, but the phase angle of $\Re_{RR}$ has a cycle jump under certain satellite elevation angle and soil humidity conditions.

In fact, soil is a multi-component mixed medium with a rough interface with the air, and its interior is a complex porous system. After GNSS signals incident on the interface, part of the signals scatter towards the receiving antenna, and part transmit into the soil, scattering multiple times before returning to the air to be received by the antenna. Obviously, the actual phenomenon occurring at the soil-air interface is volume scattering rather than surface reflection, and "reflection coefficient" no longer strictly follows Eqs. (4.28) and (4.29). However, the concept of "reflection coefficient" can still be used in the case of a flat soil surface to describe the changes in signal characteristics of GNSS signals before and after reflecting off a relatively flat soil surface. From current research, the observed modulus of the reflection coefficient is smaller than the theoretical value due to the effect of soil roughness, while observed phase angle changes of the reflection coefficient are slightly larger than theoretical values due to signal transmission. Extracting features related to soil reflection coefficient from Reflected GNSS Signals and constructing a relationship model between soil moisture and these features using theoretical or empirical methods can achieve Land Soil Moisture Retrieval.

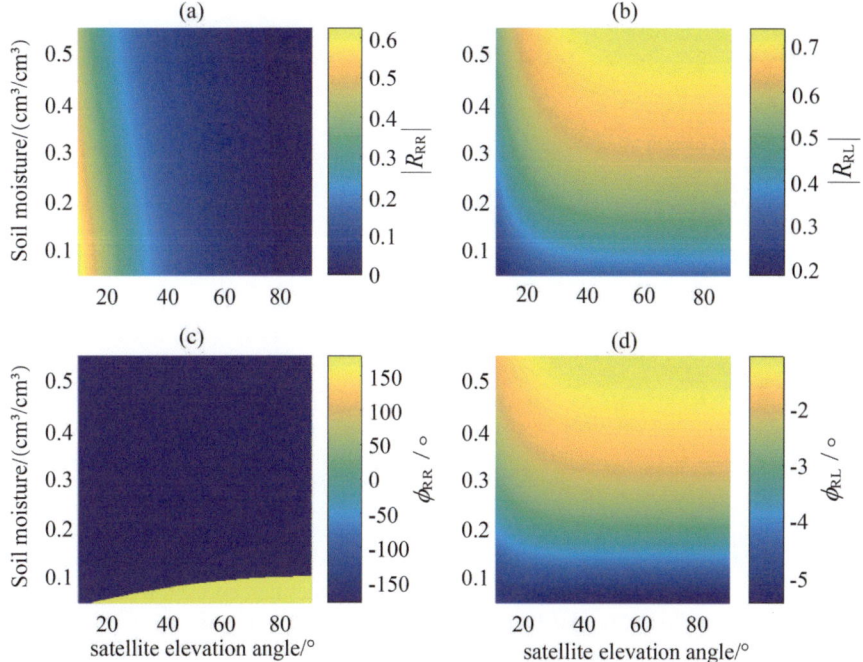

Fig. 7.3 Soil reflection coefficients at different satellite elevation angles and soil moisture levels

7.2 Dual Antenna Soil Moisture retrieval

Dual antenna soil moisture retrieval, similar to GNSS-R ocean remote sensing, involves receiving direct signals with an RHCP antenna and land surface reflected signals with an LHCP antenna. Co-processing these signals estimates the reflection coefficient to invert soil moisture. GNSS-R receivers can be installed on various platforms such as the ground, air, low-Earth orbit (LEO) satellites to measure soil moisture at different scales.

7.2.1 Extraction of Observables

As described in Chap. 4, the primary observables in GNSS reflection signal processing are their correlation power at different delays and Doppler shifts, namely the delay-Doppler Map (DDM). For GNSS-R waveform measurement applications (such as the retrieval of sea surface wind speed), in most cases, the correlation power of the specular reflection signal's Doppler and various delays, also known as the delay waveform, is considered. This is a typical observable in waveform measurement applications. Meanwhile, in GNSS-R power measurement applications (such as

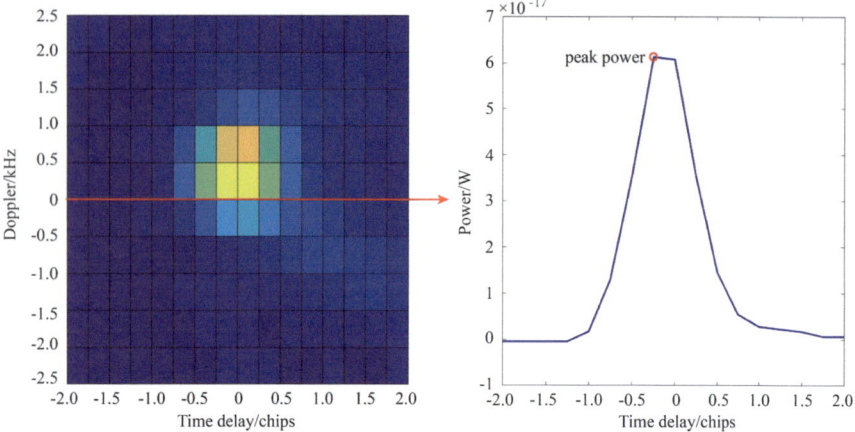

Fig. 7.4 Schematic of typical GNSS-R observables relationship

soil moisture retrieval), the correlation power of reflected signals within the specular reflection area is often considered. This is both a degenerate expression of the Delay-Doppler Map in the delay and Doppler dimensions and a degenerate expression of the delay waveform in the delay dimension, serving as a typical observable for power measurement applications. The relationship among these three observables is shown in Fig. 7.4.

In addition, the reflected signals correlation power contains both coherent and incoherent components. For relatively flat terrestrial surfaces, the coherent component dominates. When neglecting the incoherent component, the correlation power of reflected signals is given by

$$\langle |Y_r(\tau,f)|^2 \rangle = \frac{\lambda^2 P_t G_t G_r}{16\pi^2 (R_{t,sp} + R_{r,sp})^2} |\Re|^2 \chi^2(\tau,f) \tag{7.5}$$

where $R_{t,sp}$ and $R_{r,sp}$ respectively represent the distances from the GNSS satellite and GNSS-R receiver to the specular reflection point.

To reduce the impact of direct signal power variations on the measurement of reflected signal power, the correlation power of direct signals is used to normalize the correlation power of reflected signals. In reality, both direct and reflected signals contain thermal noise. The normalized ratio after excluding thermal noise is expressed as follows [9]:

$$K = \frac{\langle |Y_{r,sp}|^2 \rangle - \sigma_r^2}{\langle |Y_d|^2 \rangle - \sigma_d^2} \tag{7.6}$$

where $Y_{r,sp}$ represents the reflected signal complex correlation value, Y_d is the direct signal complex correlation value, σ_r^2 is the noise power in the reflected signal complex

7.2 Dual Antenna Soil Moisture retrieval

correlation value, σ_d^2 is the noise power in the direct signal complex correlation value, and

$$\left\langle |Y_{r,sp}|^2 \right\rangle = \frac{\lambda^2 P_t G_t^r G_r^r}{16\pi^2 (R_{t,sp} + R_{r,sp})^2} |\Re|^2 + \sigma_r^2 + \Delta_r$$

$$\left\langle |Y_d|^2 \right\rangle = \frac{\lambda^2 P_t G_t^d G_r^d}{16\pi^2 R_d^2} + \sigma_d^2 + \Delta_d$$
(7.7)

where Δ_d represents the random error in the correlation power of the direct signal and Δ_r is the random error in the correlation power of the reflected signal. For a specific satellite, the noise power in the direct-reflected signal correlation powers can be replaced either by the average correlation power outside one chip before the peak of the direct-reflected signal delay one-dimensional correlation power curve or by the average cross-correlation power with the PRN code sequence of other invisible satellites.

By substituting Eq. (7.7) into Eq. (7.6), we obtain

$$K = \frac{G_t^r}{G_t^d} \cdot \frac{G_r^r}{G_r^d} \cdot \frac{R_d^2}{(R_{t,sp} + R_{r,sp})^2} \cdot |\Re|^2 \qquad (7.8)$$

where G_r^d represents the gain of the direct signal receiving antenna, G_r^r represents the gain of the reflected signal receiving antenna, and R_d is the distance between the GNSS satellite transmitting antenna and the direct signal receiving antenna of the GNSS-R device. Note that the effect of random errors is ignored here.

7.2.2 Estimation of Soil Reflection Coefficients

In ground-based observation, the distance between the direct antenna and the specular reflection point is relatively short, and the gains of the transmission antenna in the direction of the specular reflection point and the receiving antenna direction can be regarded as approximately the same, namely

$$\frac{G_t^r}{G_t^d} \approx 1, \quad \frac{R_d}{R_{t,sp} + R_{r,sp}} \approx 1 \qquad (7.9)$$

Then, the normalized correlation power of the reflected signal is

$$K = \frac{G_r^r}{G_r^d} \cdot |\Re|^2 \qquad (7.10)$$

Define the antenna gain correction factor as

$$F_g = \sqrt{\frac{G_r^d}{G_r^r}} \qquad (7.11)$$

The modulus of the reflection coefficient for a flat bare surface can be obtained as

$$|\Re| = \sqrt{K} \cdot F_g \qquad (7.12)$$

When the soil surface is not smooth enough, with the increase of roughness, the coherent component in the reflected signal decreases and the incoherent component increases, leading to a certain attenuation in the received reflected signal strength. Define the roughness compensation factor as [10]

$$F_r = \exp(2k\sigma_s \sin\theta) \qquad (7.13)$$

where $k = 2\pi/\lambda$ is the wave number of the GNSS signal and σ_s is the root mean square height of surface undulations.

Thus, soil reflection coefficient for rough bare surface was estimated by the following equation:

$$|\Re| = \sqrt{K} \cdot F_g \cdot F_r \qquad (7.14)$$

When the soil surface is covered with vegetation, the signal scattering situation become more complex, potentially involving single scattering signals from vegetation, single scattering signals from soil, double/multiple scattering signals from vegetation, and coupling double/multiple scattering signals from vegetation and soil. The primary effect can be equivalently considered as attenuation of the signal. In this case, Eq. (7.14) is further corrected as [11]

$$|\Re| = \sqrt{K} \cdot F_g \cdot F_r \cdot F_v \qquad (7.15)$$

where, F_v is the vegetation attenuation compensation factor, which is obtained from the optical thickness of the vegetation γ and the satellite altitude angle θ

$$F_v = \exp\left(2\frac{\gamma}{\sin\theta}\right) \qquad (7.16)$$

It should be noted that an implicit premise of the above derivation is that the direct reflection signal processing channels of the GNSS-R device have the same gain. When the consistency of the GNSS-R receiving channels is poor, separate corrections are also needed.

7.2.3 Soil Moisture retrieval Tests

After obtaining the soil reflection coefficient, the soil moisture value is solved using theoretical or empirical models. Based on the soil dielectric constant model and the corrected reflection coefficient, the mathematical expression for the retrieval of soil moisture (here referring to volumetric moisture content) can be derived, as shown in Eq. (7.17).

$$m_v = \frac{-b + \sqrt{b^2 - 4ac}}{2a} \tag{7.17}$$

where,

$$\begin{aligned} a &= 2.862 - 0.012S + 0.001C \\ b &= 3.803 + 0.462S - 0.341C \\ c &= 119.006 - 0.500S + 0.633C - \varepsilon^{-1}(|\Re_{RL}|)\big|_\theta \end{aligned} \tag{7.18}$$

where $\varepsilon^{-1}(\cdot)\big|_\theta$ is the inverse function of the relationship modeled between the reflection coefficient and the dielectric constant at a satellite elevation angle of θ and the effect of the imaginary part of the soil dielectric constant is neglected here.

Below, an analysis is conducted on the soil moisture retrieval experiments carried out by the authors' research group in 2014−2015.

1. **Experimental Scene Description**

Three independent data collection experiments were carried out in November 2014, April and May 2015 at the experimental field of Daiyue District Meteorological Station, Tai'an City, Shandong Province, China (36.16°N, 117.15°E), which were recorded as Experiments 1, 2 and 3, respectively. During the experiments, the field surface was leveled, and the winter wheat was planted. A satellite view of the field is shown in Fig. 7.5.

During the period, November 26–28, 2014, was the tillering stage of winter wheat, April 14–16, 2015 was the jointing stage, and May 19–21 was the filling stage. The growth conditions of winter wheat during these stages were shown in Fig. 7.6a−c, respectively. The water content and density information of winter wheat were collected during the last two stages, as shown in Table 7.1.

The setup of the data collection equipment during the experiments is shown in Fig. 7.7. The antenna was installed at a height of about 5 m, with the RHCP antenna's maximum gain direction pointing to the zenith and the LHCP antenna's maximum gain direction obliquely pointing to the ground, at an azimuth of about 116° and an elevation angle of about 45°.

At every whole hour during the GNSS direct and reflected signal collection process, a multi-point soil sample collection was conducted. Then, using the oven-drying weighing method, the soil weight moisture of each sample point was measured and averaged. Based on the predetermined bulk density of the soil, this was converted

Fig. 7.5 Satellite view of experimental site in Tai'an, Shandong

Fig. 7.6 Growth conditions of winter wheat during the experimental period

Table 7.1 Parameters of winter wheat collected during the experimental period

Vegetation parameters	Experiment 1	Experiment 2	Experiment 3
Number of sample plants	N/A	20 plants	20 plants
Sample wet weight	N/A	90.97 g	175.75 g
Sample dry weight	N/A	16.61 g	60.54 g
Plant density	N/A	1078 plants/m^2	477 plants/m^2

into the soil volumetric moisture content, serving as the true value of soil moisture at the time of soil sample collection. Figure 7.8 shows the soil sampling and oven-drying process during the measurement of soil moisture.

2. **Results Analysis of Experiment 1**

In Experiment 1, as winter wheat was in the tillering stage, the impact of wheat seedlings on the soil reflection signal could be neglected. Therefore, the obtained soil moisture served as the measurement result under bare soil conditions. Conversely,

7.2 Dual Antenna Soil Moisture retrieval

Fig. 7.7 Installation of the signal collection equipment

Fig. 7.8 Soil sampling and oven-druing weighing

during Experiments 2 and 3, the impact of wheat plants could not be ignored and was considered as measurements under vegetation-covered conditions, requiring correction.

During the three experiments, the specular reflection point trajectories of visible stars and the corresponding spatial distribution of antenna gain were essentially the same. Figure 7.9 shows the results of a particular moment during Experiment 1. The pentagram marks the ground projection position of the antenna, with BeiDou PRN01, PRN04, and PRN08 satellites 'reflected signals' specular reflection points within the experimental area. PRN01 and PRN04 are GEO satellites, and PRN08 is an IGSO satellite.

The soil moisture retrieval results and error analysis for Experiment 1 are given in Fig. 7.10. The x-axis from 0–1800 represents each minute from 8:00–18:00 during

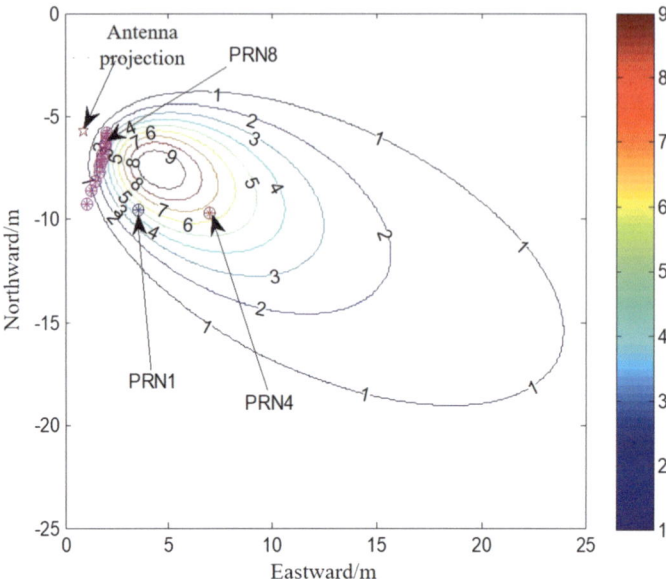

Fig. 7.9 Field reflection antenna gain distribution and specular reflection points of some visible stars

the experiment period (the results of the other two experiments are described in the same way), with precipitation occurring during some of the experiment periods. It can be observed that the precipitation process is quite evident in the inverted soil moisture results. Among the three observation satellites, the results from the two GEO satellites, PRN 01 and PRN 04, were more continuous than those from the IGSO satellite PRN08, due to the motion of the IGSO satellite causing continual changes in the observation area. The overall root mean square error of the three satellites' measurement results was about $0.083 cm^3/cm^3$, with the average absolute errors being $0.033 \ cm^3/cm^3$, $0.046 \ cm^3/cm^3$, and $0.137 \ cm^3/cm^3$, respectively.

Further linear correlation analysis between GEO satellite observations and in situ measurements is further carried out, as shown in Fig. 7.11. The correlation coefficients between the observation results of PRN01 and PRN04 and the actual measurement results were 0.88 and 0.68, respectively, with measurement root mean square errors of 0.040 and 0.055 cm^3/cm^3. Evidently, the observation results of PRN01 were slightly better than those of PRN04, due to PRN01 having a higher elevation angle, stronger reflected signal, and smaller retrieval error. After averaging the observation results of the two GEO satellites, the correlation coefficient with the in-situ measurement results was 0.90, with a root mean square error of 0.030 cm^3/cm^3, clearly showing that averaging improved the soil moisture retrieval performance.

3. **Analysis of the results of tests 2 to 3**

Since Experiments 2 and 3 were conducted after the jointing stage of winter wheat, the influence of wheat plants must be considered. At this point, it is first necessary to

7.2 Dual Antenna Soil Moisture retrieval

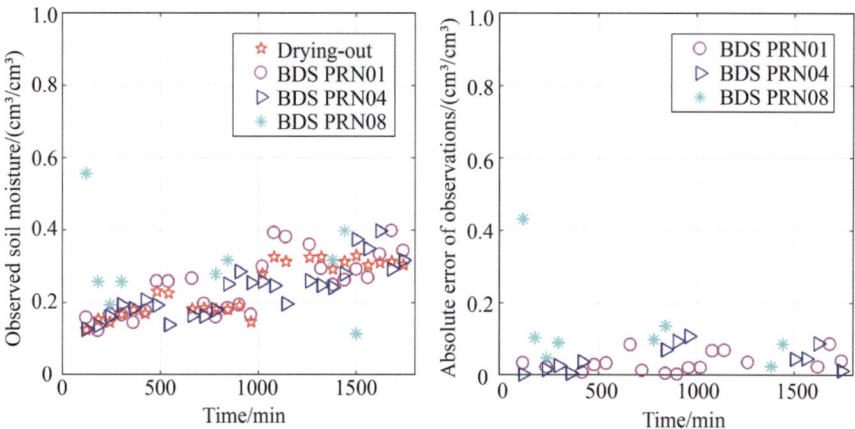

Fig. 7.10 Soil moisture measurement results and absolute error during experiment 1

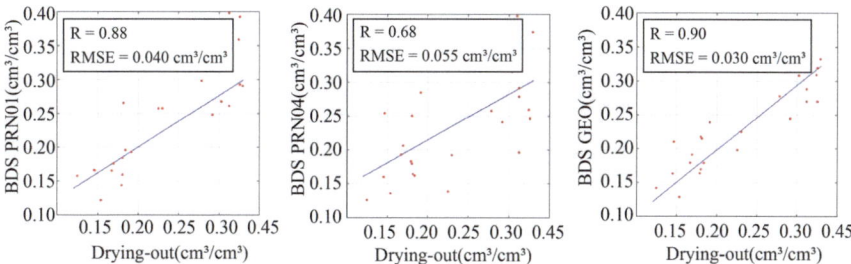

Fig. 7.11 Correlation analysis of BDS GEO satellite measurement results

calculate the optical thickness of the winter wheat vegetation, as shown in Eq. (7.19).

$$\gamma = b \cdot PWC \tag{7.19}$$

where, b is the empirical coefficient, with the usual range of values for wheat crops being 0.12 ± 0.03, and the middle value of 0.12 is taken in this paper, and PWC is the vegetation water content per unit of land area, which is calculated by Eq. (7.20):

$$PWC = \frac{m_{wet} - m_{dry}}{N_{sample}} N_{total} \tag{7.20}$$

where, N_{sample} is the number of sample plants, m_{wet} is the total wet weight of sample plants, m_{dry} is the total dry weight of sample plants and N_{total} is the plant density. According to the vegetation parameters given in Table 7.1, the water content of vegetation per unit land area during Experiment 2 and 3 were 4.008 kg/m² and 2.748 kg/m², respectively.

The soil moisture observations results for Experiments 2 and 3 are given in Figs. 7.12 and 7.13, respectively. Note that due to the influence of vegetation, the observation data from the PRN08 satellite is almost unavailable, and the focus here is on analyzing the measurement results from the two GEO satellites, PRN01 and PRN04.

The measurement results of Experiments 2 and 3 showed more fluctuations compared to Experiment 1, with relatively larger absolute errors. In Experiment 2, the absolute average errors of the measurement results from PRN01 and PRN04 were 0.073 cm^3/cm^3 and 0.054 cm^3/cm^3, respectively, while in Experiment 3, the average absolute errors were 0.056 cm^3/cm^3 and 0.155 cm^3/cm^3, respectively. Table 7.2 shows the average deviations and standard deviations of PRN01, PRN04, and their

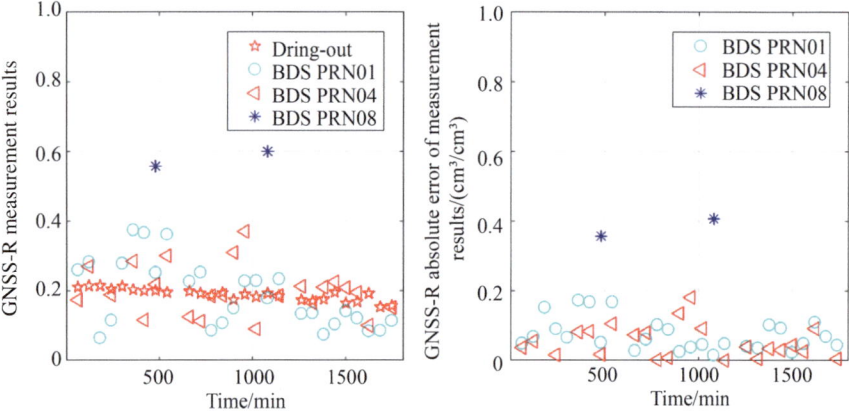

Fig. 7.12 Soil moisture measurement results and absolute error during experiment 2

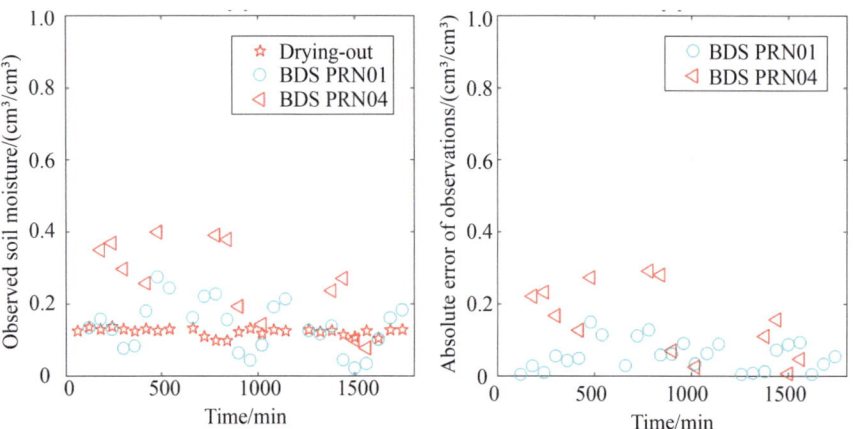

Fig. 7.13 Soil moisture measurement results and absolute error during experiment 3

7.3 Single Antenna Soil Moisture retrieval

Table 7.2 Statistics analysis of soil moisture measurements errors for Experiments 2 and 3

Observations	Test 2		Test 3	
	Mean deviation $(cm^3/cm)^3$	Standard deviation $(cm^3/cm^3)(cm^3/cm)^3$	Mean deviation $(cm^3/cm)^3$	Standard deviation $(cm^3/cm)^3$
PRN 01	−0.003	0.095	0.013	0.069
PRN 04	0.010	0.071	0.144	0.112
Average of PRN01 and PRN04	≈0	0.070	0.050	0.078

Table 7.3 HD8030 chip main technical parameters

Parameter name	Parameter value
Navigation system	GPS, BDS
Frequency	L1, B1
RF front-end quantization bits	3 bit
Capture sensitivity	−148 dBm
Tracking sensitivity	−165 dBm
Operating voltage	2.5~3.6V
Operating current	25 mA
Main frequency	100 MHz
On-chip SQI FLASH (for firmware storage)	512KB
On-chip SRAM	256 KB
On-chip real-time clock (RTC)	32,768 Hz
Interfaces/peripherals	USB × 1,UART × 2,SPI × 3,I2C × 1,PWM × 2, SDRAM × 1,TFTC × 1,GPIO × 56
Package	BGA100
Sizes	7.0 mm × 7.0 mm

average values; the average deviation of observation results in Experiment 2 was smaller, especially evident for PRN04.

7.3 Single Antenna Soil Moisture retrieval

Single-antenna soil moisture retrieval is to the dual-antenna approach. Its basic principle is similar to the interferometry method for sea level measurement discussed in Sect. 6.2.5, which also involves a single antenna receiving a direct-inverse signal simultaneously and extracts the characteristic quantity resulting from the interference of the two signals. A mapping relationship between soil moisture and the characteristic quantity is established, thereby applying it to soil moisture retrieval. This chapter distinguishes between single antenna and dual antenna methods from the system perspective. In the previous chapter, the distinction between the collaborative method (corresponding to the dual-antenna method here) and the interferometry method (corresponding to the single antenna here) was made from the signal processing perspective.

Generally, remote sensing applications of navigation satellite reflectance signals are abbreviated as GNSS-R. Correspondingly, single-antenna interferometry applications are abbreviated as GNSS-IR (Interferometric Reflectometry) or GNSS-MR (Multipath Reflectometry). Since the prerequisite for interference is that the direct and reflected signals have the same frequency, such applications are only suitable for ground-based observation scenarios where the receiving antenna is installed at a low height. Additionally, utilizing existing Global Continuous Operating Reference Stations can also achieve soil moisture measurement and information services for the site's region.

7.3.1 Extraction of Observables

Consider the ground-based GNSS-IR soil moisture observation scenario shown in Fig. 7.14, where the vertical height from the RHCP antenna phase center to the ground is H, and the GNSS satellite elevation angle is θ.

Ignoring the effect of antenna cross-polarization gain, the signal received by the RHCP antenna can be expressed as

$$s(t) = s_d(t) + s_r(t) + n(t) \tag{7.21}$$

Fig. 7.14 Schematic of GNSS-IR observation scenario

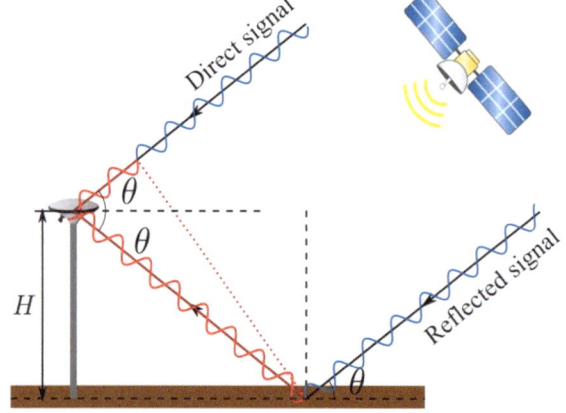

7.3 Single Antenna Soil Moisture retrieval

where s_d is the direct signal, s_r is the reflected signal, and n is the system noise.

In ground-based observation applications, the antenna phase center height is generally low, allowing for the following approximations [15]: (1) the GNSS direct signal arriving at the RHCP antenna has the same power density as the direct signal arriving at the specular reflection point;(2) the GNSS direct signal and the reflected signal reaching the RHCP antenna have the same carrier frequency. Assuming that the reflecting surface is horizontal, the GNSS direct and reflected signals are represented as

$$s_d(t) = A_d(t)D(t)C(t)e^{-j(2\pi f_0 t + \varphi_0)} \\ s_r(t) = A_r(t)D(t-\tau)C(t-\tau)e^{-j(2\pi f_0 t + \varphi_0 - \varphi_{RR})} \tag{7.22}$$

where A_d is the amplitude of the direct signal, A_r is the amplitude of the reflected signal, $D(\cdot)$ is navigation message data bits, $C(\cdot)$ is the pseudo-random noise code sequence, f_0 is the carrier frequency, φ_0 is the carrier initial phase, and φ_{RR} is the phase angle of the signal reflection coefficient $\Re_{RR}$.

The amplitude of the GNSS direct and reflected signals is [12]

$$A_d(t) = \sqrt{\zeta(t) \frac{\lambda^2 G[+\theta(t)]}{4\pi}} \\ A_r(t) = \sqrt{\zeta(t) \frac{\lambda^2 G[-\theta(t)]}{4\pi}} |\Re_{RR}(t)| \tag{7.23}$$

where, ζ is the power density of the GNSS signal at the antenna and reflection point, λ is the signal wavelength, G is the RHCP gain distribution, $|\cdot|$ represents the modulus operator, and "+"and "−" represent the forward and backward incidences by the antenna. The signal-to-noise ratio data series of the GNSS interferometric signal is given by [13]

$$SNR(i) = SNR_d(i) + SNR_r(i) + 2\sqrt{SNR_d(i)SNR_r(i)} \cos[2kH \sin \theta(i) + \varphi_{RR}(i)] \tag{7.24}$$

$$SNR_d(i) = \frac{\lambda^2 \psi(i)}{4\pi \sigma^2} G[+\theta(i)] \\ SNR_r(i) = \frac{\lambda^2 \psi(i)}{4\pi \sigma^2} G[-\theta(i)] \cdot |\Re_{RR}(i)|^2 \tag{7.25}$$

where, σ^2 is the noise power; $k = 2\pi/\lambda$ is the GNSS signal wave number.

The GNSS interferometric Signal-to-Noise Ratio(SNR) contains two types of components, the trend term and the oscillatory term, where the former contains the sum of the direct reflection signals SNRs, and the latter represents the cross-term of SNRs of the direct and reflected signals. Changes in soil moisture affect the characteristics of the reflected signal, thus both the trend and oscillatory components of the interferometric SNR are related to soil moisture. However, in practice, the direct signal received by the antenna is much stronger than the reflected signal, so the oscillatory term is more sensitive to soil moisture than the trend term, and usually only the oscillatory component is used to invert the soil moisture.

The oscillatory component, mainly dominated by the relatively weaker reflected signal, has a cosine-like waveform. When the soil surface is rough, the cosine-like oscillatory component produces irregular distortions, and a constant amplitude standard cosine function cannot describe the additional information brought by corresponding distortions. Reference[14] introduced a waveform reconstruction method based on an adaptive filtering algorithm. This method first fits the SNR data with a low-order polynomial to obtain the trend component (denoted as P_0), and the SNR data after removing the trend component is

$$SNR_c(i) = P_1(i) + P_2(i) \cos\left[2\pi f(i) \sin \theta(i) + \varphi_0(i)\right] \tag{7.26}$$

where P_1 is the fitting error of the trend component, P_2 is the time-varying oscillation amplitude; f is the time-varying oscillation frequency; and φ_0 is the initial phase.

7.3.2 Estimation of Soil Reflection Coefficients

To obtain the ratio of the direct to the reflected signal power of the soil surface, i.e., the soil reflection coefficient, further processing of the SNR data with the trend component removed is required. Equation (7.26) is rewritten in the form of vector multiplication as follows:

$$SNR_c(i) = \begin{bmatrix} P_1(i) & P_2(i) \cos \varphi_0 & P_2(i) \sin \varphi_0 \end{bmatrix} \begin{bmatrix} 1 \\ \cos\left[2\pi f(i) \sin \theta(i)\right] \\ -\sin\left[2\pi f(i) \sin \theta(i)\right] \end{bmatrix} \tag{7.27}$$

where the row vector contains the unknown parameters to be estimated P_1, the P_2 and φ_0; the column vectors contain the quantities estimated in the previous section. Using SNR_f as the filter output and c_0, c_1 and c_2 as the filter coefficients, a filter of the following form is constructed, i.e.

$$SNR_f(i) = \begin{bmatrix} c_0 & c_1 & c_2 \end{bmatrix} \begin{bmatrix} 1 \\ \cos\left[2\pi f(i) \sin \theta(i)\right] \\ -\sin\left[2\pi f(n) \sin \theta(i)\right] \end{bmatrix} \tag{7.28}$$

The coefficients are adjusted to minimize the root-mean-square error between the filter output SNR_f and the input SNR_c, and the corresponding coefficients c_0, c_1 and c_2 are the valuations of P_1, $P_2(i) \cos \varphi_0$ and $P_2(i) \sin \varphi_0$. Figure 7.15 shows the working principle of the constructed filter.

7.3 Single Antenna Soil Moisture retrieval

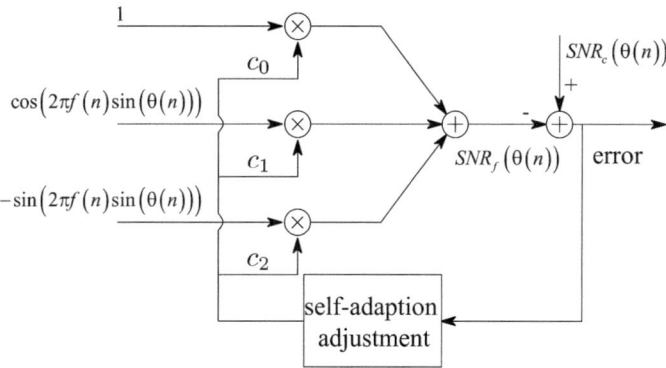

Fig. 7.15 Schematic diagram of the adaptive filter's working principle

Based on the trend term P_0 and the filter coefficients c_0, c_1 and c_2, the following set of equations is constructed:

$$\begin{cases} S_d(i) + S_r(i) = P_0(i) + c_0(i) \\ 2\sqrt{S_d(i) \cdot S_r(i)} = \sqrt{c_1^2(i) + c_2^2(i)} \end{cases} \quad (7.29)$$

The solution is

$$\begin{cases} S_r(i) = \frac{P_0(i)+c_0(i)-\sqrt{[P_0(i)+c_0(i)]^2-[c_1^2(i)+c_2^2(i)]}}{2} \\ S_d(i) = \frac{P_0(i)+c_0(i)+\sqrt{[P_0(i)+c_0(i)]^2-[c_1^2(i)+c_2^2(i)]}}{2} \end{cases} \quad (7.30)$$

Thus, the modulus of the soil reflection coefficient is then

$$|\Re(i)| = \sqrt{S_r(i)/S_d(i)} \quad (7.31)$$

Once the soil reflectance coefficient is obtained, the soil moisture can be obtained using the same theoretical model as in the previous Sect. 7.2.3 Dual antenna method, and can also be inverted by constructing an empirical model between soil reflectance coefficient and soil moisture.

7.3.3 Soil Moisture retrieval Experiment

1. Experimental Scene Description

The single antenna ground-based soil moisture retrieval experiment was conducted from September 10th to November 9th, 2018, lasting two months, at the experimental field of the National Vegetable Engineering Technology Center in Tongzhou District,

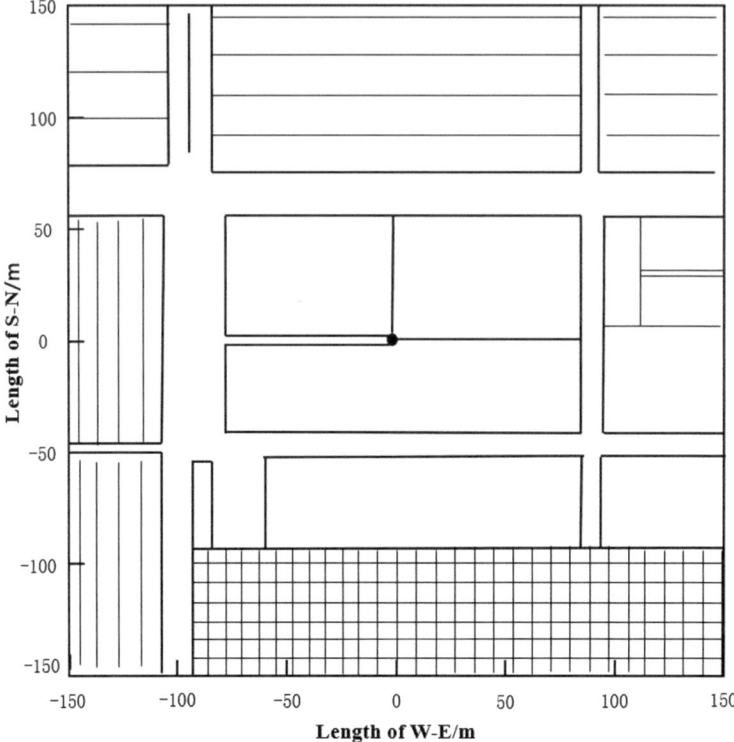

Fig. 7.16 Satellite view of the experimental area

Beijing. The experimental area was about 200 m long east to west and about 50 m wide north to south, with its satellite map shown in Fig. 7.16.

The experimental equipment is a Huace N72 reference station type receiver, with the antenna model Antcom G5Ant-52AT1. A small weather station is also equipped to real-time collect the average soil moisture at 0~6 cm depth as comparative data. The equipment installation is shown in Fig. 7.17.

The plot underwent plowing on September 19, 2018, which made the soil surface very rough and remained this condition until the end of the experiment. For this purpose, a 20 m long rope was stretched horizontally across the plot, points were evenly marked on the rope, and the vertical distance from the soil surface to the rope was measured, as shown in Fig. 7.18. A total of 109 sample points were collected, with an average height of this horizontal rope from the soil surface being 9.2 cm, the root mean square height being 5.1 cm, and the maximum soil surface height being 32.8 cm.

2. Experimental Results Analysis

During the two months of the experiment, interferometric signals from GPS and BDS navigation constellations were collected, with the minimum elevation angle

7.3 Single Antenna Soil Moisture retrieval

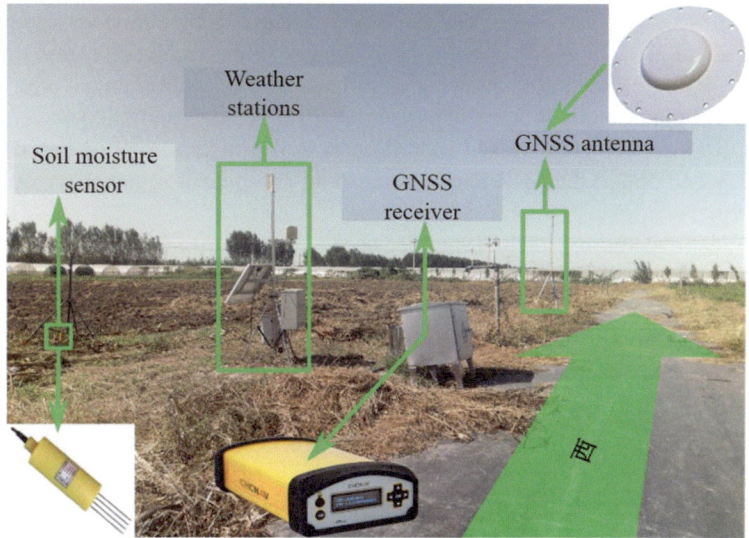

Fig. 7.17 Experimental equipment installation

Fig. 7.18 Soil surface roughness measurement scene

of satellites set to 10° to ensure that the first Fresnel reflection zone was within the experimental area. Figure 7.19 shows the changes in the receiver's signal-to-noise ratio (SNR) before and after plowing as a function of elevation angle, with larger elevation angles leading to more apparent distortions in the oscillation waveform.

The SNR data was processed according to the extraction method described in Sect. 7.3.1 to obtain the SNR results after trend component removal, as shown in Fig. 7.20.

The adaptive filtering method was then used to solve the direct and reflected signal power, as shown in Fig. 7.21, and subsequently, the modulus of the soil reflection coefficient was obtained according to Eq. (7.29).

Through data analysis, a second-order empirical model relating soil moisture to the soil reflection coefficient was established:

$$y = -44.06x^2 + 14.16x - 0.79 \tag{7.32}$$

where y is soil moisture and x is the reflection coefficient.

Using this model to invert soil moisture, the root mean square error of the retrieval results was 0.023 cm^3/cm^3, with an average error of 0.085 cm^3/cm^3, and the correlation coefficient between the retrieval results and the in-situ measurements was 0.70.

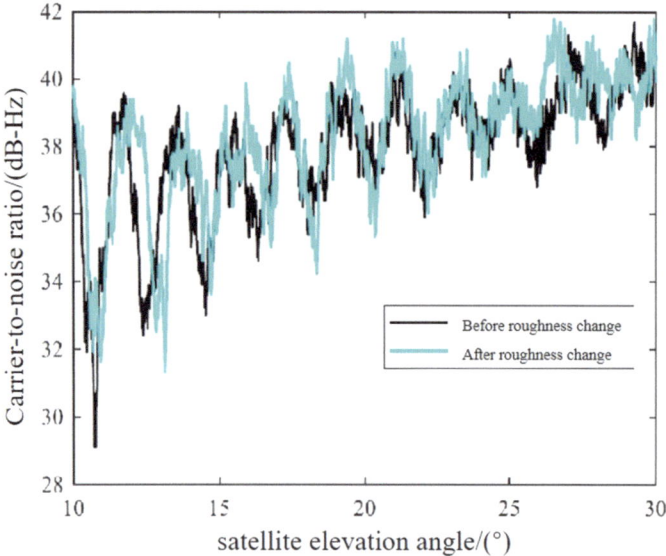

Fig. 7.19 Changes in SNR before and after changes in roughness (*Note* GPS PRN32 satellite)

7.4 Example of a Soil Moisture Monitoring System

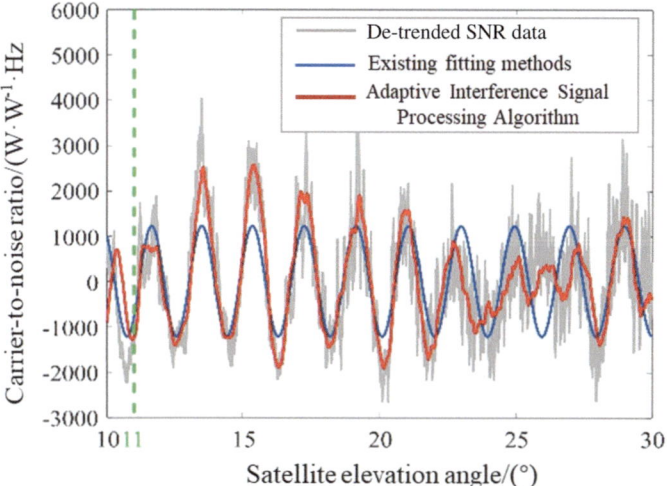

Fig. 7.20 SNR data tracking results (*Note* GPS PRN32 satellite)

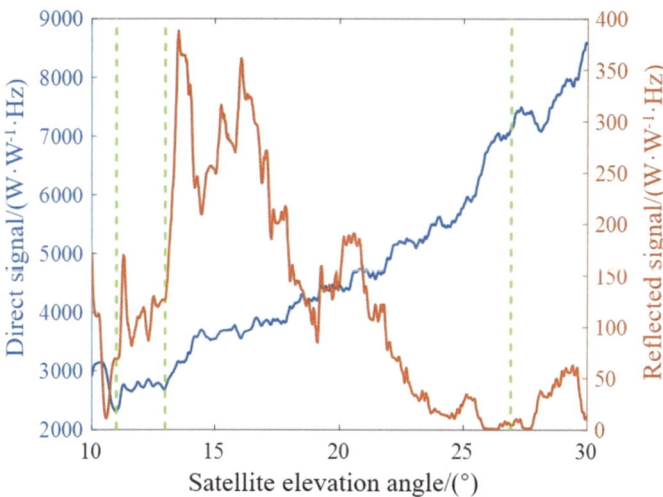

Fig. 7.21 Direct and reflected signal SNR reconstruction results (*Note* GPS PRN32 satellite)

7.4 Example of a Soil Moisture Monitoring System

The dual-antenna soil moisture monitoring system is not fundamentally different from the ocean remote sensing system. The front-end data collection part can be said to be exactly the same, with only slight differences in the backend algorithms. This chapter introduces an example of a single antenna soil moisture monitoring system, including the corresponding hardware design, software design, and field test results, etc.

7.4.1 Overall Design

The process of measuring soil moisture by single antenna method is roughly divided into two stages: the first stage involves extracting the interference features sensitive to soil moisture from the Signal-to-Noise Ratio (SNR) data, such as oscillating waveform amplitude, frequency, initial phase, etc. The second stage involves utilizing the interference features to invert the soil moisture. The main features of the first stage are the larger data volume, high computational intensity, and a relatively fixed processing flow, which can be carried out in an embedded microprocessor, paired with a low-cost, customizable, and integrable GNSS module as the interference data collection frontend, achieving an "on-the-go" online working mode. The main features of the second stage are relatively high computational intensity, and relatively fixed processing flow, which can be carried out in an embedded microprocessor. The second stage is characterized by relatively lower computational intensity and higher flexibility, which can be conducted on a user-friendly, high-performance general-purpose computer, using either an offline or online processing method. This way, a highly flexible and scalable GNSS interference signal processing system can be formed, whose overall design schematic is shown in Fig. 7.22.

The entire system is divided into front-end and back-end parts. The front-end includes the hardware and software required for the first stage of work, also referred to as the GNSS interference signal processing terminal, with the overall design goal

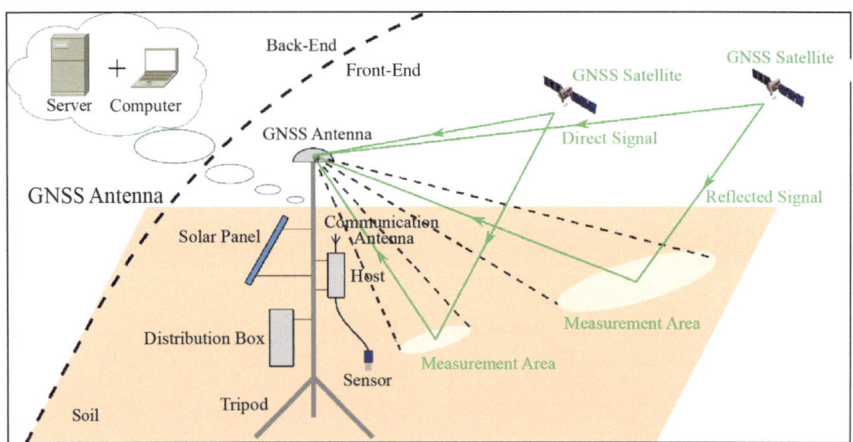

Fig. 7.22 Schematic of the overall design of the GNSS interference signal processing system

7.4 Example of a Soil Moisture Monitoring System 247

of forming a low-cost device capable of online extraction of interference characteristics while simultaneously collecting comparative data. This device can independently carry out observations in outdoor environments such as fields without human supervision, remotely transmit observation data wirelessly, and possess certain networking capabilities. The back-end includes the hardware and software required for the second stage of work, with the overall design goal of managing and monitoring multiple terminals, storing, visualizing, processing, and inverting soil moisture from terminal-transmitted data.

7.4.2 Terminal Hardware Design

The terminal hardware includes the host system, the power supply system, and the equipment mount, and this book only focuses on the host system. Figure 7.23 gives a block diagram of the host system design, including the GNSS module, core board, sensor module, wireless communication module, and power supply module.

1. **GNSS module**

The GNSS module functions to receive and process navigation satellite signals and output the observables required for interferometric measurements. The chosen module is the highly integrated RF baseband navigation chip HD8030 from Shenzhen Huada Beidou Technology Co., Ltd. This chip integrates an ARM-Cortex M3 processor, supports secondary development, and can obtain navigation data in real-time through the chip's Software Development Toolkit (SDK), achieving navigation positioning and interference signal processing, with the results output through a serial port. Table 7.1 provides the main technical parameters of this chip.

2. **Core boards**

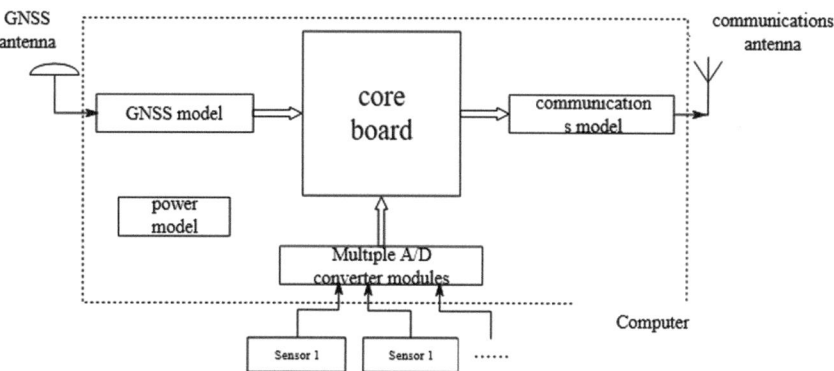

Fig. 7.23 Block diagram of host system components

The core board is the computation and control center, with functions including: (1) receiving and parsing the observational data output by the GNSS module, extracting interferometric measurements such as SNR, satellite elevation angles, azimuth, etc., and then performing interferometric processing on the qualifying observables; (2) collecting comparative data from environmental sensors at fixed time intervals, such as soil temperature and moisture, air temperature and humidity, etc.; (3) integrating, packaging interferometric measurement results with comparative data, and then transmitting to the wireless communication module for remote data transmission. An embedded microcontroller board equipped with an ARM-Cortex M4 core STM32F407ZGT6 is selected, which offers rich peripheral resources, strong expandability, numerous interfaces, and low cost.

3. **Sensor modules**

The sensor module collects environmental data, including soil temperature and moisture sensors and air temperature and humidity sensors, providing comparative and auxiliary data for establishing and validating the single antenna soil moisture retrieval model. Sensors come in digital and analog output types, the former directly outputs binary digital values representing measurements, while the latter outputs voltage or current signals proportional to the measurements, requiring an additional Analog-to-Digital Converter (ADC) to obtain digital measurements, allowing for higher resolution and flexibility with the appropriate choice of A/D converter bits.

4. **Wireless communication module**

The wireless communication module is responsible for receiving the data from the core board and transmitting it to the remote host at the specified Internet Protocol (IP) address through the wireless communication network. An industrial-grade 4G universal data transmission module is used for wireless communication, integrating a TCP/IP protocol stack, supporting bidirectional conversion between serial data and TCP/IP data, allowing terminal devices to be permanently online, with flexible parameter configuration and storage.

7.4.3 Terminal Software Design

The software in the GNSS interference signal processing terminal is implemented as multitasking in parallel based on a Real-Time Operating System (RTOS), with its processing flow shown in Fig. 7.24.

In Fig. 7.24, the "Interrupt Service Routine" mainly responds to interrupt requests from the GNSS module to complete navigation data reception and verification and stores the data in a "temporary data area" in memory. The GNSS module updates data at a frequency of 1 Hz, generating an interrupt every second; the "Real-Time Acquisition of Low-Elevation Satellites" retrieves the current navigation data from the "temporary data area" and extracts satellite elevation angles, SNR, etc., from it, then filters and stores data with low satellite elevation angles (ranging from 5° to

7.4 Example of a Soil Moisture Monitoring System

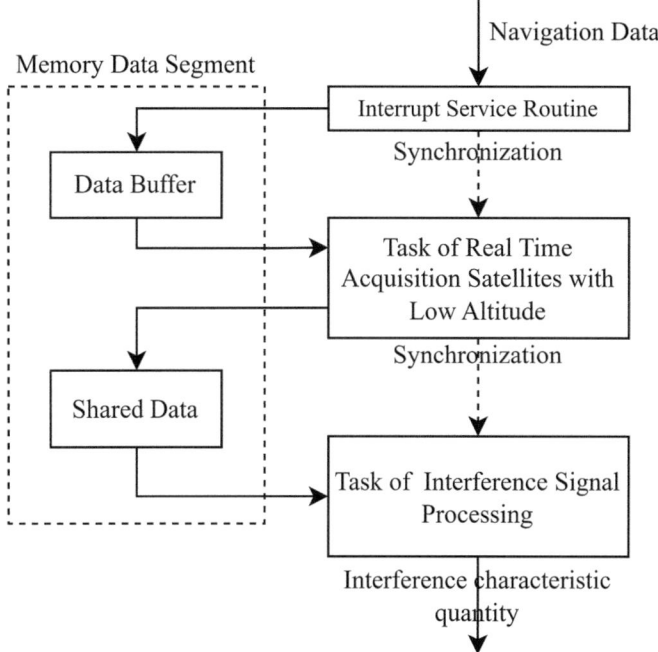

Fig. 7.24 GNSS interference signal processing flow

30°) into a "shared data area"; the "Interference Signal Processing" retrieves data for a satellite once its low elevation angle data is collected and executes interference signal processing algorithms to extract interference features for output. The coordination among modules relies on the RTOS's task scheduling mechanism, interrupt management, synchronization mechanisms, etc.

In the operating system, each task has two states: "running" and "not running." When a task is in the "running" state, the processor executes that task's code; if in the "not running" state, the task temporarily sleeps, preserving its state. The task scheduling mechanism determines which state a task is in, implemented by the task scheduling algorithm. RTOS typically uses a preemptive priority scheduling algorithm to ensure system response speed. All tasks are assigned priorities, and a task in the "running" state can be interrupted at any time by a higher priority task. In this system, the "Low Satellite Elevation Angle Data Real-time Capture Task" has a higher priority than the "Interference Signal Processing Task" to ensure new navigation data reception, filtering, and storage are not lost. Both are event-driven tasks, with the former always waiting for synchronization events generated by the interrupt service routine to notify it to retrieve the latest navigation data from the temporary data area.

The GNSS interference signal processing time allocation diagram is shown in Fig. 7.25.

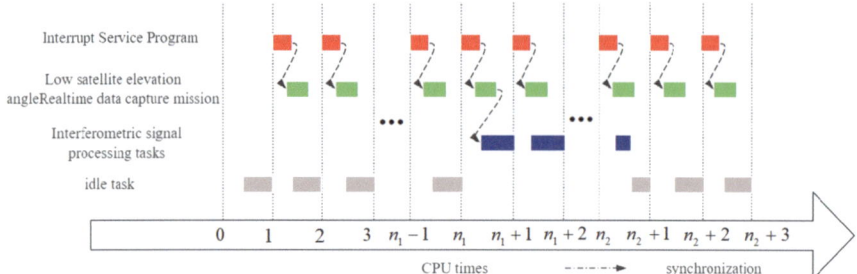

Fig. 7.25 GNSS interference signal processing time allocation diagram

In Fig. 7.25, tasks' priority increases from bottom to top, with the "Interrupt Service Routine" triggered by hardware interrupts, hence its default priority is higher than other tasks. Different lengths of horizontal lines indicate computation time. Assuming the "Interrupt Service Routine" and "Low Satellite Elevation Angle Data Real-time Capture Task" take milliseconds, then the "Interference Signal Processing Task" takes seconds.

The "Low Satellite Elevation Angle Data Real-time Capture Task" closely follows the "Interrupt Service Routine," forming a "deferred interrupt handling" mechanism. To quickly respond to external events, real-time systems require interrupt service routines to be as short as possible, such as only completing interrupt source identification, data verification, hardware flag bit resetting, etc., with the remainder of the work handled by system tasks.

7.4.4 Experimental Results

The GNSS interference signal processing terminal was deployed at the Beijing Academy of Agriculture and Forestry Sciences, National Engineering Research Center for Vegetables (116.68972° E, 39.69767° N) on October 10, 2021, as shown in Fig. 7.26. At that time, the site was temporarily in a fallow state, with flat bare soil, and the antenna installation height was approximately 2.1 m.

Terminal data was transmitted in real-time to the server for storage, and users could remotely access the terminal data in a visual manner, as shown in Fig. 7.27, which is one of the server interfaces.

Using the measurement values from contact soil moisture sensors as benchmark comparison data, the server extracts GNSS interference signal features, randomly divides them into two groups for training the retrieval model and testing. Figure 7.28 shows the soil moisture retrieval results based on the empirical model, with the data overall performing well throughout the experimental period.

7.4 Example of a Soil Moisture Monitoring System

Fig. 7.26 Deployment of the GNSS interference signal processing terminal

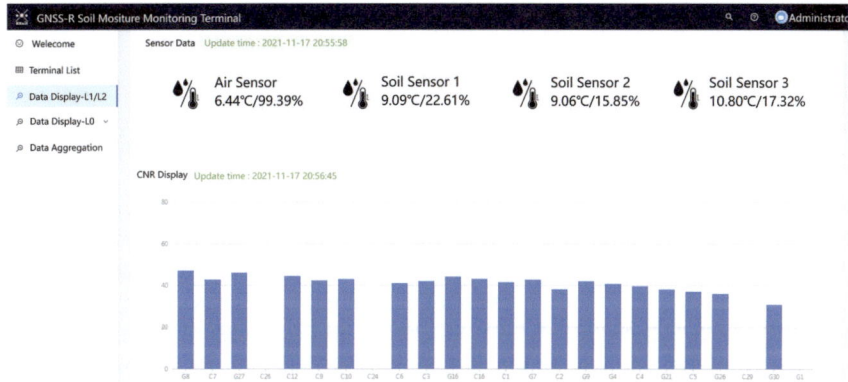

Fig. 7.27 Server-side data visualization

Table 7.4 summarizes the statistical average results of multiple random groupings, training, and retrieval, including average Root Mean Square Error ($\overline{RMSE}$), average Maximum Absolute Error ($\overline{MAE}$), and average correlation coefficient ($\overline{R}$).

The time for the terminal to process the interferometric signals and extract the interferometric eigenquantities on-line will fundamentally determine the optimum time for soil moisture measurements, and here two indicators, "measurement delay" and "measurement interval", are used to measure the time for the terminal to process the signals on-line. "Measurement delay" pertains to an individual satellite, that is, the time required from the receiving the first set of low-satellite elevation angle data for that satellite to the terminal outputting the observation results of the interference features for that satellite. It consists of two parts: one part is the time for the accumulation of low satellite elevation angle data, which is dependent on the design of

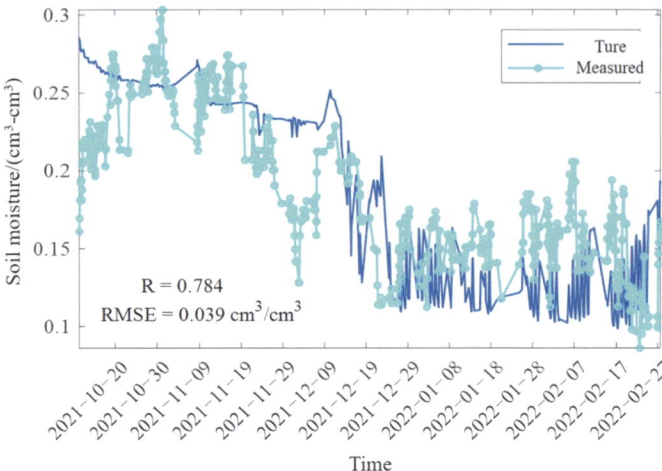

Fig. 7.28 Soil moisture retrieval results based on the empirical model

Table 7.4 Statistics of soil retrieval results based on empirical model

$\overline{\text{RMSE}}$ (cm³/cm)³	$\overline{\text{MAE}}$ (cm³/cm)³	$\overline{\text{R}}$
0.039	0.118	0.785

the orbit of the navigation satellite and the location of the terminal, It consists of two parts. One part is the time for the accumulation of low satellite altitude angle data, which is dependent on the design of the orbit of the navigation satellite and the location of the terminal, but is independent of the terminal's design. This part represents the shortest delay achievable under ideal conditions. The other part is the terminal data processing time of low satellite elevation angle data, dependent on the navigation satellite's orbital design and the terminal's location, unrelated to the terminal design, representing the shortest delay achievable under ideal conditions; the other part is the terminal data processing time, dependent on the terminal's hardware and software design. "Measurement interval" is the time difference between the terminal outputting observation results for two consecutive satellites, also consisting of two parts: one part is the time difference between two consecutive satellites appearing, dependent on the navigation satellite's orbital design and the terminal's location, unrelated to the terminal design, representing the shortest interval achievable under ideal conditions; the other part is the difference in time taken by the terminal to process data for the two satellites, dependent not only on the terminal design but also on the data volume of the two satellites.

Since the visible duration of the same satellite at different positions on the Earth's surface is different, and the time difference between two satellites appearing one after another is also different, both of the above indicators are non-uniform maximum, minimum and average values are used here. for statistical analysis, with results shown in Table 7.5. Under ideal conditions, the average measurement delay is around 80 min

Table 7.5 Analysis results of terminal online processing timeliness (satellite elevation angle 5° to 30°)

	Minimum (minutes)	Maximum (minutes)	Average (minutes)
Measurement delay (ideal)	23.17	146.00	77.78
Measurement delay (actual)	39.02	177.69	102.87
Measurement interval (ideal)	0	101.75	19.66
Measurement interval (actual)	0	203.29	30.08

and the average measurement interval is around 20 min; the average measurement delay achieved by the terminal is around 102 min and the average measurement interval is around 30 min.

7.5 Summary

This chapter introduced the content related to terrestrial soil moisture detection based on Reflected GNSS Signals, covering the basic principles, typical methods, and an example of a soil moisture observation system. The typical detection methods introduced mainly include single antenna detection and dual antenna detection, both of which involve extracting features from the observation sequence that are sensitive to soil moisture and using the relationship model between the feature observables and soil moisture to invert soil moisture. The required feature observables mainly include observables related to the amplitude of the reflected signal and observables related to the phase of the reflected signal, with retrieval models including theoretical and empirical models. Results from ground-based experiments indicate that both methods possess the capability for quantitative retrieval of soil moisture. The example of a single antenna soil moisture observation system provided insights into the system architecture, functional implementation, and field testing situation, offering some reference value for the application of GNSS-R in soil moisture retrieval.

References

1. Hegarat-Mascle SL, Zribi M, Alem F, et al. Soil moisture estimation from ERS/SAR data: toward an operational methodology [J]. IEEE Trans Geosci Remote Sens. 2002;40(12):2647–58.
2. Schmugge TJ, Jackson TJ. Passive microwave remote sensing of soil moisture[C]. Berlin: Land Surface Processes in Hydrology;1997. p. 135–51.
3. Njoku EG, Entekhabi D. Passive microwave remote sensing of soil moisture[J]. J Hydrol. 1996;184:101–29.
4. Zavorotny V, Voronovich AG. Bistatic GPS signal reflections at various polarizations from rough land surface with moisture content[C]. In: IEEE international geoscience and remote sensing symposium. Honolulu; 2000. p. 2852–4.

5. Zavorotny VU, Masters D, Gasiewski A, et al. Seasonal polarimetric measurements of soil moisture using tower-based GPS bistatic radar[C]. IEEE geoscience and remote sensing symposium. Toulouse;2004. p. 781–3.
6. Wang JR, Schmugge TJ. An empirical model for the complex dielectric permittivity of soils as a function of water content[J]. IEEE Trans Geosci Remote Sens. 1980; GE-18(4):288–95.
7. Dobson MC. Microwave dielectric behavior of wet soil-part II: dielectric mixing models[J]. IEEE Trans Geosci Remote Sens. 1985;23(1):35–46.
8. Hallikainen MT, Ulaby FT, Dobson MC, et al. Microwave dielectric behavior of wet soil-part I: empirical models and experimental observations[J]. IEEE Trans Geosci Remote Sens. 1985;23(1):25–34.
9. Larson KM, Small EE, Gutmann E, et al. Using GPS multipath to measure soil moisture fluctuations: initial results[J]. GPS Solut. 2007;12(3):173–7.
10. Arroyo AA, Camps A, Aguasca A, et al. Dual-polarization GNSS-R interference pattern technique for soil moisture mapping[J]. IEEE J Sel Top Appl Earth Obs Remote Sens. 2014;7(5):1533–44.
11. Mo T, Choudhury BJ, Schmugge TJ, et al. A model for microwave emission from vegetation-covered fields[J]. J Geophys Res: Ocean. 1982;87(C13):11229–37.
12. de Roo RD, Ulaby FT. Bistatic specular scattering from rough dielectric surfaces[J]. IEEE Trans Antennas Propagation. 1994;42(2):220–31.
13. Nievinski FG, Larson KM. Forward modeling of GPS multipath for near-surface reflectometry and positioning applications[J]. GPS Solut. 2014;18(2):309–22.
14. Han M, Zhu Y, Yang D, et al. Soil moisture monitoring using GNSS interference signal: proposing a signal reconstruction method[J]. Remote Sens Lett. 2020;11(4):373–82.

Open Access This chapter is licensed under the terms of the Creative Commons Attribution-NonCommercial-NoDerivatives 4.0 International License (http://creativecommons.org/licenses/by-nc-nd/4.0/), which permits any noncommercial use, sharing, distribution and reproduction in any medium or format, as long as you give appropriate credit to the original author(s) and the source, provide a link to the Creative Commons license and indicate if you modified the licensed material. You do not have permission under this license to share adapted material derived from this chapter or parts of it.

The images or other third party material in this chapter are included in the chapter's Creative Commons license, unless indicated otherwise in a credit line to the material. If material is not included in the chapter's Creative Commons license and your intended use is not permitted by statutory regulation or exceeds the permitted use, you will need to obtain permission directly from the copyright holder.

Chapter 8
GNSS-R Imaging

Global Navigation Satellite System-Reflectometry Synthetic Aperture Radar (GNSS-R SAR) is an integrated space-air-ground bi/multi-static SAR system comprising navigation satellites and low-orbit satellites or airborne/ground-based receivers, which can make full use of the existing navigation satellite resources, and has the advantages of a large number of satellites, flexible and diversified geometrical configurations, strong concealment, short revisit cycle, long observation time, etc. It is one of the important directions for the future development of air-ground and space-ground SAR radar networks. This chapter focuses on the geometric configuration, signal model, spatial resolution and imaging algorithms of the GNSS-R imaging system, and on this basis, the results of simulation and experimental verification are analyzed.

8.1 System Configuration and Signal Modeling

8.1.1 System Configuration

Synthetic Aperture Radar (SAR) can achieve high-resolution microwave imaging, with various features such as all-time, all-weather capabilities, and wide swath among other characteristics. A bi-static SAR refers to a system where the transmitter and receiver are placed on two different platforms, transmitter and receiver antennas' phase centers are in different spatial positions for the same pulse.

Bi-static SAR can be categorized into several different modes based on the different modes of motion of the transmitting and receiving platforms. Prof. Ender of the German Aerospace Center (DLR) has proposed a classical classification method which classifies the different modes of bi-static SAR into a hierarchy [1]. As the classification level increases, the geometric complexity of the bi-static SAR system increases, making imaging more challenging, but the system's flexibility improves and its applicability becomes more extensive. Table 8.1 shows the classification of bi-static SAR systems.

GNSS-R SAR is a typical passive bi/multi-static SAR system in which the transmitter is a navigation satellite and the receiver is located at a fixed position on a low-orbiting satellite, an airborne platform or on the surface. It can be seen from the topological relationship that the direct beam coverage of navigation satellites on the ground is very large, while the echo beam coverage of the receiver is relatively small when close to the observation area, meaning the geometrical configuration of the system is highly asymmetric. Typically, a GNSS-R SAR system consists of a direct signal channel and an echo signal channel. The direct signal channel receives the direct signals from navigation satellites through an RHCP omnidirectional antenna and obtains accurate carrier phase, code phase and positioning results, providing precise reference information for signal synchronization in the echo channel. The echo channel usually adopts a high-gain circularly or linearly polarized antenna to receive the reflected echo signals from the observation area for imaging.

As Fig. 8.1a shows, the receiver is a stationary ground station, hovering UAV, or stationary near-space floaters, etc., referred to here as a fixed station mode and categorized as a special stationary mode. For example Fig. 8.1b, placing the receiver on a flying carrier is called airborne mode. In this case, there is a significant difference in altitude between the navigation satellite and the airborne platform, and they do not satisfy the conditions of parallel paths and uniform linear motion. If the navigation satellite's motion during the synthetic aperture time is equivalent to uniform linear motion, when the aircraft's trajectory is parallel to the navigation satellite's trajectory, it can be categorized as constant velocity mode; when the aircraft's trajectory is not parallel to the satellite's, it can be categorized as arbitrary mode.

Table 8.1 Classification of Bi-static SAR systems

Hierarchy	Paradigm	Define
1	Tandem	the transmitter and receiver are placed on two different platforms moving along the same track at equal speeds in a straight line
2	Translational invariant	the transmitter and receiver are placed on two different platforms moving along two parallel tracks at equal speeds in a uniform linear motion
3	Constant velocity	the transmitter and receiver move with constant velocities, but their tracks are not parallel and/or their speeds are not equal
4	Arbitrary motion	the transmitter and receiver are placed on two different platforms each moving along an arbitrary path (e.g., non-linear or accelerated motion)

8.1 System Configuration and Signal Modeling

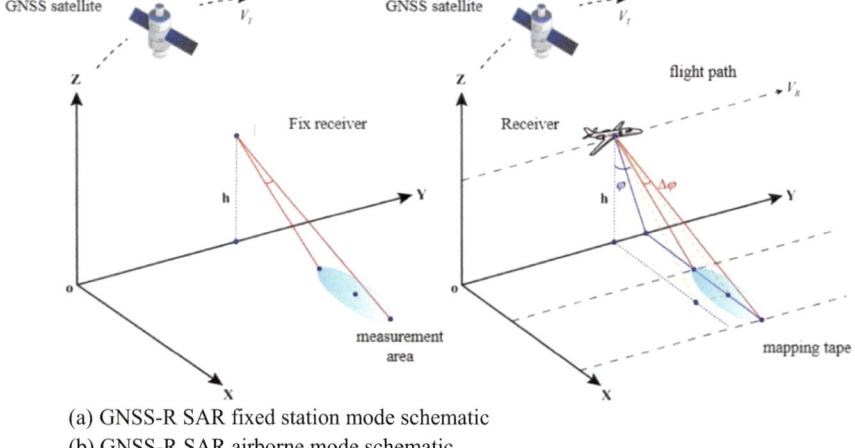

(a) GNSS-R SAR fixed station mode schematic
(b) GNSS-R SAR airborne mode schematic

Fig. 8.1 Schematic of GNSS-R SAR modes

8.1.2 Two-Dimensional Signal Model

The signal transmitted by the navigation satellite can be represented in complex form as

$$s_t(t) = A_e c(t) \exp(-j2\pi f_c t - j\varphi - j\pi d(t)) \tag{8.1}$$

where, A_e denotes the amplitude of the navigation signal, $c(t)$ denotes the pseudo-random code, f_c is the carrier frequency, φ denotes the initial phase, and $d(t)$ denotes the navigation data code, taking the value of ± 1. From Eq. (8.1), it can be seen that changes in the polarity of the navigation data code will lead to phase jumps in the echo signal. To maintain continuity in the Doppler phase, it is necessary to mitigate the effects of the navigation data. In the signal reconstruction process, synchronization is achieved through the direct signal, the navigation data code is stripped, and the amplitude and initial phase, which do not affect the imaging algorithm, are omitted, resulting in the signal taking the following form:

$$s_d(t) = c(t - \tau_d(t)) \exp(-j2\pi f_c \tau_d(t)) \tag{8.2}$$

where $\tau_d(t)$ is the propagation delay of the direct wave. For a target with a position vector $\mathbf{P}$ in the observation scene, the echo signal can be expressed as

$$s_r(t) = c(t - \tau_r(t)) \exp(-j2\pi f_c \tau_r(t)) \tag{8.3}$$

where $\tau_r(t)$ is the propagation delay of the echo. The mathematical models of the delay of the direct wave $\tau_d(t)$ and the delay of the return wave $\tau_r(t)$ are denoted respectively as

$$\tau_d(t) = \frac{|\mathbf{R_t}(t) - \mathbf{R_r}(t)|}{v_c} \tag{8.4}$$

$$\tau_r(t) = \frac{|\mathbf{R_t}(t) - \mathbf{P}|}{v_c} + \frac{|\mathbf{R_r}(t) - \mathbf{P}|}{v_c} \tag{8.5}$$

where $\mathbf{R_t}(t)$ denotes the position vector of the navigation satellite, $\mathbf{R_r}(t)$ denotes the position vector of the receiver, v_c denotes the speed of light, and |•|denotes the Euclidean norm. The signal models in Eqs. (8.2) and (8.3) are a one-dimensional model, while a two-dimensional signal model is required for the SAR system to achieve two-dimensional resolution of scene targets. Based on the characteristics of the continuous wave nature of navigation signals, the one-dimensional signal can be processed into two dimensions, using a pseudo-random code period (e.g., 1 ms for the GPS L1 C/A signal) as the width of the equivalent pulse in the range direction to generate a two-dimensional echo signal, i.e., the system's Pulse Repetition Frequency (PRF) is 1000 Hz. The echo signal is divided according to azimuth time and range time, and Eq. (8.3) can be rewritten as follows:

$$s_r(t, \tau) = w(t) c \left[\tau - \frac{R(t)}{v_c} \right] \exp\left[-j2\pi f_c \frac{R(t)}{v_c} \right] \tag{8.6}$$

where t is the azimuthal time, also called slow time; τ is the range time, also called fast time; $w(t)$ is the rectangular envelope of the echo signal in the azimuth direction; and $R(t)$ is the total path length of the reflected signal.

As shown in Fig. 8.2 shown, $R_T(t)$ indicates the instantaneous distance from the navigation satellite to the target area, and $R_B(t)$ indicates the instantaneous distance from the navigation satellite to the receiver. $R_R(t)$ denotes the instantaneous distance from the receiver to the target area, then

$$R_B(t) = |\mathbf{R_t}(t) - \mathbf{R_r}(t)| \tag{8.7}$$

$$R_T(t) = |\mathbf{R_t}(t) - \mathbf{P}| \tag{8.8}$$

$$R_R(t) = |\mathbf{R_r}(t) - \mathbf{P}| \tag{8.9}$$

$$R(t) = R_T(t) + R_R(t) \tag{8.10}$$

The Doppler frequency variation causes changes in the phase of the signal within each pulse duration, then Eqs. (8.2) and (8.3) are represented as

$$s_d(t, \tau) = c \left[\tau - \frac{R_B(t)}{v_c} \right] \exp\left[-j2\pi \frac{R_B(t)}{\lambda} - j2\pi f_{dB} t \right] \tag{8.11}$$

8.1 System Configuration and Signal Modeling

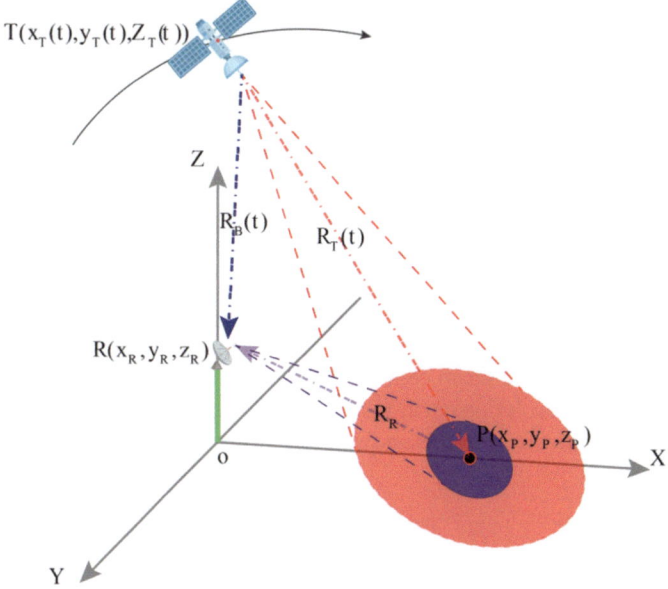

Fig. 8.2 Geometric configuration of the GNSS-R SAR fixed ground station mode

$$s_r(t, \tau) = c\left[\tau - \frac{R(t)}{v_c}\right] \exp\left[-j2\pi \frac{R(t)}{\lambda} - j2\pi (f_{dT} + f_{dR})t\right] \quad (8.12)$$

where f_{dB} and $(f_{dT} + f_{dR})$ are the Doppler shifts generated by the variation of $R_B(t)$ and $R(t)$, respectively, within a pseudo-random code period, which can be expressed as

$$\begin{cases} f_{dB} = d[(2\pi/\lambda)R_B(t)]dt \\ f_{dT} + f_{dR} = d[(2\pi/\lambda)R(t)]dt \end{cases} \quad (8.13)$$

The actual received navigation signal also contains many error factors, such as local oscillator drift and atmospheric interference caused delay and phase error. Define τ_e and φ_e as the combined delay and phase errors caused by various interference factors, then Eqs. (8.11) and (8.12) become:

$$s_d(t, \tau) = c\left[\tau - \frac{R_B(t)}{v_c} - \tau_e\right] \exp\left[-j2\pi \frac{R_B(t)}{\lambda} - j2\pi f_{dB}t - j\varphi_e\right] \quad (8.14)$$

$$s_r(t, \tau) = c\left[\tau - \frac{R(t)}{v_c} - \tau_e\right] \exp\left[-j2\pi \frac{R(t)}{\lambda} - j2\pi (f_{dT} + f_{dR})t - j\varphi_e\right] \quad (8.15)$$

8.2 Spatial Resolution

Spatial resolution is an important indicator of the overall performance of the SAR system, representing the two-dimensional capability of a SAR system to resolve targets adjacent in the target area, including azimuth resolution and range resolution, which are mainly related to signal parameters and geometric configurations. In monostatic SAR system, the range direction is defined as the beam center of the return antenna, and the azimuth direction is defined as the motion direction of the radar platform, and its slant distance course is only related to the motion parameters of the radar platform. GNSS-R SAR is a bi-static SAR system with separate transmitters and receivers, involving two directions of motion for the transmitter and receiver and two directions for the beam centers of the transmitting and echo antennas. The definition of spatial resolution needs to consider the geometric configuration, i.e., the relationship between the positions and velocity vectors of the two platforms relative to the target [2].

8.2.1 Resolution Definition

Azimuth resolution ρ_a reflects the Doppler frequency resolution capability of the SAR system and is defined as the distance change mapped by a unit Doppler resolution cell [3]:

$$\rho_a = \frac{\partial r_a}{\partial f_d} df_d = \frac{1}{\partial f_d / \partial r_a} df_d \qquad (8.16)$$

where r_a is the azimuth coordinate characterized by distance, and f_d denotes the Doppler frequency of the echo signal. Range resolution ρ_r reflects the delay resolution capability of the SAR system and is defined as the change in distance mapped in a unit delay resolution cell:

$$\rho_r = \frac{\partial r_g}{\partial \tau} d\tau = \frac{1}{\partial \tau / \partial r_g} d\tau \qquad (8.17)$$

where r_g is the range coordinate characterized by distance. The SAR system employs a gradient method to separately analyze the range and azimuth directions, where the azimuth direction is the direction of maximum change in Doppler frequency and the range direction is the direction of maximum change in the time delay. The two-dimensional resolution of a bi-static SAR system can be obtained by solving for the maximum change rate of the Doppler frequency f_d and time delay τ in their corresponding directions.

Figure 8.3 shows the geometric configuration of GNSS-R SAR in airborne mode, where, $\mathbf{V_{Tx}}$ is the velocity vector of the navigation satellite, $\mathbf{V_{Rx}}$ is the velocity vector

8.2 Spatial Resolution

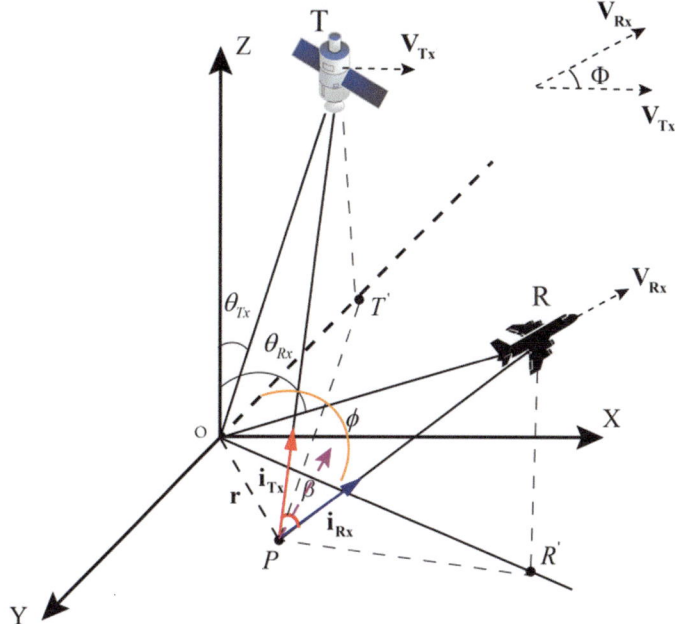

Fig. 8.3 Geometric configuration of GNSS-R SAR in airborne mode

of the receiver, the angle between the two is Φ, θ_{T_x} and θ_{R_x} are the ground incidence angles of the navigation satellite and receiver, and the center of the target area O, $T\prime$ and $R\prime$ are the nadir and flight points, respectively. The observation angle of the aircraft ϕ is the angle between $\mathbf{OT'}$ and $\mathbf{OR'}$, with P being any point target in the target area, and $\mathbf{i_{T_x}}$ and $\mathbf{i_{R_x}}$ are the unit vectors pointing from the point target P towards the navigation satellite and receiver. The unit vector of P is defined as $\mathbf{OP} = \mathbf{r}$.

The echo delay information at the point target P can be expressed as

$$t(\tau, \mathbf{r}) = \frac{1}{v_c}[|\mathbf{R_{Tx}}(\tau, \mathbf{r})| + |\mathbf{R_{Rx}}(\tau, \mathbf{r})|] \tag{8.18}$$

When the vector $\mathbf{r}$ changes along any direction, the change in time delay of the echo can be expressed as

$$dt = \nabla \mathbf{t} \cdot d\mathbf{r} \tag{8.19}$$

$$\nabla \mathbf{t} = \frac{1}{v_c}(\mathbf{i_{Tx}}(\tau, \mathbf{r}) + \mathbf{i_{Rx}}(\tau, \mathbf{r})) \tag{8.20}$$

where $\nabla \mathbf{t}$ is the gradient of $t(\tau, \mathbf{r})$ in the $\mathbf{r}$ direction. The bandwidth of the satellite navigation signal is B and its minimum time-resolved interval in the distance direction is

$$T_r = \frac{1}{B} \tag{8.21}$$

From Eqs. (8.19) and (8.20), we have

$$|d\mathbf{r}| = \frac{dt}{|\nabla \mathbf{t}| \cdot \cos \theta_{rt}} \geq \frac{dt}{|\nabla \mathbf{t}|} = \frac{T_r}{|\nabla \mathbf{t}|} \tag{8.22}$$

where θ_{rt} is the angle between $\nabla \mathbf{t}$ and $d\mathbf{r}$, which takes the value of 0°~90°, and the equality in Eq. (8.22) holds if and only if $\theta_{rt} = 0°$ ($\nabla \mathbf{t}$ is in the same direction as $d\mathbf{r}$). Typically, both the distance and azimuthal resolution directions are not in the imaging plane. Assuming that the vector of $\nabla \mathbf{t}$ after projecting it to the imaging plane (usually the surface plane is chosen) is $\nabla \mathbf{t_g}$, then the range-oriented resolution vector of the GNSS-R SAR imaging plane is

$$d\mathbf{r}_g = \frac{1/B}{|\nabla \mathbf{t_g}|} \cdot \mathbf{i}_{tg} \tag{8.23}$$

where $\mathbf{i}_{tg}$ is the unit vector of $\nabla \mathbf{t_g}$. Due to the small imaging scene of GNSS-R SAR in airborne or fixed station mode, using the center point O of the target area in Fig. 8.3 as a reference to calculate the ground imaging plane range resolution yields

$$|\nabla \mathbf{t_g}| = \frac{1}{v_c} \sqrt{\sin^2 \theta_{Tx} + \sin^2 \theta_{Rx} + 2 \sin \theta_{Tx} \sin \theta_{Rx} \cos \phi} \tag{8.24}$$

$$\mathbf{i}_{tg} = \begin{bmatrix} \frac{-\sin \theta_{Rx} \cos \phi - \sin \theta_{Tx}}{\sqrt{\sin^2 \theta_{Tx} + \sin^2 \theta_{Rx} + 2 \sin \theta_{Tx} \sin \theta_{Rx} \cos \phi}} \\ \frac{\sin \theta_{Rx} \sin \phi}{\sqrt{\sin^2 \theta_{Tx} + \sin^2 \theta_{Rx} + 2 \sin \theta_{Tx} \sin \theta_{Rx} \cos \phi}} \end{bmatrix} \tag{8.25}$$

Similarly, the azimuth resolution vector for the GNSS-R SAR can be derived as

$$d\mathbf{r}_a = \frac{1/T_s}{|\nabla \mathbf{f_{dg}}|} \cdot \mathbf{i}_{fg} \tag{8.26}$$

where T_s is the minimum time interval in the azimuth direction, $\mathbf{i}_{fg}$ is the unit vector of $\nabla \mathbf{f_{dg}}$, and $\nabla \mathbf{f_{dg}}$ is the projection of the Doppler gradient of the echo signal $\nabla \mathbf{f_d}$ onto the surface imaging plane, denoted by

$$\nabla \mathbf{f_{dg}} = \frac{1}{\lambda} \left[\frac{1}{|\mathbf{R_{Tx}}|} (\mathbf{V_{Tx}} - (\mathbf{V_{Tx}} \cdot \mathbf{i_{Tx}}) \mathbf{i_{Tx}}) + \frac{1}{|\mathbf{R_{Rx}}|} (\mathbf{V_{Rx}} - (\mathbf{V_{Rx}} \cdot \mathbf{i_{Rx}}) \mathbf{i_{Rx}}) \right] \tag{8.27}$$

Typically, the Doppler gradien $\nabla \mathbf{f_d}$ changes little during the synthetic aperture time in fixed station or airborne mode GNSS-R SAR, and therefore, one can choose to calculate $\nabla \mathbf{f_d}$ at the center moment of the synthetic aperture.

8.2.2 Relationship Between Resolution and Geometric Configuration

The resolution vectors of GNSS-R SAR in the distance and azimuth directions are related to its bi-static geometric configuration, with the two-dimensional resolution for the same echo parameter differing under various geometric conditions. From the formulas for range and azimuth resolution, it can be seen that the resolving capability of GNSS-R SAR is mainly determined by the range resolution vector, azimuth resolution vector, and the angle between them.

1. **Resolution range**

From Eqs. (8.23), (8.24) and (8.25), it can be seen that the three variables $\theta_{T_x}, \theta_{R_x}$ and ϕ together constrain the range resolution of GNSS-R SAR, which is independent of the velocities of the navigation satellite and receiver. In practical application scenarios, the incidence angles of the navigation satellite and receiver can be calculated by observing the scenarios, orbital information of the navigation satellite and receiver's motion trajectory, θ_{T_x} and θ_{R_x}, which both take the value range of $0° \sim 90°$. The value range of ϕ is $0° \sim 360°$. When $\cos\phi = 1$, namely, when, i.e. $\phi = 0°$ or $\phi = 360°$, the optimal range resolution is obtained, and the worst range resolution is obtained when $\cos\phi = -1$, i.e. $\phi = 180°$. The optimal and worst range resolution can be expressed as

$$|d\mathbf{r_g}|_{min} = \frac{v_c}{B} \frac{1}{|\sin\theta_{Tx} + \sin\theta_{Rx}|} \quad (8.28)$$

$$|d\mathbf{r_g}|_{max} = \frac{v_c}{B} \frac{1}{|\sin\theta_{Tx} - \sin\theta_{Rx}|} \quad (8.29)$$

From Eq. (8.29), when $\theta_{Tx} = \theta_{Rx}$ and $\phi = 180°$, the time delay gradient ∇t is perpendicular to the imaging plane, making $\nabla \mathbf{t_g}$ a zero vector, and the GNSS-R SAR ground range resolution tends towards infinity, meaning there is no ground range resolving capability. Taking the BDS-B3 signal as an example, Fig. 8.4 shows the relationship between range resolution and observation angle for different receiver incidence angles, with parameter settings as shown in Table 8.2. The incidence angle of the navigation satellite is 45°, and the incidence angles of the receiver are selected as 30°, 40° and 50°. When the observation angle is $0° \leq \phi \leq 180°$, the range resolution decreases with an increase in the observation angle, when the observation angle is $180° \leq \phi \leq 360°$, the range resolution increases with the increase in the observation angle, and the resolution is the worst when $\phi = 180°$. For receivers with varying angles of incidence, the range resolution becomes worse with increasing angle of incidence.

2. **Azimuth resolution**

From Eqs. (8.26) and (8.27), the azimuth resolution of the GNSS-R SAR is related to the positions and velocities of the navigation satellites and receiver. For the airborne

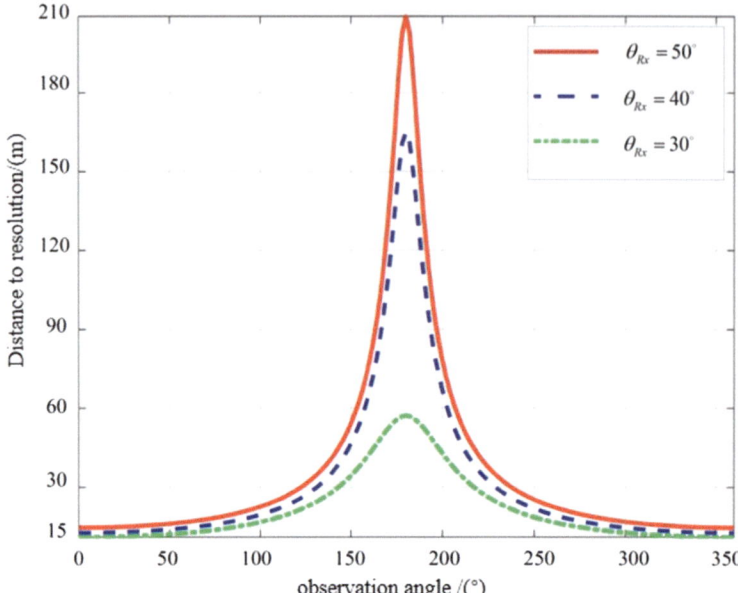

Fig. 8.4 Change in range resolution with observation angle under different receiver incidence angles

Table 8.2 GNSS-R SAR system parameters based on the BDS-B3 signal

Parameters	Navigation satellite	TV or radio receiver
Code	BDS-B3	
Bandwidths	20.46 MHz	
High degree	20000 km	1000m
Movement speed	3000 m/s	65m/s
Angle of incidence	45°	30°/40°/50°

receivers, the Doppler gradient of the echo signal $\nabla \mathbf{f_{dg}}$ in Eq. (8.27) mainly contains the navigation satellite motion and the airborne receiver. Although the aircraft motion speed is smaller than that of the navigation satellite ($|\mathbf{V_{Tx}}| \gg |\mathbf{V_{Rx}}|$), the Doppler gradient of the echo signal in Eq. (8.27) mainly originates from the second term because the distance from the target scene to the navigation satellite is much larger than that to the airborne receiver ($|\mathbf{R_{Tx}}| \gg |\mathbf{R_{Rx}}|$). In a fixed station mode, the Doppler gradient of the echo signal $\nabla \mathbf{f_{dg}}$ is mainly derived from the first term of Eq. (8.27). Typically, the synthetic aperture time is increased to accumulate a larger Doppler bandwidth to obtain higher azimuth resolution. Figure 8.5 presents the GNSS-R SAR azimuth resolution results based on the conditions in Table 8.2. With different observation angles and angles between velocity vectors, the azimuth resolution varies symmetrically. With a synthetic aperture time of 3 s, the optimal azimuth resolution

8.3 Backward-Projection Imaging Algorithm

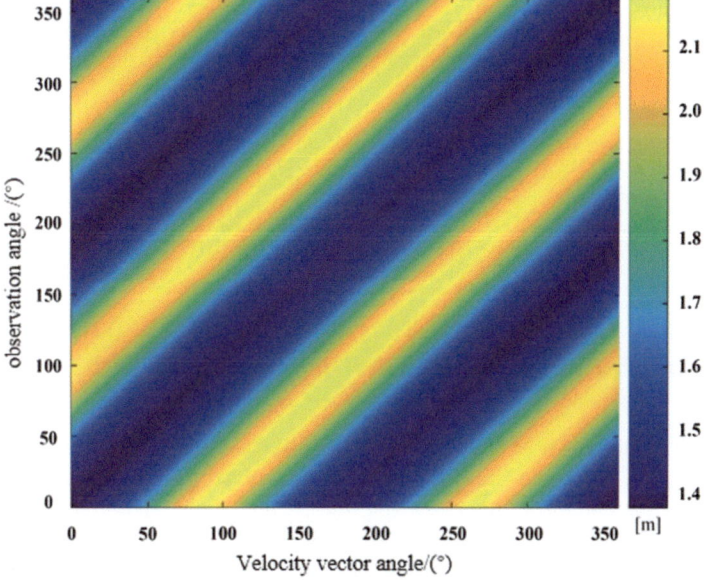

Fig. 8.5 Relationship between azimuth resolution and observation angle and the angle between velocity vectors (with a synthetic aperture time of 3 s)

value is about 1.31 m. As the synthetic aperture time increases or the speed of the airborne platform increases, the azimuth resolution will further improve.

8.3 Backward-Projection Imaging Algorithm

The BP (Back Projection) algorithm is an imaging algorithm based on time domain processing. It calculates the distance delay between the antenna and the image pixels based on their locations, back-projects the echo data according to the delay information into the image domain, and coherently accumulates at each pixel point to produce a two-dimensional image [4]. The BP (Back Projection) algorithm is a point-by-point processing method that can achieve the optimal matching for all targets as well as high image resolution.

The GNSS-R SAR Back-projection imaging algorithm, as shown in Fig. 8.6, use delay, phase, Doppler, and navigation data code information obtained from the direct signal capturing and tracking process to compensate the echo signal. The distance-oriented compression is realized by the matched filter method, after which the output correlation values are back-projected to realize the azimuthal focusing, and finally the images of each azimuthal moment are superimposed to produce the final image.

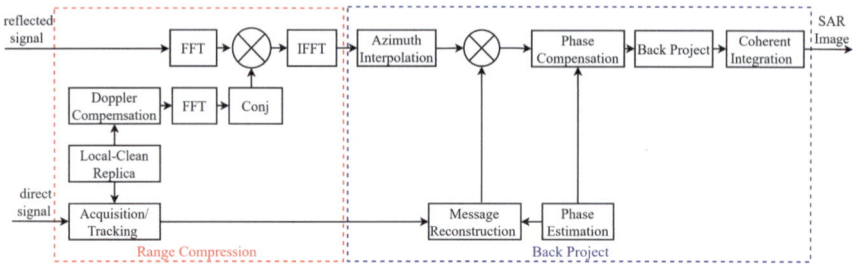

Fig. 8.6 GNSS-R SAR back-projection imaging algorithm

8.3.1 Range Compression and Synchronization

The locally generated reference signal $h(t)$ is used as the impulse response function of the matched filter, and the echo signal is convolved in the time domain with it to complete matched filtering, extracting the corresponding delay and phase information, or converted to the frequency domain using a Fourier transform. The echo signal carries errors produced by Doppler modulation, navigation data code phase, and interference phase, which need to be eliminated one by one to enhance the focusing accuracy in both range and azimuth directions.

1. **Intrapulse Doppler effect elimination**

The Doppler offset introduced in Eqs. (8.12) and (8.15) $(f_{dT} + f_{dR})$ is estimated using the navigation satellite orbit data, receiver track data, and scene information, defining as $(f_{dT} + f_{dR})\prime$. The reference signal is then reconstructed

$$h\prime(t) = h(t) \exp(j2\pi (f_{dT} + f_{dR})\prime t) \qquad (8.30)$$

The estimated deviation $\Delta f = (f_{dT} + f_{dR})\prime - (f_{dT} + f_{dR})$ is obtained by a two-step search of the echo signal. The Doppler frequency generated by the navigation satellite's motion relative to a fixed station on the ground is generally within the range of $-5 \sim 5\,kHz$. A rough estimate of the Doppler of the received signal is obtained by traversing the search at intervals of 500Hz, then searching at intervals of 100Hz in the range of ± 400Hz centered on the rough estimate.

2. **Navigation data code impact elimination**

The effect of the navigation data code is not considered in Eq. (8.15), but the phase information of the navigation data code $d(t)\varphi_d$ is included in the results after range compression, as shown in Eq. (8.31).

$$F(t, \tau) = P[\tau - \frac{R(t)}{v_c}] \exp[-j2\pi \frac{R(t)}{\lambda} - j2\pi \Delta f t - j\varphi_d - j\varphi_e] \qquad (8.31)$$

where $P(\tau)$ is the autocorrelation function of the pseudo-random code, Δf is the Doppler frequency offset between the reference signal and the echo signal, and φ_d

8.3 Backward-Projection Imaging Algorithm

takes the value of 0 or π, with this phase jumping when the data changes. Therefore, multiplying the corresponding navigation data code by the phase change of the range compression signal peak can eliminate its impact.

Figure 8.7 shows the impact results of a navigation data code. Figure 8.7a shows a fragment of azimuth peak phase history after range compression, with phase jumps at five sample points. Figure 8.7b shows the result after differential processing, with a total of five jumps in phase at π. Figure 8.7c shows the detected navigation data code, with data jump positions matching phase jump positions. Figure 8.7d shows the Doppler phase of the azimuth peak signal after stripping the navigation data code, and the value changes continuously after the effect of the navigation data code is removed.

3. **Interference phase error elimination**

Under fixed station or airborne mode conditions, the direct and echo signals can be regarded as passing through the same atmospheric layer, and both have the same atmospheric interference phase error φ_e, mainly composed of higher-order terms

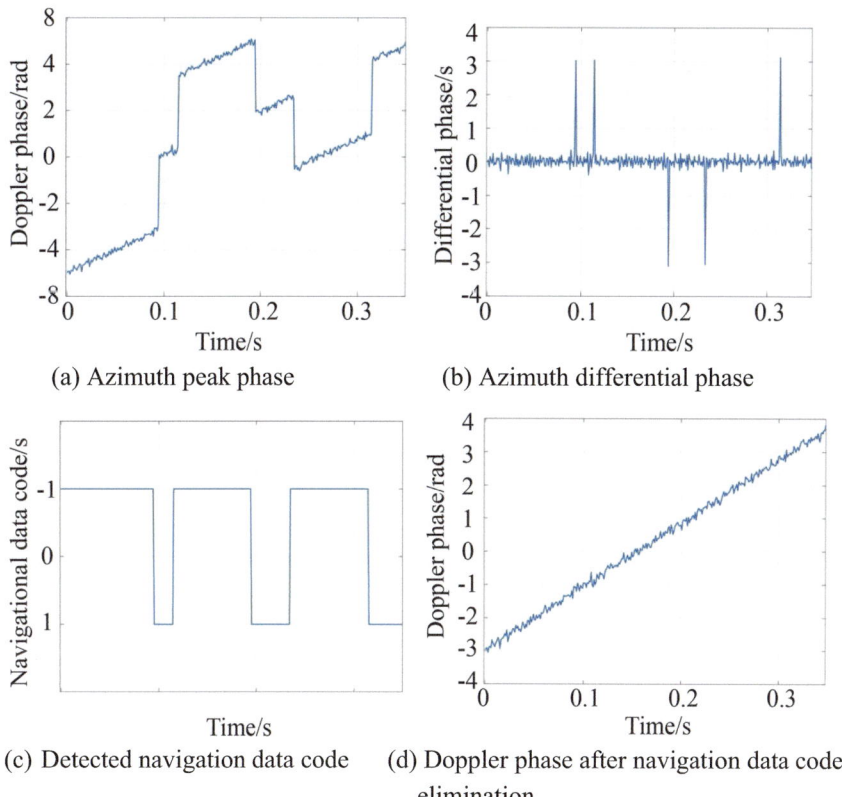

Fig. 8.7 Impact results of the navigation data code

caused by local oscillator drift, represented by polynomial fitting. After removing the effects of Doppler and navigation data code from the range compression results, subtracting the phase information at the correlation peak from the polynomial fitting results obtains the higher-order phase residuals $\varphi_e\prime$ and the phase compensation term $\exp(-j\varphi_e\prime)$.

The range compression result after synchronization, the echo signal is

$$F(t, \tau) = P[\tau - \frac{R(t)}{v_c}] \exp[-j2\pi \frac{R(t)}{\lambda}] \quad (8.32)$$

In the actual discrete signal processing, when the time delay value of the echo signal does not coincide with sampling points, interpolation is necessary to improve the azimuth focusing accuracy. In general, the sampling rate in the navigation signal reception processing exceeds 6–8 times the data code rate, meeting the requirements for the imaging algorithm back-projection process on the sampling points without the need for interpolation.

8.3.2 Back-Projection

Back-Projection maps the range-compressed signal to each pixel in the imaging area, specifically, to the center points of grids after gridding. Figure 8.8 shows a schematic of the geometric configuration of the back-projection imaging algorithm. The size of the grid, related to the resolution of the imaging, should be carefully selected considering both image quality and computational complexity. Usually, the grid is divided by the azimuth ground range resolution with better resolution, and the spacing is less than or equal to $\frac{1}{2}|d\mathbf{r_a}|$. At a certain azimuth time, the positions of the navigation satellite, pixel point and receiver are fixed and known, and the time delay of the echo signal at the pixel point (x_m, y_n) is calculated based on the signal propagation distance, as shown by

$$\tau_{mn}(t) = \frac{R_T(t, m, n) + R_R(t, m, n)}{v_c} \quad (8.33)$$

$R_T(t, m, n)$ is the instantaneous distance from the navigation satellite to the pixel (x_m, y_n, z), $R_B(t)$ is the instantaneous distance from the navigation satellite to the receiver, and $R_R(t, m, n)$ is the instantaneous distance from the receiver to the pixel point (x_m, y_n, z).

$$R_T(t, m, n) = \sqrt{[x_T(t) - x_m]^2 + [y_T(t) - y_n]^2 + [z_T(t) - z]^2} \quad (8.34)$$

$$R_R(t, m, n) = \sqrt{[x_R(t) - x_m]^2 + [y_R(t) - y_n]^2 + [z_R(t) - z]^2} \quad (8.35)$$

8.3 Backward-Projection Imaging Algorithm

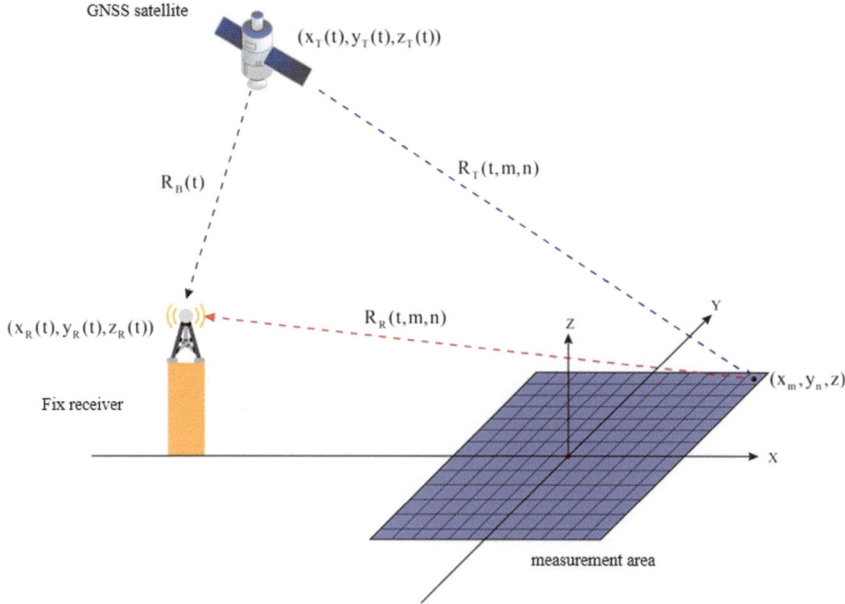

Fig. 8.8 Schematic of the back-projection imaging algorithm geometry

$$R_B(t) = \sqrt{[x_T(t) - x_R(t)]^2 + [y_T(t) - y_R(t)]^2 + [z_T(t) - z_R(t)]^2} \quad (8.36)$$

Substituting Eqs. (8.34) and (8.35) into (8.32) gives the correlation value of the pixel point (x_m, y_n) as

$$F(t, \tau_{mn}(t)) = P[\tau - \tau_{mn}(t)] \exp[-j2\pi \frac{R_T(t, m, n) + R_R(t, m, n)}{\lambda}] \quad (8.37)$$

The receiver receives the echo signal as well as the direct signal, and the correlation values for each pixel point in the imaging region and the propagation delay difference between the two signals are inextricably linked, which is compensated for in the algorithm to avoid azimuthal defocusing [5].

$$h(t, m, n) = \exp[j2\pi \frac{R_T(t, m, n) + R_R(t, m, n) - R_B(t)}{\lambda}] \quad (8.38)$$

The range-compressed result after phase compensation becomes the grayscale value for that pixel, generating an image at each azimuth time.

$$S(t, m, n) = F(t, \tau_{mn}(t)) \cdot h(t, m, n) \quad (8.39)$$

Finally, the images from each azimuth moment within the synthetic aperture time $[-\frac{T_p}{2}, \frac{T_p}{2}]$ are integrated to produce a complete image.

$$S(m, n) = \int_{-\frac{T_p}{2}}^{\frac{T_p}{2}} S(t, m, n) dt \qquad (8.40)$$

8.4 Imaging Algorithm Implementation

In recent years in the field of graphics and image processing, the traditional "single" processing approach of Central Processing Units (CPUs) is gradually evolving towards a "co-processing" approach utilizing both CPUs and Graphics Processing Units (GPUs). NVIDIA Corporation introduced the Compute Unified Device Architecture (CUDA) programming model, which takes full advantage of both CPU and GPU strengths. The CPU is seen as the "host" carrying out serial data computations, while the GPU, as the "device," performs parallel data computations by calling kernels. Due to the parallel nature of the Back-Projection algorithm, its implementation in the CUDA programming model shows good acceleration effects, reducing algorithm run times without compromising image quality, meeting the requirements for rapid imaging processing [6–11].

8.4.1 Heterogeneous Architecture Parallel Platforms

When the target area increases or the synthetic aperture time extends, the computational cost of the Back-Projection algorithm dramatically rises. To ensure imaging quality and efficiency, both range compression and Back-Projection processing of echo signals can be performed in parallel. Figure 8.9 shows a heterogeneous parallel processing platform, where the CPU primarily handles tasks such as direct signal capture, tracking, positioning, and synchronization processing of the echo signal, while the GPU performs Back-Projection imaging and sends the output to the CPU for display and storage.

8.4.2 Range Compression

Range compression primarily involves FFT, IFFT, and complex multiplication operations, where FFT and IFFT are based on cuFFT functions provided in CUDA, and complex multiplication is achieved through constructing kernel functions. Memory overflow is a common error in GPU applications, and FFT and IFFT require high

8.4 Imaging Algorithm Implementation

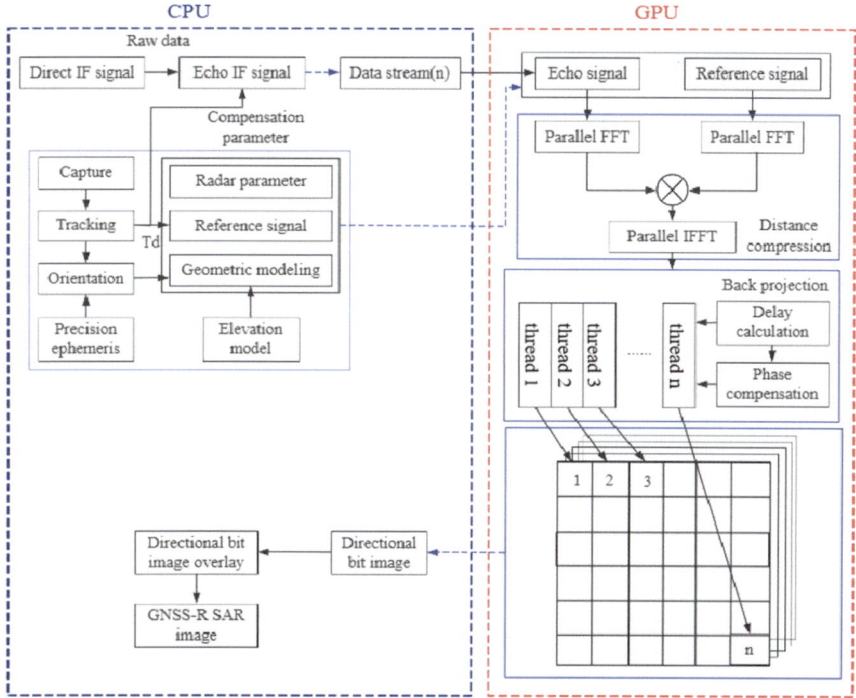

Fig. 8.9 Heterogeneous parallel processing platform

memory demands for large data lengths, especially with large volumes of Reflected GNSS Signal data. It is essential to match echo data and storage space sizes to prevent memory overflow. Figure 8.10 shows an FFT/IFFT execution strategy, where 1, 2, …, n denote different range gate numbers, allocated independent threads for performing FFT and IFFT operations on their data, with a single thread handling one or more range gate data processing, determined by considering the GPU's memory space size.

The range compression process is as follows:

(1) Determine the number of Fourier transform points based on the width of one equivalent pulse of the navigation signal and the sampling rate, as well as data storage space in CPU and GPU, transferring the reference signal and echo signal from CPU to GPU.
(2) Design and execute a range Fourier transform scheme for the reference signal and echo signal based on the number of range sampling points.
(3) Construct parallel kernel functions for the reference signal's complex conjugate matrix, echo signal matrix, and complex multiplication, allocating threads according to the number of range gates to complete parallel computations.
(4) Design and implement a range inverse Fourier transform scheme for the reference signal and echo signal based on the number of range sampling points.

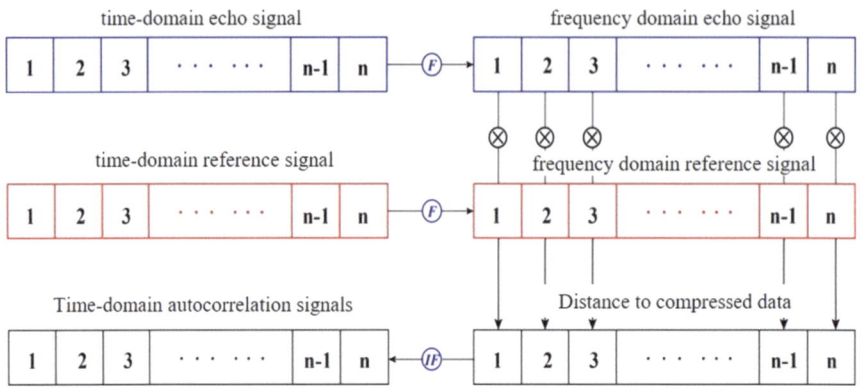

Fig. 8.10 An FFT/IFFT execution strategy

8.4.3 Back Projection

Back-Projection primarily involves time delay calculation and phase compensation, including complex multiplication and addition operations within the kernel function. There are two different implementation methods: based on pixel point mapping and based on Pulse Repetition Interval (PRI) mapping.

(1) When implementing Back-Projection based on pixel point mapping, each thread corresponds to each pixel in the image. Each kernel function call for complex addition and multiplication operations targets different pixel points within the current PRI, and the kernel function is called repeatedly for each PRI to obtain the imaging result.

(2) When implementing Back-Projection based on PRI mapping, each thread corresponds to a PRI. In a single kernel function call, it coherently accumulates echo signals for the same pixel point, and then the kernel function is called repeatedly for each pixel point to obtain the imaging result.

580. Main frequency

Comparing the two methods, the former processes all data across various PRIs, which involves less memory access than the latter that repeats for each pixel point, potentially reducing or avoiding memory access conflicts. To further reduce resources consumed during kernel function calls, "batch" processing can be adopted, i.e., a single kernel function call processes multiple PRIs' echo data. Figure 8.11 shows "batch" processing in parallel based on pixel point mapping, where the number of PRIs processed each time is referred to as a "batch".

8.4 Imaging Algorithm Implementation

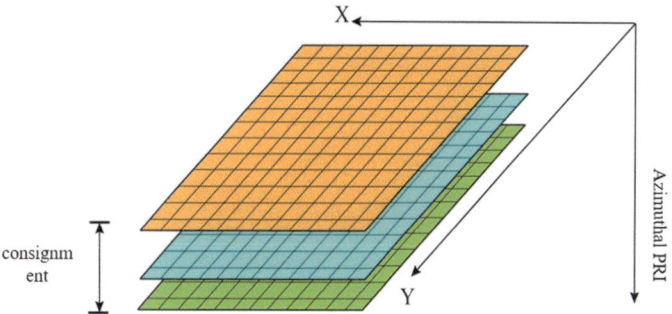

Fig. 8.11 Schematic of "batch" processing in parallel based on pixel point mapping

8.4.4 Stream Processing Structure Optimization

In the CUDA programming model, the data transfer interface is a high-speed serial computer expansion bus standard (Peripheral Component Interconnect-Express, PCI-E), with a maximum transmission rate of PCI-E 4.0 up to 8GB/s, used for transferring processing data from CPU memory to GPU memory and then back from GPU memory to CPU memory after processing. If the interface data is restricted, the performance improvement of GPU for data processing acceleration will not be fully guaranteed, especially in cases of extremely large data volumes. Segmenting data for serial sequential execution of imaging for each segment, despite parallel processing for each segment, cannot maximally exploit the parallel processing capability of the CPU+GPU architecture.

Asynchronous multi-level stream processing adopts a pipeline structure, alternating between transmission and processing using different engines, as shown in Fig. 8.12. Each level of stream processing structure is responsible for processing a segment of signal data, with asynchronous parallel execution between structures, significantly reducing delays caused by interface transmission. The computing engine of Level 1 stream processing structure and the duplication engine of Level 2 stream processing structure work in parallel; the computing engine of the previous level and the duplication engine of the next level work alternately, thereby maximizing the use of GPU work units. Data transfer delays between the CPU and GPU occur in parallel, shortening overall delays and improving computational efficiency.

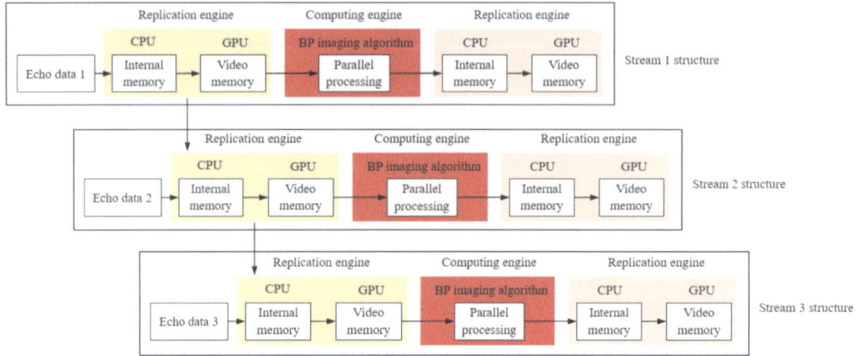

Fig. 8.12 Asynchronous multi-level stream processing structure

8.5 Simulation Analysis and Experimental Verification

8.5.1 Simulation Analysis

1. **Simulation Scenario Description**

Using a navigation satellite as the transmitter, with the receiver fixed at a certain location, 25 point targets are uniformly distributed over a 4 × 4km plane, with their relative geometric relationships as shown in Fig. 8.13, numbered 1~25. With a Pulse Repetition Interval (PRI) of 1ms (i.e., one code cycle), the Doppler frequency offset estimation error Δf should be less than 500 Hz (to satisfy the Nyquist Sampling Theorem), otherwise, aliasing phenomena will appear in the range compression results, as shown in Fig. 8.14a, showing ($\Delta f = 800$Hz) the result for target at point 13 (with a synthetic aperture time of 300s). When $\Delta f = 0$Hz, the aliasing phenomenon disappears, as shown in Fig. 8.14b shows.

Following the implementation process depicted in Fig. 8.9, stacking azimuth images yields all imaging results for the 25 point targets within the area, with the image for target number 13 located at (0,0) as shown in Fig. 8.15. The relative geometric relationships of the point targets and the simulation settings are entirely consistent. Taking typical targets number 5, 10, and 13 as examples, their PSLR (peak side lobe ratio) and ISLR (integration side lobe ratio) values are calculated, consistent with theoretical values in both range and azimuth directions, without significant deviation due to changes in geometric configuration, with specific values as shown in Table 8.3.

2. **Parallel acceleration of imaging algorithms**

Based on the CPU+GPU parallel hardware platform described in Sect. 8.4 and programmed in C language, the imaging data processing is implemented. The main device models and parameters are as shown in Table 8.4.

8.5 Simulation Analysis and Experimental Verification

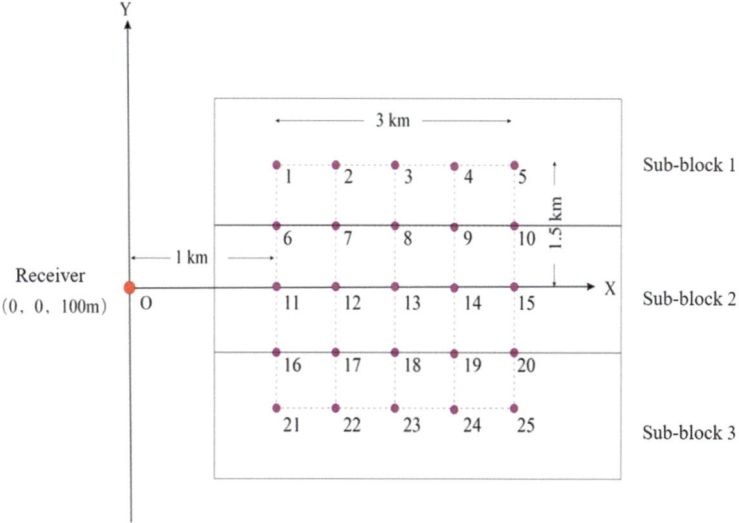

Fig. 8.13 Distribution of point targets in the imaging area

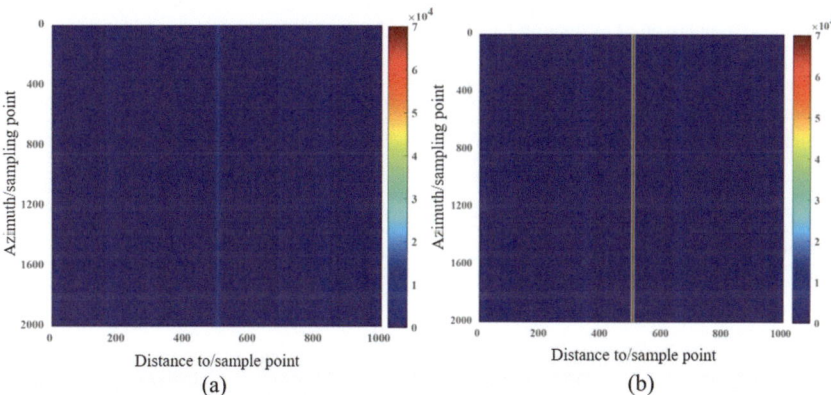

Fig. 8.14 Effect of different doppler frequency offset estimation errors on range compression

To compare the role of the GPU in imaging processing, imaging is implemented using the CPU alone and in tandem with the GPU. The introduction of the GPU does not alter the algorithm's corrections, errors, etc., hence the imaging performance in both cases is equivalent, with identical range and azimuth resolution, PSLR, and ISLR values. Table 8.5 presents the evaluation parameters for target number 25, with results under both running conditions being exactly the same.

Although the use of the GPU does not improve imaging quality, it significantly enhances imaging efficiency. The speed-up ratios for FFT/IFFT operations and

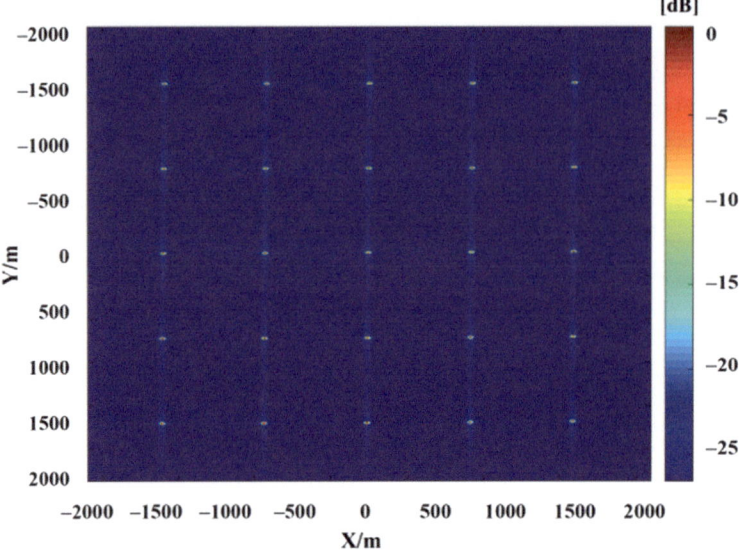

Fig. 8.15 Imaging results for point targets (with target number 13 at position (0,0))

Table 8.3 Comparison of evaluation parameters and theoretical values for typical point targets

	Away from			Azimuthally		
	PSLR/dB	ISLR/dB	Resolution/m	PSLR/dB	ISLR/dB	Resolution/m
Target 5	−34.80	−13.45	15.9	−12.25	−10.90	5.6
Target 10	−34.80	−13.45	15.9	−13.25	−10.90	5.6
Target 13	−34.80	−13.45	15.9	−13.25	−10.90	5.6
Theoretical value	−35.00	−13.45	15.9	−13.27	−10.90	5.6

Table 8.4 Physical parameters of the hardware platform

Hardware name	Model number	Capacity	Clock Frequency(GHz)
CPU	AMD ryzen 3800X	~	3.9
GPUs	Nvidia GeForce GTX 3090	24GB	1.95
Random access memory (RAM)	DDR4	64GB	3

Table 8.5 Evaluation parameters for target number 25

8.5 Simulation Analysis and Experimental Verification

Table 8.6 Time comparison for imaging 25 point targets

Parameters	FFT&IFFT	Complex multiplication	Backward projection	Time consumption
CPU(s)	74.4	20.88	17,361.6	17,456.88
GPU(s)	0.636	0.08	105.752	106.468
Acceleration ratio	117	261	164.17	163.96

Table 8.7 Analysis of integrated BP (Back Projection) algorithm acceleration performance for different scene sizes

Area size	128 × 128	256 × 256	512 × 512	1024 × 1024	2048 × 2048	4096 × 4096	8192 × 8192
Unit	s	s	s	s	s	s	s
CPU	123.1	206.4	507	2087.4	7611.9	30,543.3	68,871.9
CPU + GPU	1.1	1.73	3.43	13.38	45.47	180	407.26
Acceleration ratio	111.6	119.07	147.6	156	167.4	168.9	169.11

complex multiplication involved in range compression are 117 and 261, respectively. The overall speed-up ratio for the back-projection process, mainly involving complex multiplication and addition, can reach 164. Considering only the back-projection and range compression processes, the former accounts for over 99% of the time, so the GPU's computational efficiency improvement is very beneficial for the engineering implementation of GNSS reflectometry signal imaging applications. Table 8.6 provides a time comparison for imaging 25 point targets.

To further evaluate the efficiency of the CPU+GPU parallel platform for imaging different scenarios, the back-projection imaging algorithm was executed for seven different-sized scenes with a grid spacing of 1m. Imaging times and speed-up ratios are listed in Table 8.7. It can be seen that when the imaging area exceeds 2048× 2048, the speed-up ratio of the CPU+GPU parallel platform compared to a single CPU platform no longer shows significant changes. This is because the average imaging time per point in large scenes tends to stabilize, with all of the GPU's stream processors working, leading to synchronous increases in total time spent by both the CPU and CPU+GPU platforms, hence the change in speed-up ratio becomes less noticeable [15].

8.5.2 Test Validation

The imaging area selected for the test scene is the Beihang Stadium, as shown in Fig. 8.16. Data collection took place from 9:50 to 10:20 on March 25, 2021, and the receiver is placed in the stadium grandstand. The imaging scene is divided into 0.4

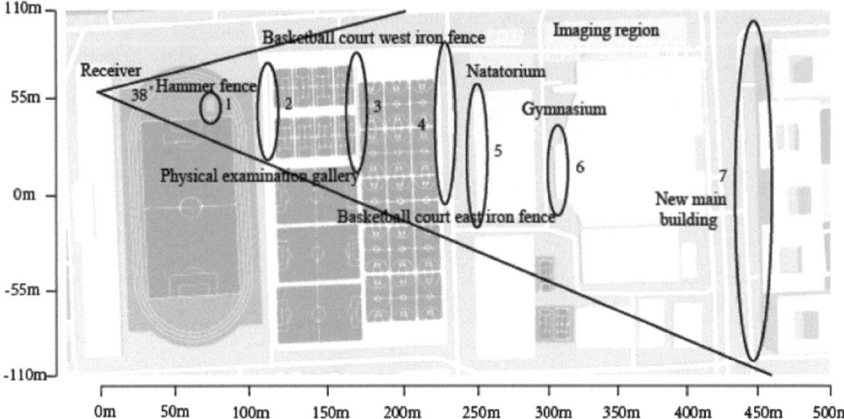

Fig. 8.16 Schematic of the navigation satellite position, receiver position and building distribution in the target area

× 0.4 m grids, and the imaging area is set as a square of 600 × 600 m, with a grid number of 1500 × 1500 accordingly.

The receiver position coordinates are set to (0,0), the synthetic aperture time is 1800s (i.e., the entire data acquisition time period of 30min), and the backward projection imaging results are shown in 8.17, all buildings in the actual scene are focused into the image.

Due to the height limitation of the receiver, echoes primarily originated from the west side of buildings, appearing as strong scattering points or surfaces in the image. The imaging result's pixel location information was matched with the optical image's location information, as shown in Fig. 8.18. The strongest echo signal position is approximately at point (315m, 10m), corresponding to target number 6 (the west edge of the gymnasium) in Fig. 8.16.

From the range profile in Fig. 8.19, it can also be seen that the power peak points of targets number 2, 3, 4, 5, 6, and 7 in the imaging area correspond one-to-one with the locations of each building. The range measurement for target number 6 is approximately 18 m, slightly larger than the theoretical range resolution value of 16.8 m (i.e., slightly reduced resolution). The azimuth profile measurement for the west edge of the gymnasium is 40m, comparable to the results of optical imaging.

Similar to the simulation experiment, the actual data was processed both with the CPU alone and with integrated CPU+GPU processing, taking approximately 5.5 h and 2 min, respectively, a nearly 165-fold efficiency improvement, consistent with the simulation's speed-up ratio results.

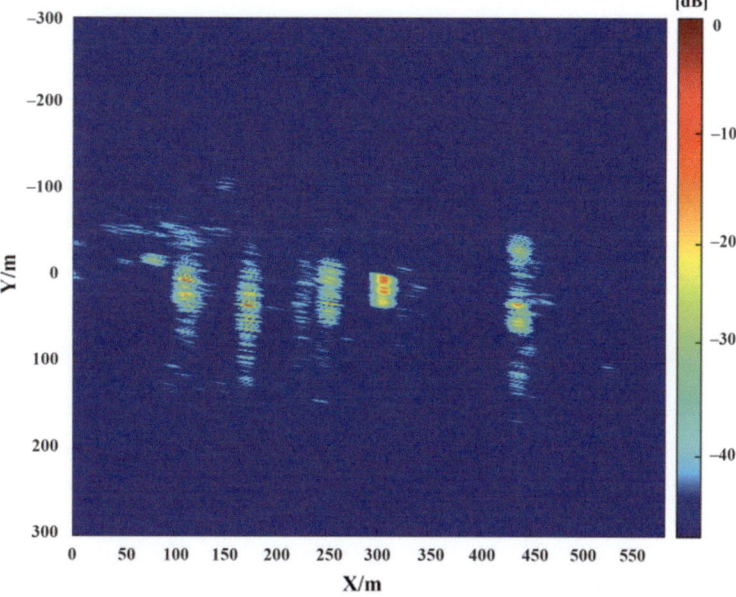

Fig. 8.17 Imaging results

8.6 Summary

As the radiation source of the GNSS-R SAR system, the global coverage and signal characteristics of navigation satellites ensure that the system has all-weather and all-day capability, and the images of the observation area can be acquired at any time. This chapter focuses on analyzing the influence of geometric configuration on the spatial resolution of GNSS-R SAR, and researches the implementation method for GNSS-R SAR system imaging, both simulation analysis and actual experiments have verified its feasibility.

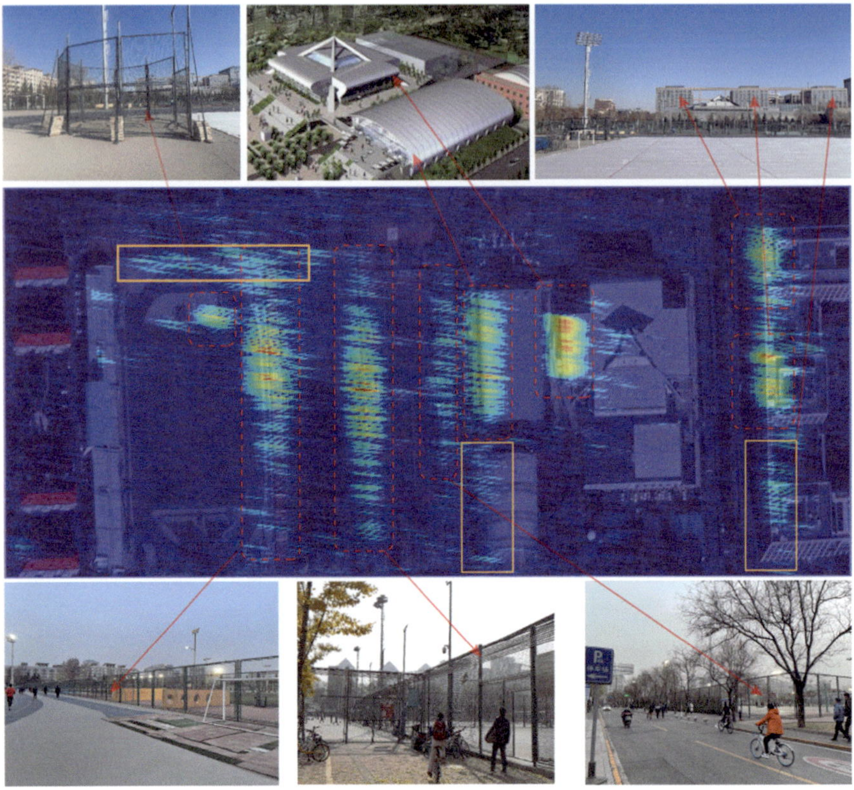

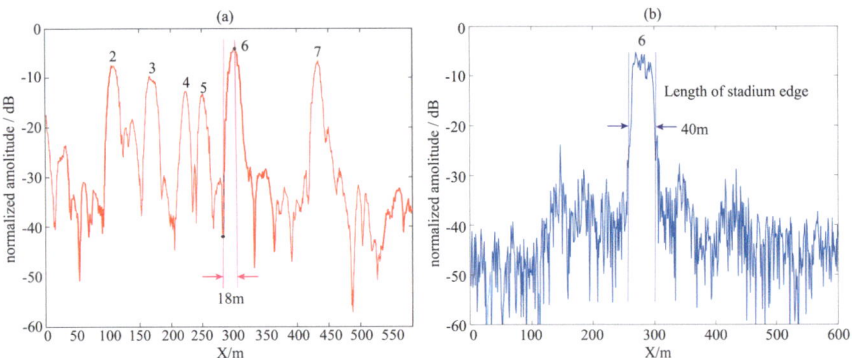

Fig. 8.18 GNSS-R SAR image and optical image matching results

Fig. 8.19 SAR image profile analysis

References

1. Ender J. A step to bistatic SAR processing. EUSAR, 2004: 356–9.
2. Cardillo GP. On the use of the gradient to determine bistatic SAR resolution[A]. In: International symposium on antennas and propagation society, merging technologies for the 90's[C], vol. 2. Dallas: IEEE; 1990. p. 1032–5.
3. Moccia A, Renga A. Spatial resolution of bistatic synthetic aperture radar: impact of acquisition geometry on imaging performance[J]. IEEE Trans Geosci Remote Sens. 2011;49(10):3487–503.
4. Munson DC, O'Brien JD, Jenkins WK. A tomographic formulation of spotlight-mode synthetic aperture radar. Proc IEEE. 1983;71(8):917–25.
5. Shao YF, Wang R, Deng YK, et al. Fast back projection algorithm for bistatic SAR imaging[J]. IEEE Geosci Remote Sens Lett. 2013;10(5):1080–4.
6. Zhang F, Yao X, Tang H, et al. Multiple mode SAR raw data simulation and parallel acceleration for Gaofen-3 mission[J]. IEEE J Sel Top Appl Earth Obs Remote Sens. 2018;11(6):2115–26.
7. Liu Y, Zhou Y, Zhou Y, et al. Accelerating SAR image registration using swarm-intelligent GPU parallelization[J]. IEEE J Sel Top Appl Earth Obs Remote Sens. 2020;13:5694–703.
8. Ming-Cong S, Ya-Bo L, Feng-Jun Z, et al. Processing of SAR data based on the heterogeneous architecture of GPU and CPU[A]. In: IET international radar conference 2013[C]. XIAN: IET; 2013. p 1–5.
9. Fasih A, Hartley T. GPU-accelerated synthetic aperture radar back projection in CUDA[A]. In: 2010 IEEE radar conference[C]. Arlington: IEEE; 2010. p. 1408–13.
10. Balz T, Stilla U. Hybrid GPU-based single- and double-bounce SAR simulation[J]. IEEE Trans Geosci Remote Sens. 2009;47(10):3519–29.
11. Devadithya S, Pedross-Engel A, Watts CM, et al. GPU-accelerated enhanced resolution 3-D SAR imaging with dynamic metamaterial antennas[J]. IEEE Trans Microw Theory Tech. 2017;65(12):5096–103.

Open Access This chapter is licensed under the terms of the Creative Commons Attribution-NonCommercial-NoDerivatives 4.0 International License (http://creativecommons.org/licenses/by-nc-nd/4.0/), which permits any noncommercial use, sharing, distribution and reproduction in any medium or format, as long as you give appropriate credit to the original author(s) and the source, provide a link to the Creative Commons license and indicate if you modified the licensed material. You do not have permission under this license to share adapted material derived from this chapter or parts of it.

The images or other third party material in this chapter are included in the chapter's Creative Commons license, unless indicated otherwise in a credit line to the material. If material is not included in the chapter's Creative Commons license and your intended use is not permitted by statutory regulation or exceeds the permitted use, you will need to obtain permission directly from the copyright holder.

Chapter 9
Exploring New Applications of GNSS Reflectometry

Since the late 1980s, GNSS-R has been developed over 30-years, it has gone through stages of concept proposition, theoretical innovation, technical demonstration and application promotion. Currently, with the launch of the American CYGNSS, Chinese Fengyun 3-E satellites and the construction of shore-based observation stations, sea surface wind speed inversion has begun to enter the stage of operational application and promotion; soil moisture and sea ice detection have also entered the stage of technological demonstration; GNSS Bi-SAR technology is gradually entering the stage of experimental validation from conceptual research. In this process, with the deepening understanding of GNSS-R technology and the continuous progress in signal processing technology, new concepts and applications have been put forward one after another in the fields of earth observation and radar detection, including mobile target detection, inland water body detection and river identification. Considering the current limitations in the coverage area, narrow bandwidth, and high sidelobes of conventional external radiation source signals such as Global System for Mobile communication (GSM) and Frequency Modulation (FM) signals, using GNSS signals as external radiation sources for target detection represents an emerging field with robust vitality. Inland water body detection technology is a further extension of GNSS-R applications beyond soil moisture, marking another new direction in spaceborne GNSS-R Earth observation. Rivers, as a special type of water body on Earth's surface, have characteristics such as water body width, river boundaries, and river surface height that significantly impact regional climate and agricultural safety production. GNSS reflection signals can also play a precise monitoring role.

This chapter provides a brief description for ground-based airborne target detection, ground-based river boundary detection and spaceborne water body identification, including the theoretical simulation and experimental validation results from the author's research group, trying to provide some new ideas and methods for subsequent research in this field.

9.1 Mobile Target Detection

Mobile target detection has a strong background in military and defense applications, and is also a necessary cutting-edge technology in modern smart cities and security management. According to the range of target movement, it can be divided into two categories: ground/sea surface mobile target detection and airborne mobile target detection. The former can be achieved by the imaging technical methods in Chap. 8, and this chapter focuses on the latter. For airborne mobile target detection, existing radar products have been applied for many years, and have played an important role in many scenarios at home and abroad. Active radar systems can easily expose themselves due to their electromagnetic signals emission, and there are still different degrees of blind spots in detecting low, slow-moving airborne targets. The carrier frequencies of GNSS signals are within the L-band (1.1–1.6 GHz), and receivers can be configured in bistatic or multistatic modes, fully utilizing navigation satellites from all spatial directions as radiation sources to create flexible and variable geometric configurations for airborne mobile target detection. This can compensate for the deficiencies of other types of radar and represents a new direction for future development.

9.1.1 Introduction to Detection Mechanisms

There are various configurations for bistatic airborne mobile target detection, with the receiver placed on the ground being the simplest and most common. In this mode, the target to be tested is located within the space between the navigation satellite and the receiver. Figure 9.1 shows the geometric relationship schematic for airborne mobile target detection. The coordinate origin is the location of the receiver, the Y-axis is the projection of the receiver antenna beam on the ground plane, the Z-axis points to the zenith, and XYZ forms a right-handed orthogonal coordinate system, with the tested target and receiver located within the same YOZ plane.

Fig. 9.1 Geometric relationship schematic of airborne mobile target detection

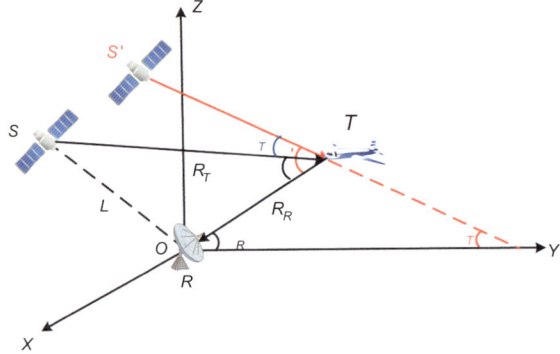

9.1 Mobile Target Detection

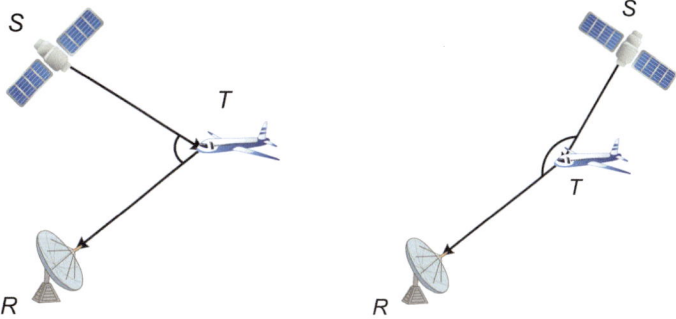

(a) Backward scattering configuration   (b) Forward scattering configuration

Fig. 9.2 Two geometrical configurations of bistatic radar

In the figure, S denotes a single navigation satellite, T denotes an airborne target to be measured, R denotes a <u>receiver</u> located at a fixed point on the ground, and S' is the projection of S on the YOZ plane. β is called the double base angle, which varies with the movement of the navigation satellite and the aerial target under the condition that the receiver is fixed. The target scattering signal received when its value is the signals received when $0° < \beta < 135°$ and $135° < \beta < 180°$ are called backward and forward scattering, respectively. Backward scattering and forward scattering are two geometric configurations of a bistatic radar system, as shown in Fig. 9.2 [1, 2].

1. **Backscatter configuration**

The backscattering configuration is a common type in bistatic radar systems. Since GNSS signals are not specifically designed for target detection, utilizing GNSS reflectometry signals for airborne target detection can be considered a specific case of a non-cooperative bistatic detection system. The process of airborne target detection based on the backscattering configuration is shown in Fig. 9.3, where the satellite selection module primarily selects the satellite with the optimal geometric configuration based on the detection area, the ephemeris of navigation satellites, the fixed position of the receiver, and the scattering characteristics of the target. The direct and reflected antennas refer to the antennas receiving direct signals from navigation satellites and reflected signals from targets, respectively, each with their own processing channels. The direct signal is captured and tracked to obtain a reference signal, which is correlated with the echo signal received and processed by the reflected antenna to obtain time-delay and Doppler observables for target detection.

The reception of the direct signal is handled in the same way as in a normal navigation and positioning receiver, where signal capture can be considered as a search process in three dimensions: pseudo-random code, time delay and Doppler. If the navigation satellites used are selected and determined in advance, only the latter two are searched, and the occurrence time of the correlation peak determines the parameters of the reference signal. The two-dimensional multichannel correlation

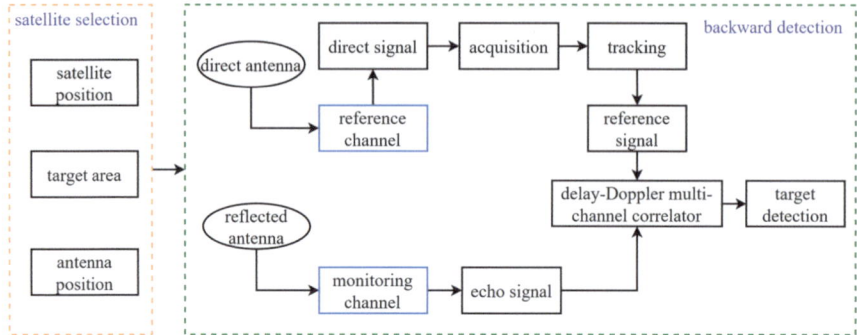

Fig. 9.3 Backscatter target detection flow

data processing for delay and Doppler is similar to what is described in Chap. 5, where all chips and frequency bins to be searched need processing. Considering the complexity of implementation,accomplished by processing delay in parallel and Doppler serially as shown in Fig. 9.4.

2. **Forward scattering configuration**

The geometric relationship of forward scattering corresponds to the diffraction phenomenon of wireless signal propagation. When electromagnetic waves propagates through the air and encounters an obstacle (herein referred to as the airborne target),new secondary waves are generated, bypassing the obstacle and entering the shadow zone, also known as the forward scattering region Ground receivers receive electromagnetic waves that have bypassed the obstacle, and by identifying the differences with unobstructed signals, airborne targets can be detected.

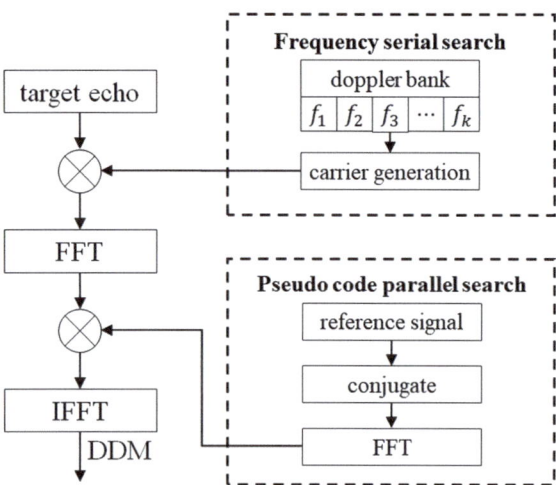

Fig. 9.4 Schematic of the parallel time-delay matched filtering process

9.1 Mobile Target Detection

Considering the actual physical size of an airborne moving target, The changes in received signals are analyzed using the degree of the target's entry into the diffraction region as the independent variable.

A set of concentric ellipses spaced half a wavelength apart with the transmitter S and receiver R as the foci point is known as the Fresnel zone, as shown in Fig. 9.5 [3]. For a target T_n located on the *nth concentric* ellipse, according to the geometric relationship, we have

$$|ST_n| + |T_nR| - |SR| = n\lambda/2 \tag{9.1}$$

where λ is the wavelength of the signal carrier. The area enclosed within the first concentric ellipse is called the first Fresnel zone, and the area between the (n-1)th and nth concentric ellipses is called the nth Fresnel zone. The distance between the target

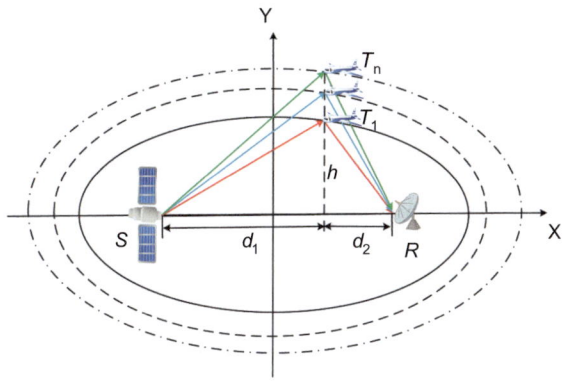

Fig. 9.5 Schematic of Fresnel zone

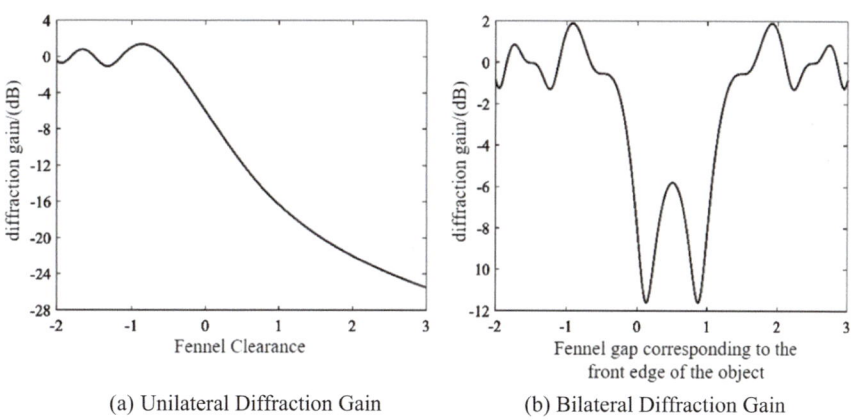

(a) Unilateral Diffraction Gain

(b) Bilateral Diffraction Gain

Fig. 9.6 Examples of changes in diffraction gain with fresnel gap

point T_n and the line-of-sight connection of the transmitter and receiver is denoted by h. The distance between its projection on the line-of-sight (i.e., the direct path) and the distances to the transmitter and receiver is given as d_1 and d_2, respectively.

After the navigation satellite signal passes through any point on the illustrated ellipse to reach the receiver, the difference in propagation path distance compared to the direct path distance to the receiver is denoted by Δd, which yields

$$\Delta d = |\mathbf{ST_n}| + |\mathbf{T_nR}| - |\mathbf{SR}|$$
$$= d_1\sqrt{1 + \frac{h}{d_1}} + d_2\sqrt{1 + \frac{h}{d_2}} - (d_1 + d_2) \tag{9.2}$$

The phase difference between the two is

$$\Delta\varphi = \frac{2\pi \Delta d}{\lambda} \tag{9.3}$$

When the target is within the first Fresnel zone, i.e., $n = 1$, $h \ll d_1$, then the following approximate expression is available [4]:

$$\Delta d \approx \frac{h^2(d_1 + d_2)}{2d_1 d_2} \tag{9.4}$$

$$\Delta\varphi \approx \frac{\pi h^2(d_1 + d_2)}{\lambda d_1 d_2} \tag{9.5}$$

The size of the extent of the first Fresnel zone is defined by the radius r_1, which is

$$r_1 = \sqrt{\frac{\lambda d_1 d_2}{d_1 + d_2}} \tag{9.6}$$

The proportion of an airborne moving target that enters the area from outside the first Fresnel zone is called the Fresnel gap u. The value of u is negative when the target is outside the first Fresnel zone, zero for u when the target is in the center of the Fresnel zone, and positive for u when the target leaves the Fresnel zone and thereafter. The numerical size of the Fresnel gap u is defined as

$$u = \frac{h}{r_1} \tag{9.7}$$

Based on the target type and its motion process, the diffraction transmission of navigation satellite signals can be analyzed in two models: unilateral and bilateral.

Assuming that E_o is the field strength of the navigation satellite signal in free space, the unilateral model is

9.1 Mobile Target Detection

$$\frac{E_d}{E_o} = F(v) = \frac{1+j}{2} \int_v^\infty \exp\left(\frac{-j\pi z^2}{2}\right) dz \tag{9.8}$$

where $F(v)$ is the Fresnel integral that varies with the Fresnel diffraction parameter v, which is often solved by looking up tables or approximate calculations. At this point, the gain from diffraction (expressed in dB units) is

$$G_{single} = 20 \lg |F(v)| \tag{9.9}$$

As the target crosses the first Fresnel zone, diffraction occurs at both the front and rear edges, and the received signal is the sum of the diffraction signals from both ends, thus forming the bilateral model. The field strength of the diffraction signal from the target's front edge is

$$F(v_{front}) = \frac{1+j}{2} \int_{v_{front}}^\infty \exp\left(\frac{-j\pi z^2}{2}\right) dz \tag{9.10}$$

The field strength of the diffracted signal from the target's rear edge is

$$F(v_{back}) = \frac{1+j}{2} \int_{-\infty}^{v_{back}} \exp\left(\frac{-j\pi z^2}{2}\right) dz \tag{9.11}$$

According to Eqs. (9.10) and (9.11), the gain produced by bilateral diffraction is

$$G_{double} = 20 \lg |F(v_{front}) + F(v_{back})| \tag{9.12}$$

Examples of the changes in diffraction gain with the position of the target's front edge in both unilateral and bilateral diffraction modes are shown in Fig. 9.6a and b. In the case of unilateral diffraction, an infinite plate enters the first Fresnel zone from the position of $u < 0$, and the front end of the flat plate moves from the position of $u = -2$ to the position of $u = 3$, with the diffraction gain gradually decaying as the target enters; in the case of bilateral diffraction, a sphere is selected as the target, and the length of its crossing the Fresnel zone is the radius of the first Fresnel zone, and the front end of the target moves from the position of $u = -2$ to the position of $u = 3$. When the target completely enters the first Fresnel zone, the diffraction gain will have a brief peak, and the received signal power will show a "W" type change.

9.1.2 Field Test Analysis and Validation

1. **Backscatter configuration experiment**

For the airborne moving target detection experiment based on the backscattering configuration, commercial airplanes near the Capital Airport were selected as the objects. The experiment was conducted on the morning of April 24, 2022, at the West Lake Garden Bagua Platform near Beijing Capital Airport (40°02′55″N, 116°02′56″E), as shown in Fig. 9.7.

The antenna parameters used for data collection are shown in Table 9.1. A right-hand circular polarization antenna was used to receive direct signals from BeiDou navigation satellites, while a left-hand circular polarization antenna received signals reflected off airplanes from BeiDou satellites. After down-conversion to intermediate frequency, the signals were sampled, quantized, and stored for data processing as depicted in Figs. 9.3 and 9.4.

During the experiment, the flight path of the commercial airplane and the antenna coverage area are illustrated in Fig. 9.8. Data collection started at 12:42:29, with Air China flight CA1712 departing around 12:41, flying from north to south. It entered the coverage area of antenna 3 at 12:42:48, while the distribution of BeiDou navigation satellites during this period is shown in Fig. 9.9. Satellites No. 2, No. 7, No. 10, No. 30, and No. 36 all formed backscattering geometric relationships.

Flight CA1712 entered the antenna's coverage area (satellite No. 36) 16.8 s after the receiver was activated and left the coverage area (satellite No. 2) after 21.6 s. The data processing results showed that compared to the signal received

Fig. 9.7 Backscattering configuration experiment site

Table 9.1 Antennas used in the experiment

Antenna number	Antenna type	Maximum gain	Installation	Polarization mode
1	Omnidirectional antenna	2.8dBi	Set rigidly in place	RHCP
2	Omnidirectional antenna	5.5dBi	Set rigidly in place	RHCP
3	Directional antenna	10dBi	Set rigidly in place	LHCP
4	Directional antenna	13dBi	Hand-held aligned aircraft	LHCP

9.1 Mobile Target Detection

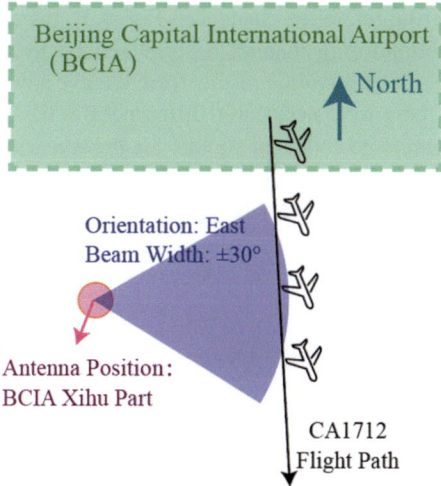

Fig. 9.8 Flight trajectory and antenna coverage of the civil aircraft during the experiment

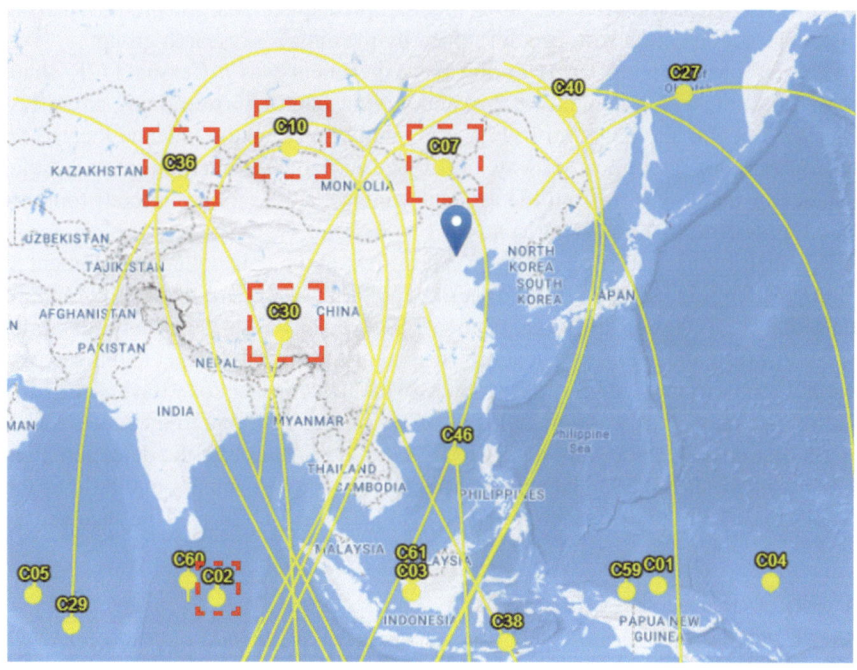

Fig. 9.9 Distribution of Beidou navigation satellites during the experiment

by the direct antenna (the reference channel in Fig. 9.3), the signal received by the reflected antenna (the monitoring channel in Fig. 9.3) had a frequency offset range of $-5\,Hz \sim -20\,Hz$, and the observation results from the five navigation satellites were consistent. However, there were noticeable differences in the delay after the correlative processing of direct and reflected signals for the five satellites, with coherent integration time results for 20 ms as shown in Table 9.2.

Figure 9.10 presents the time-delay Doppler correlation results for satellites No. 36, No. 10, No. 30, and No. 2 in chronological order, clearly demonstrating that BeiDou navigation satellite signals can effectively detect airborne moving targets under this geometric configuration.

2. Forward scattering configuration tests

The analysis in this section is based on experimental data conducted on November 23, 2020, at the Shengli Airport in Dongying, Shandong, focusing on the changes in navigation satellite signals under the Fresnel zone diffraction conditions introduced in the previous section. Two types of antennas, omnidirectional and directional, were used, with their specifications listed in Table 9.3. Signals were down-converted to intermediate frequency after reception, then sampled, quantized, and processed using custom target detection software developed by the author's research group.

The airborne moving target during the experiment was a Cessna 172R small trainer aircraft, with a length of 8.2 m and a wingspan of 11 m training aircraft of the experimental site is shown in Fig. 9.11, with the reception antenna located on the west side of the airport runway, positioned horizontally facing east at a radial distance of about 450 m from the airplane runway, as the trainer aircraft took off heading north across the detection area.

The distribution of visible navigation satellites at the data collection moment at 13:20 is shown in Fig. 9.12. Based on the double-base angle range of the forward scattering configuration, the signal from satellite No. 24 was selected for observation.

By inserting the actual physical dimensions of the airborne target and the experimental scenario into Eq. (9.6), the radius of the first Fresnel zone traversed by the aircraft during takeoff was calculated to be 10.7 m, comparable to the target's size. In the first data set, with only a single target taking off through the detection area, a "V" shaped attenuation change occurred in the received signal amplitude as the target crossed the first Fresnel zone between satellite No. 24 and the receiver, as

Table 9.2 Detection results for each satellite

Satellite PRN	Start time/s	Termination time/s	Latency/us	Doppler/Hz
30	19.1	19.5	0.68	−10
7	17.88	18	0.23	−5
36	16.8	17.3	0.71	−10
10	17.4	17.7	0.45	−10
2	21.24	21.6	1.03	−20

9.1 Mobile Target Detection

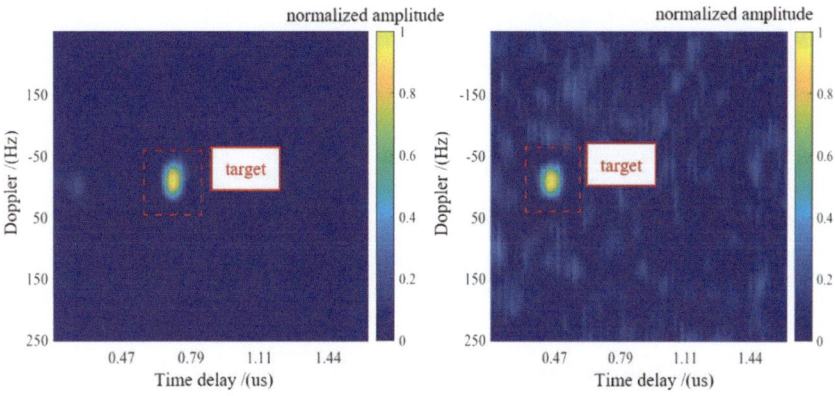

(a) Satellite No. 36 (b) Satellite No. 10

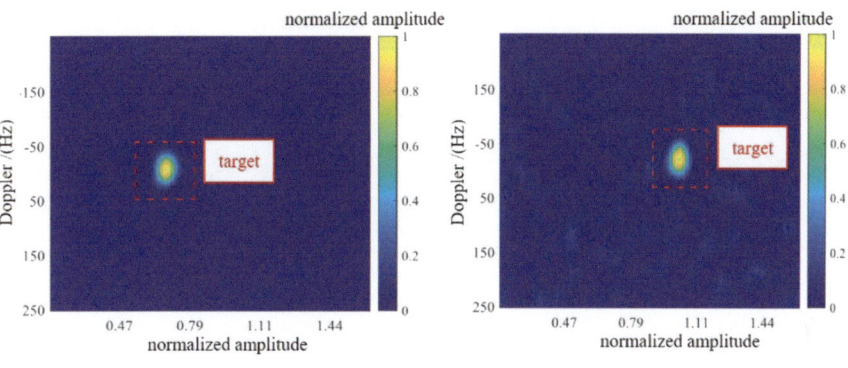

(c) Satellite No. 30 (d) Satellite No. 2

Fig. 9.10 Time-delay doppler correlation values for each BeiDou satellite

Table 9.3 Parameters of forward scattering reception antennas

Antenna type	Antenna usage	Operating frequency	Antenna gain	Polarization mode
Omnidirectional antenna	Receive direct signal	GPS L1 (supports B1 signal)	3dBi	RHCP
Directional antenna (±20°)	Receive diffraction signals	GPS L1	10dBi	RHCP

shown in Fig. 9.13. The duration of this change approximates the time the aircraft stayed within the first Fresnel zone.

In the second data set, the aircraft took off and circled above the airport, crossing the first Fresnel zone multiple times, resulting in several "V" shaped attenuation changes in the received signal, as shown in Fig. 9.14. As the aircraft circled back and forth, its radial distance from the ground-based receiver increased, thus enlarging the

Fig. 9.11 Schematic of the experimental site

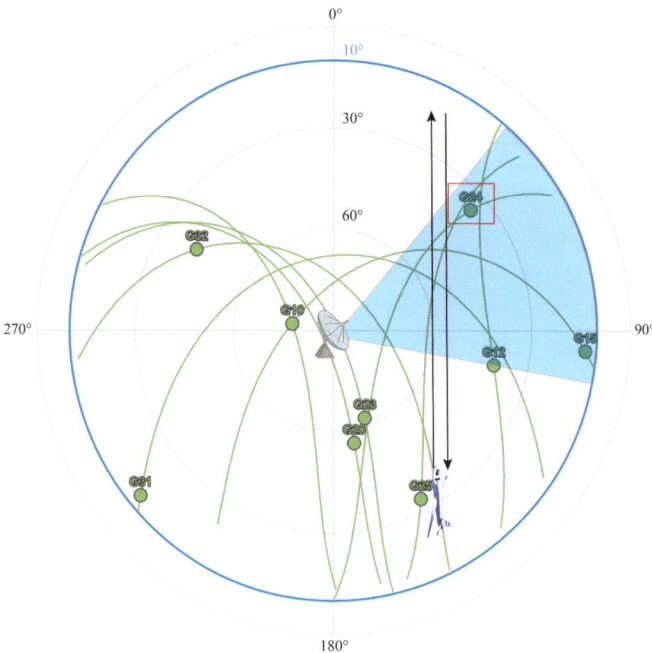

Fig. 9.12 Visible satellite distribution at the time of data collection

radius of the first Fresnel zone. During the experiment, situations occurred where the aircraft completely entered the first Fresnel zone, causing "W" shaped changes in the received signal attenuation (Target 3 in Fig. 9.14), and the time the aircraft stayed within the first Fresnel zone increased.

This experimental data processing result indicates that when an aerial mobile target appears multiple times within the Fresnel zone, it can be identified from the

9.1 Mobile Target Detection

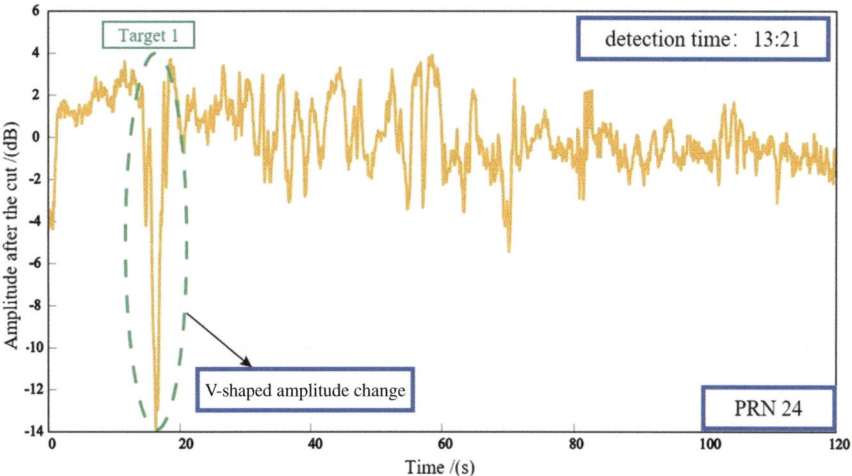

Fig. 9.13 Signal amplitude as the target crosses the first Fresnel zone once

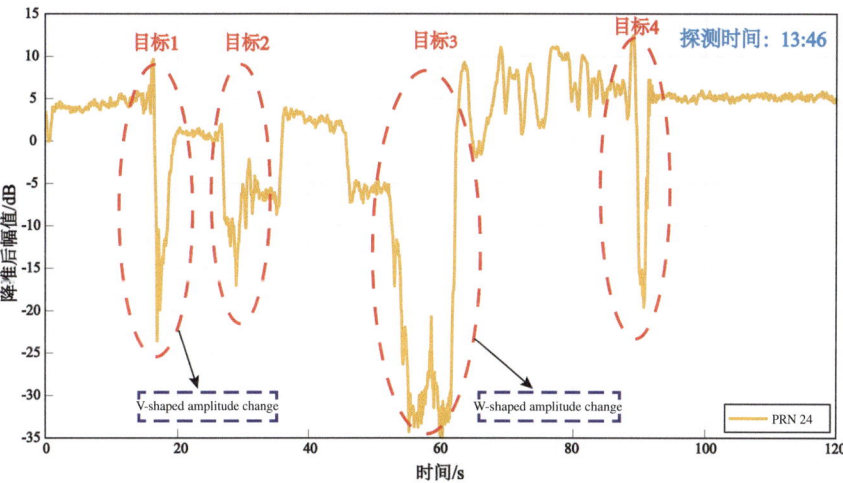

Fig. 9.14 Signal changes as the target crosses the first fresnel zone multiple times (*Note* Targets 1–4 in the figure refer to the sequence of appearances of the same target)

changes in the received signals. However, if multiple different targets appear simultaneously within the detection area, additional observational quantities are needed for identification, such as data from multiple navigation satellites, adding more antennas, or networking ground receivers.

9.2 River Boundary Detection

Rivers, defined as bodies of water that collect on the earth's surface from precipitation or ground emergence and flow along depressions under the influence of gravity, play a vital role in maintaining the earth's water cycle, energy balance, climate change, and disaster monitoring. With the intensification of human activities, rivers have evolved from a natural attribute to a "nature-human" dual attribute. Currently, contact-based cableway wireless flow measurement at water level stations is the common method for river monitoring. This method, while consuming significant financial and material resources, fails to meet the demands of complex environments such as wind, rain, flood seasons, etc., resulting in sparse or missing hydrological stations in remote and climatically harsh areas of China, and thus limited river hydrological information. The development of remote sensing technology has introduced various river monitoring methods, including photography [5], UHF ground wave radar [6], active and passive microwave radar [7], and satellite multispectral imaging [8], supplementing large-scale and underdeveloped area observations to some extent. Optical monitoring results depend on the composition of the study area, climate conditions (e.g., cloud, rain), and its resolution, and satellite optical sensors can only extract river distribution without obtaining flow velocity or water level information. Similar to optical sensors, satellite synthetic aperture radar with microwave imaging can provide river distribution information under complex climate conditions.

As an important carrier of surface water resources and a basic unit of water resource management, real-time monitoring of river boundaries, flow velocity, water levels, and flow rates is crucial. As mentioned, GNSS reflection signals can extract physical parameters such as soil moisture from the reflection ground in Earth observation. The application of this technology to detect river boundaries, differentiating between river water and banks (soil or concrete materials), and solving suitable models and observational quantities is the focus of this section.

9.2.1 Detection Principle Analysis

As previously mentioned, the detection of river boundaries using GNSS reflection signals can employ either a dual-antenna mode or a single-antenna mode. The former offers a broader application scope, utilizing not only satellite-based but also airborne and ground-based geometric configurations for data collection; the latter is limited to ground-based geometric configurations, as illustrated in Fig. 9.15, which shows a schematic of the ground-based GNSS reflection signal application geometry. With the movement of navigation satellites, the specular reflection point corresponding to the ground-fixed receiver will form a continuous trajectory of position changes on the earth's surface. When the GNSS antenna is near a river, part of the specular reflection point trajectory falls on the river surface, and another part on the riverbank. Due to the different dielectric constants of river water and riverbanks (usually non-water

9.2 River Boundary Detection

materials such as soil), the navigation satellite signals reflected by the river and the banks will differ in amplitude, phase, or frequency. This forms the physical basis for distinguishing river water from riverbanks based on GNSS reflection signals. Since river boundaries are line-shaped targets and considering the slow movement speed of the specular reflection points formed by the movement of navigation satellites, river boundaries can be viewed as a series of curves connected by nearly straight lines. Detecting river boundaries can thus be simplified to solving for each approximate straight line endpoint. This is the physical basis that allows GNSS reflection signals to detect river boundaries. Moreover, since most riverbanks have a certain slope relative to the river surface, the rate of change of time delay for GNSS signals reflected by the river and the banks, relative to the direct signal, also differs. That is, the rate of change of time delay for the direct signal can also be used as an observational quantity for feature extraction and inversion of river boundaries.

1. Solving for Reflection Coefficient Magnitude Value

Similar to Sect. 7.3, when employing a single-antenna mode with a ground-based geometric configuration, the carrier-to-noise ratio (C/N0) sequence outputted by the GNSS receiver is fitted with a polynomial, yielding:

$$\overline{C} = \sum_{i=0}^{3} a_i \theta^i \tag{9.13}$$

where the polynomial order is taken as 3rd order, $a_0 \sim a_3$ is the fitting coefficient and θ is the satellite altitude angle. Equation (9.13) is also known as the slowly varying term of the C/N0 sequence. Subtracting the fitted slowly varying term from

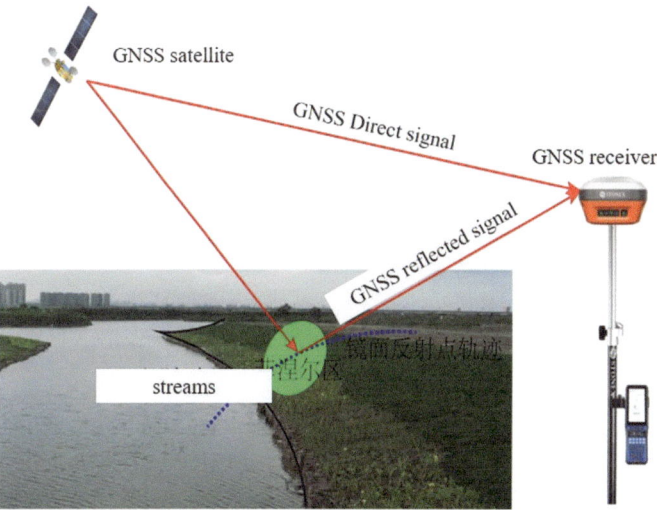

Fig. 9.15 Schematic of ground-based GNSS reflection signal application geometry

the original C/N0 sequence results in the following, also known as the oscillating term of the C/N0 sequence:

$$\tilde{C} = C - \overline{C} = 2A_d A_r \cos \varphi_{dr} \tag{9.14}$$

The definitions of A_d and A_r are the same as in Chapter 7, representing the amplitudes of the direct and reflected signals, respectively, and φ_{dr} is the phase of the oscillating term. Transforming the oscillating term using the slowly varying term yields

$$r_{rd} = \frac{\tilde{C}}{\overline{C}} = \frac{2A_d A_r}{A_d^2 + A_r^2} \cos \varphi_{dr} = \frac{2\tilde{r}_{rd}}{1 + \tilde{r}_{rd}^2} \cos \varphi_{dr} \tag{9.15}$$

where $\tilde{r}_{rd} = \frac{A_r}{A_d}$. The result after further Hilbert transform of (9.15) is written as

$$\hat{r}_{rd} = H[r_{rd}] \tag{9.16}$$

Then the envelope and phase of r_{rd} are respectively

$$|r_{rd}| = \sqrt{r_{rd}^2 + (\hat{r}_{rd})^2} \tag{9.17}$$

$$\hat{\varphi}_{dr} = \text{arctg}\left(\frac{\hat{r}_{rd}}{r_{rd}}\right) \tag{9.18}$$

Combining the above equations yields the expression $\tilde{r}_{rd}$ as

$$\tilde{r}_{rd} = \frac{1 - \sqrt{1 - |r_{rd}|^2}}{|r_{rd}|} \tag{9.19}$$

2. Grid-based Reflection Coefficient Magnitude

To match the specular reflection points with points on river boundaries as closely as possible, a grid division method is used here. That is, two parameters, the elevation angle θ and the azimuth angle ϕ, are used to generate the grid of elevation-azimuth observation units, so that they satisfy the following constraints:

$$\begin{cases} (i-1) \cdot \Delta\theta \leq \theta < i \cdot \Delta\theta \\ (j-1) \cdot \Delta\phi \leq \phi < j \cdot \Delta\phi \end{cases} \tag{9.20}$$

where $\Delta\theta$ and $\Delta\phi$ are the grid intervals for elevation angle and azimuth angle and (i, j) is the grid number. The reflection coefficient magnitude values obtained from the inversion of the navigation satellite data are mapped to the corresponding grids, and the corresponding dielectric constants $\hat{\varepsilon}_{ij}$ are obtained based on the following optimization operator, i.e.

9.2 River Boundary Detection

$$\hat{\varepsilon}_{ij} = \underset{\varepsilon_r}{\arg\min} \left\{ \sum_{n=1}^{N} ||\tilde{r}_{rdn}(i,j)| - |\tilde{r}_{rd}(\varepsilon_r)|^2| \right\} \quad (9.21)$$

where $\tilde{r}_{rd}(\varepsilon_r)$ is the theoretical model value of $\tilde{r}_{rd}$, $|\tilde{r}_{rdn}(i,j)|$ is the measured value of $\tilde{r}_{rd}$ within the grid (i,j), and N is the number of observed points in the corresponding grid (i,j).

3. Identification of River Water and Banks

Once the grid-based elevation-azimuth observation units correspond one-to-one with the dielectric constant of the reflection surface, it is possible to distinguish between river water and banks. In other words, this is a simple binary classification problem, although the feature quantity and observation data are relatively complex. The essence is to match the dielectric constant in the grid with river water/banks. To improve the accuracy of the match, the adaptive threshold segmentation method based on the maximum inter-class variance has good performance, especially suitable for application scenarios where the threshold of received data changes at any time due to the movement of navigation satellites [9]. Taking the gridded data on each azimuth angle as the segmentation object, the maximum inter-class variance of three thresholds is set as:

$$\left\{ \hat{k}_{j1}, \hat{k}_{j2}, \hat{k}_{j3} \right\} = \underset{0<k_{j1}<k_{j2}<k_{j3}<L_j}{\arg\min} \left\{ \sigma_{Bj}^2(k_{j1}, k_{j2}, k_{j3}) \right\} \quad (9.22)$$

where L_j is the maximum value of the dielectric constant at the jth azimuth; k_{j1}, k_{j2} and k_{j3} are the corresponding three segmentation thresholds; σ_{Bj}^2 is the interclass variance of the dielectric constant, defined as

$$\sigma_{Bj}^2 = \sum_{i=0}^{3} \omega_{ji}(\mu_{ji} - \mu_{jT})^2 \quad (9.23)$$

where μ_{jT} is the mean value of the dielectric constant at the jth azimuth; μ_{ji} is the intraclass mean value of class i of them; ω_{ji} is the proportion of elements in class i. When there are both river and riparian observations on the jth azimuth, the classified objects show bimodal characteristics. k_{j1} and k_{j3} are located near the clustering points of riparian and river observations, respectively, and k_{j2} is located in the fuzzy area of river and riparian observations. When there are only river or riparian observations in the jth azimuth, the classified objects show single-peak characteristics, and k_{j1} ~ k_{j3} are all located near the river or riparian observations. Therefore, the difference between k_{j1} and k_{j3} can be used as a criterion to determine whether the classified object has both river water and river bank observations. When the difference is greater than the set value, it is judged that there are both river water and river bank observations, otherwise, it is judged that there are only river water or river bank observations. k_{j2} can be used as a criterion for river water and river bank observations, i.e., if the observation value is greater than k_{j2}, the observation value is judged to be river water, and if the observation value is less than k_{j2}, the observation value is judged to be river bank. In order to reduce the influence of noise on the judgment, the number of observations judged as river water or river bank should be greater than a predefined threshold. Figure 9.16 showing the judgment process of river water and river bank, the specific steps are as follows:

(1) Read the observation value for the *jth elevation* angle.
(2) Solve for three segmentation thresholds using the maximum inter-class variance.
(3) Decide whether the absolute difference between k_{j1} and k_{j3} is greater than the set threshold T_{h1}. When it is less than T_{h1}, go to step 4); when it is greater than T_{h1}, go to step 5);
(4) Solve for the mean value of all observations on the *jth* azimuth m_j, m_j is greater than the set threshold T_{h4} Decide that all observations on the *jth* azimuth are river water, otherwise decide that all observations are river bank.
(5) Determine whether the number of observations greater than k_{j2} n_{+kj2} is greater than the set threshold T_{h2} and the number of observations less than k *is greater than the set threshold* T_{j2} n_{-kj2} is greater than the set threshold T_{h3}. If the judgment is not satisfied, return to step 4).
(6) The *ith* observation on the *jth* azimuth o_{ji} greater than k_{j2} is judged to be a river and less than k_{j2} is judged to be a riverbank.
(7) When $i \leq$ total number of jth azimuth observations N_j, jump to step 6) for the judgment of the next observation; when it is greater than, go to step 8).
(8) When $j \leq$ total number of azimuth grids N, jump to step 1) to process the next azimuth observation; when it is greater than that, jump out of the loop and the judgment is complete.

The threshold T_{h1} is set to $|o_{T1} - o_{T2}|/2$ (where o_{T1} and o_{T2} are the theoretical values of the two categorical object attributes, i.e., the theoretical values of the river and riparian observations, respectively); T_{h2} and T_{h3} are set to 10% of the total number of categorical pairs; and T_{h4} is set to $|o_{T1} + o_{T2}|/2$.

4. Determining River Boundaries

After accurately identifying river water and riverbanks, attributes of each specular reflection point are determined within the gridded elevation and azimuth observation units. To extract river boundaries, these gridded observation units, generated by the parameters elevation angle () and azimuth angle (), are remapped onto the terrestrial region where the river is located, establishing the geometric coordinates of the corresponding points. As illustrated in Fig. 9.17, using the GNSS antenna's projection point on the horizontal plane as the origin, and with the horizontal plane as the east-north plane, an east-north-sky right-angled coordinate system is established. The mapping relationship between the gridded elevation-azimuth observation units and the spatial domain is as follows

$$\begin{cases} e = h_r \sin\phi / \tan\theta \\ n = h_r \cos\phi / \tan\theta \end{cases} \quad (9.24)$$

9.2 River Boundary Detection

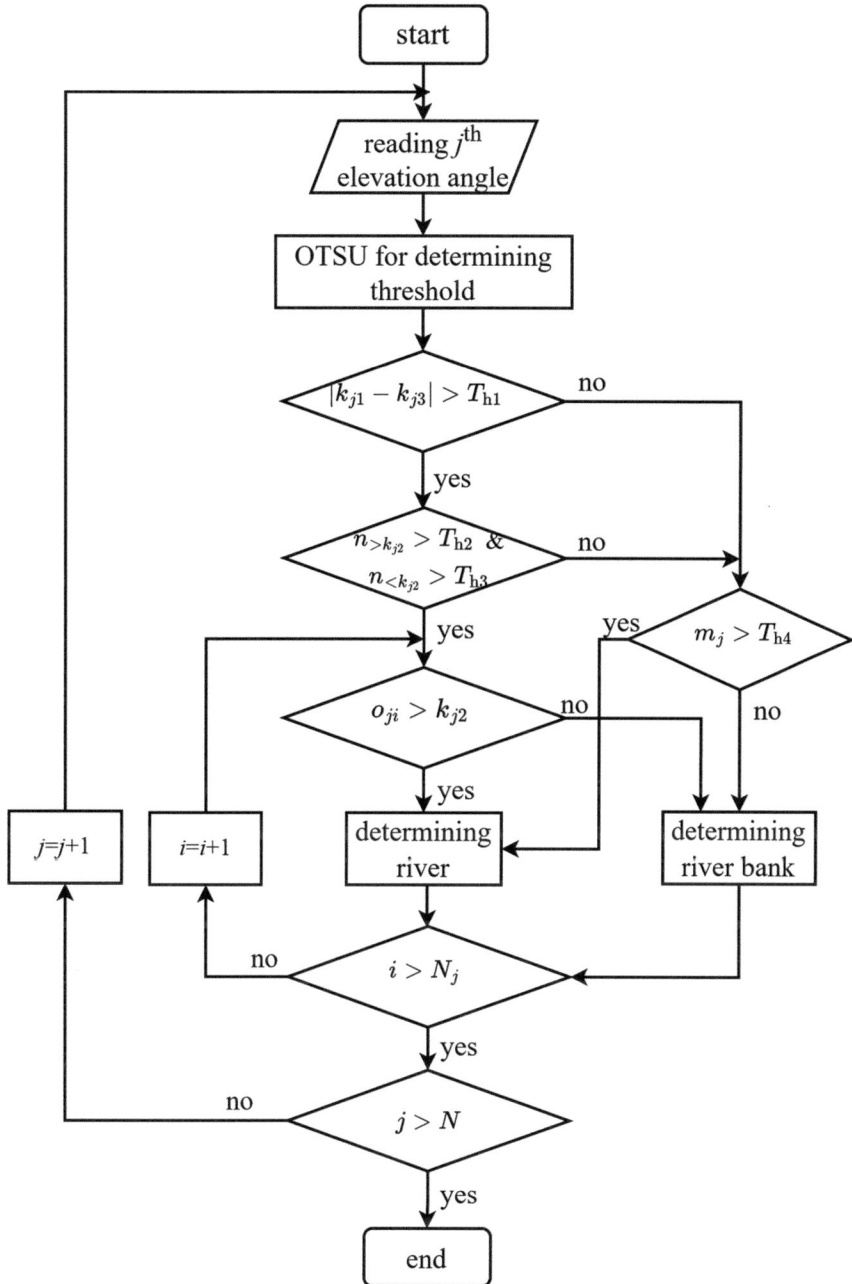

Fig. 9.16 Decision process for water and banks

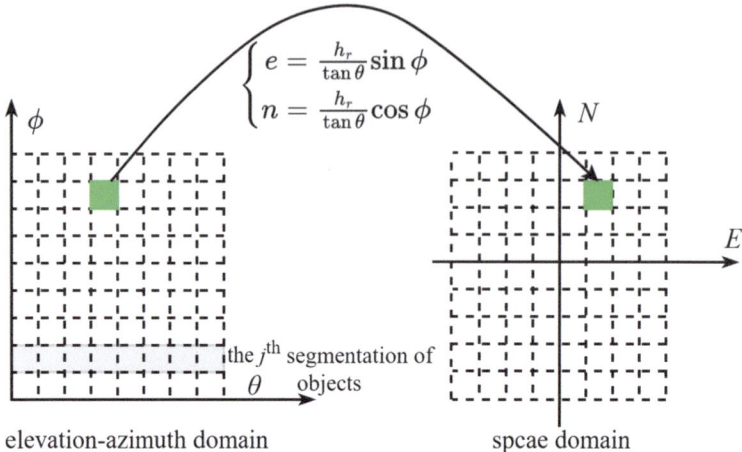

Fig. 9.17 Schematic of the mapping of the altitude angle-azimuth domain and the spatial domain (where h_r is the vertical distance between the receiver antenna and the horizontal plane)

After the river and the riverbank are identified, the points where the river is close to the riverbank are taken as the river boundary points. The altitude and azimuth angles of the detected boundary points form the set $\{\theta_1, \phi_1; \theta_2, \phi_2; \cdots ; \theta_M, \phi_M\}$. The set is mapped to the spatial domain $\{e_{b1}, n_{b1}; e_{b2}, n_{b2}; \cdots ; e_{bM}, n_{bM}\}$ using Eq. (9.24). Assume that the river boundary in the field of view is linear in the established Cartesian coordinate system, i.e.

$$n_b = p_1 \cdot e_b + p_2 \tag{9.25}$$

where e_b and n_b are the eastward and northward coordinates of the river boundary, respectively; p_1 and p_2 are the linear coefficients of the river boundary, which can be obtained by fitting of the following form.

$$\{\hat{p}_1, \hat{p}_2\} = \arg\min_{\{p_1, p_2\}} \left\{ \sum_{i=1}^{M} |n_{bi} - (p_1 e_{bi} + p_2)|^2 \right\} \tag{9.26}$$

To further improve the extraction accuracy of the river boundaries, a cyclic fitting approach can also be used, which involves:

(1) The identified river boundary points are fitted through Eq. (9.26) to obtain p_1 and p_2.
(2) Calculate the distance from each boundary point to the fitting result, i.e., Eq. (9.25).

$$d_i = \frac{|p_1 e_{bi} - n_{bi} + 1|}{\sqrt{p_1^2 + 1}} \tag{9.27}$$

9.2 River Boundary Detection

(3) If the maximum distance of the detected boundary points exceeds a predefined threshold T_{loop}, the point with the maximum distance is excluded, and the process returns to step 1); if it is below the threshold, end the loop and output p_1 and p_2.

9.2.2 Simulation Analysis and Verification

Using actual ephemerides of navigation satellites, theoretical calculations and numerical analyses were performed with different types and structural compositions of river water and banks.

1. Scenario Setup

The GNSS receiver's fixed ground position is set at (37.4475°N, 119.0100°E), with the GNSS receiving antenna 3.5 m above the river surface. The river boundary in the field of view is regarded as a perfect straight line. The linear equation of the river boundary in Scenario one is $n_b = -15$, with the river running east–west and the river boundary 15 m horizontally from the GNSS antenna. The slope of the river bank is 1.9°. Figure 9.18a and b show its side view and top view, respectively. In Scenario two, the linear equation of the river boundary is $n_b = e_b - 10$, with the azimuth angle of the river at 45°, the horizontal distance of the GNSS antenna from the river boundary is 7.07 m. The slope of the river bank is also 1.9°, as shown in the side and top views in Fig. 9.18c and d.

Data collection was done on June 19, 2021, limiting satellite elevation angles to 5°–40°. The satellite sky plot during this period, based on actual GNSS ephemerides, is shown in Fig. 9.19. This demonstrates the broader azimuthal coverage due to the higher number of GPS and BeiDou satellites in orbit compared to GLONASS and Galileo. GLONASS satellites, with their higher orbital inclinations, cover a larger azimuthal area. The plot only includes medium earth orbit navigation satellites; their inclined orbits result in poorer coverage at low northern elevation angles, affecting detection capabilities in that direction.

Gaussian white noise was added to both direct and reflected signals during data calculation, and antenna gains were based on actual directional antenna data used by the research team.

2. **Numerical Calculation Results**

The intervals for the gridded dielectric constants in elevation and azimuth angles were set as $\Delta\theta = 0.2°$ and $\Delta\phi = 1.0°$ respectively. Figures 9.20 shows the distribution of the dielectric constant in the elevation-azimuth domain for Scenario one. Despite minor oscillations in the dielectric constant due to the envelope of the reflected coefficient amplitude r_rd estimated by Hilbert transform, causing weak periodic changes with elevation angle, a clear difference exists between the dielectric constants in river water and riverbank areas.

To verify the effectiveness of the maximum inter-class variance method in distinguishing between river water and riverbanks, two datasets were analyzed: one

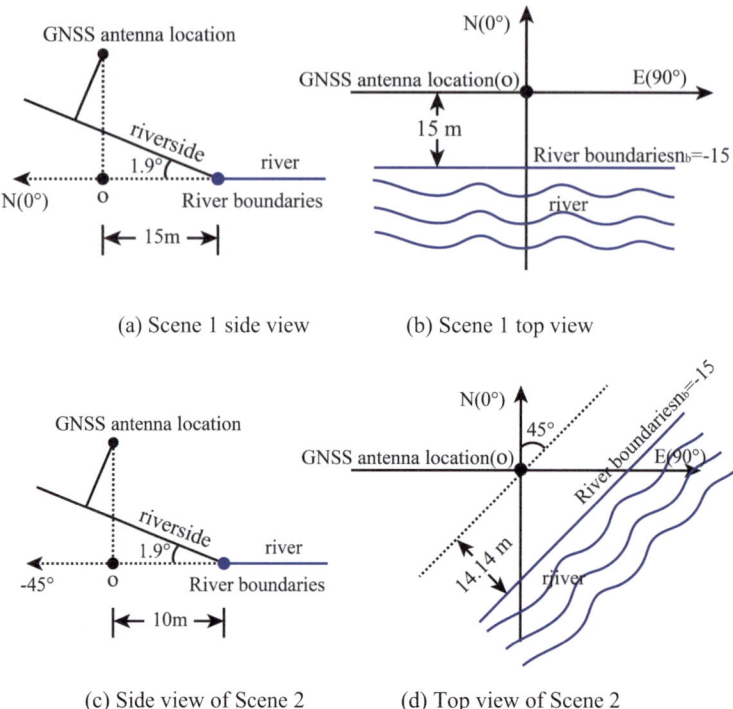

Fig. 9.18 Side and top views of the simulation scene

containing both river water and riverbanks (azimuth angles 0° ~1° in Fig. 9.20) and another consisting solely of riverbanks (azimuth angles 146° ~147° in Fig. 9.20), as shown in Fig. 9.21a and b. For the dielectric constant series at azimuth angles of 0° ~1°, the three thresholds determined by the maximum inter-class variance method are 31.28, 52.15, and 78.97. Threshold 1 (31.28) is near the riverbank's dielectric constant, and threshold 3 (78.97) is near the river water's dielectric constant, with the difference between the two thresholds being 47.69. Threshold 2 (52.15) effectively segments the inverted dielectric constants into two distinct intervals for river water and riverbanks. For the dielectric constant series at azimuth angles of 146° ~147°, the three thresholds are 17.70, 22.60, and 36.98, all located near the riverbank's dielectric constant. The difference between thresholds 1 (17.70) and 3 (36.98) is 19.28, significantly less than the difference for the first dataset at azimuth angles 0° ~1°. Hence, the difference between thresholds 1 and 3 can indicate the properties of the classified object, while threshold 2 can be used for segmentation between river water and riverbanks.

The thresholds $T_{h1} \sim T_{h4}$ set when determining the river boundary are 20, 8, 8 and 55 (the four thresholds in Fig. 9.16), and the threshold T_{loop} for the loop fitting is set to 4, then the river boundaries determined in Scenario 1 and Scenario 2 are

9.2 River Boundary Detection

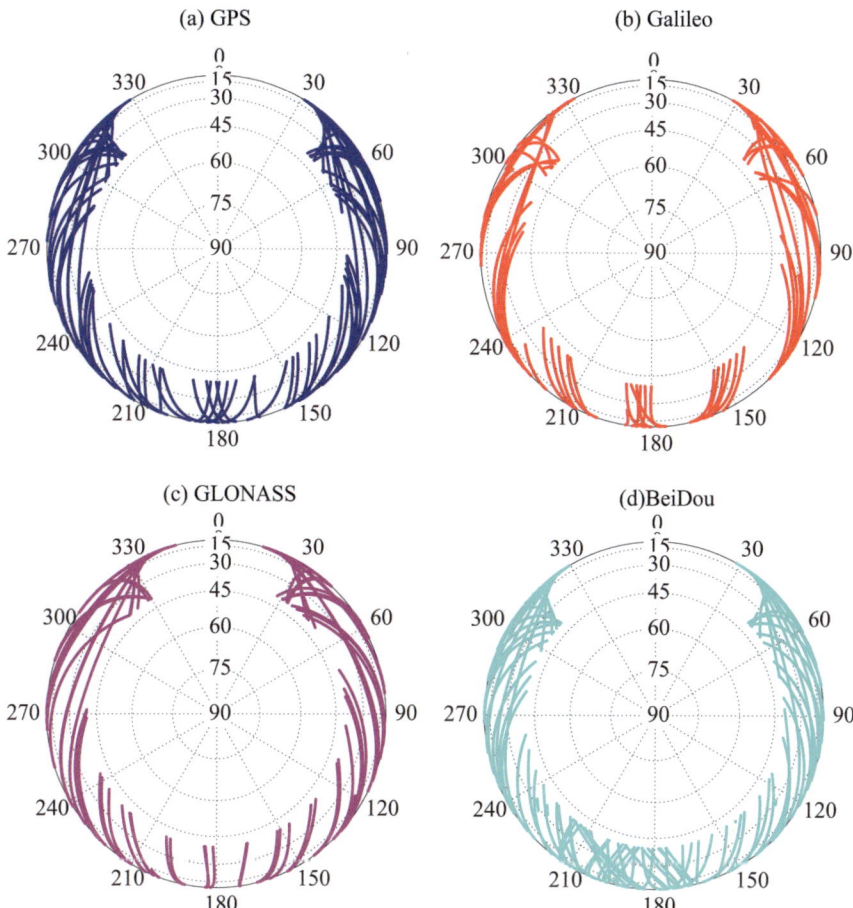

Fig. 9.19 Sky Plot of GPS, Galileo, GLONASS and BeiDou navigation satellites

$n_b = -0.05e_b - 15.81$ and $n_b = 0.99e_b - 11.06$, respectively, and the root-mean-square errors are 1.47 m and 1.13 m. The results of determining the river boundaries in Scenario 1 and Scenario 2 are shown in Fig. 9.22a and b, respectively.

Analysis of the Numerical Simulation Results has shown that the threshold value T_{loop} set during the cycle fitting process is closely related to the accuracy of river boundary determination.

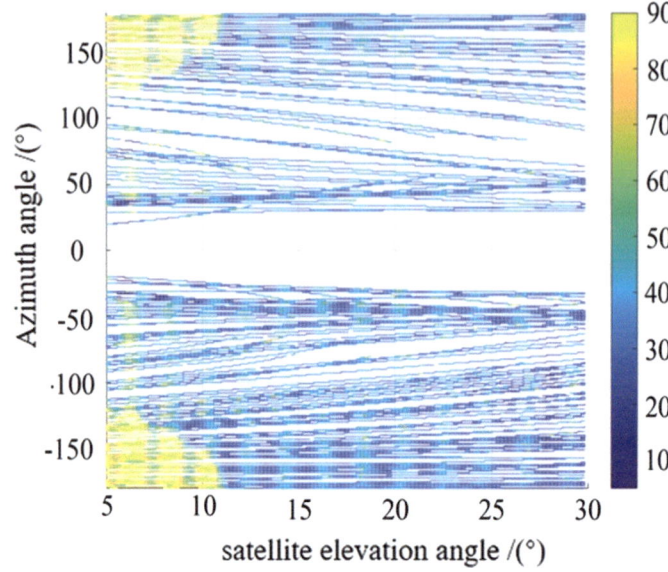

Fig. 9.20 Distribution of dielectric constant in elevation-azimuth domain for scenario one

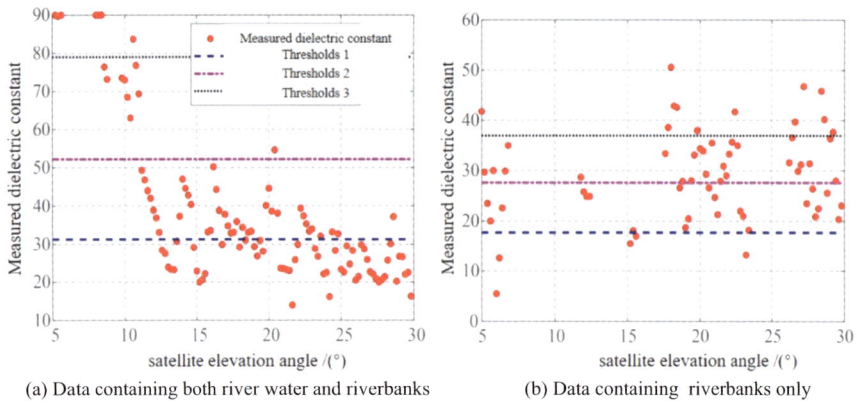

(a) Data containing both river water and riverbanks   (b) Data containing riverbanks only

Fig. 9.21 Example of inverted dielectric constant sequence data

9.3 Surface Water Body Identification

Beyond rivers, the Earth's surface features lakes, seas, glaciers, etc., collectively referred to as water bodies, crucial components of the Earth's hydrosphere. These water bodies play a significant role in climate regulation, biodiversity conservation, and human convenience. Different water bodies support agriculture, inland navigation, hydropower, flood control, and tourism, greatly advancing human civilization.

9.3 Surface Water Body Identification

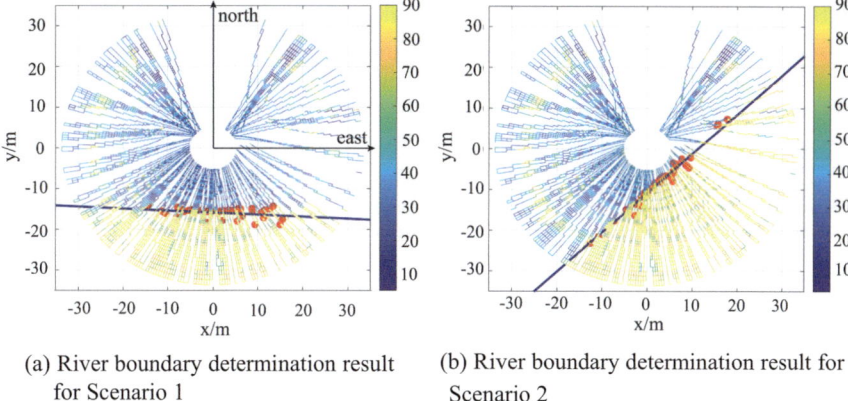

(a) River boundary determination result for Scenario 1

(b) River boundary determination result for Scenario 2

Fig. 9.22 Results of river boundary determination

Rapid changes in water and land distribution in many areas due to human activities necessitate large-scale water body observation to support global ecological protection and respond to global environmental changes. Inland water body identification methods align with those for detecting river boundaries, utilizing both satellite optical and microwave remote sensing. Using Reflected GNSS Signals sensitive to water and non-water bodies can achieve high-performance water body identification. Additionally, satellite-borne Reflected GNSS Signals offer advantages like abundant signal sources and L-band frequencies that penetrate clouds, rain, and vegetation. Research on water body identification methods using Reflected GNSS Signals can complement other satellite remote sensing methods, offering broad prospects for widespread use.

9.3.1 Water Body Identification Process

The principle of surface water body identification, similar to river water/riverbank identification, is based on the fundamental characteristic that water and non-water materials impact Reflected GNSS Signals differently. The specific remote sensing application depends on the geometric structure and receiver type. Surface water body identification differs from river boundary determination, with the former identified as a planar object (two-dimensional space) and the latter as a linear target (one-dimensional space). For linear targets, the simulation calculations considered the inclination between the target and the ground plane. For planar targets, surface roughness affects the reflected signal, with rougher surfaces causing more non-coherent components in the reflected navigation satellite signals. The amount of non-coherent components directly impacts the signal-to-noise ratio of reflected signals. Since water surfaces are smoother compared to soil, rocks, vegetation, etc., identifying water bodies using the signal-to-noise ratio of reflected signals, especially in satellite-borne Reflected GNSS Signals applications, can be highly effective. This section introduces a typical case of water body identification based on CYGNSS satellite data to illustrate the implementation method and performance of this application.

1. Extraction of Observation Quantities

For surface water body identification, the observation quantity uses the peak signal-to-noise ratio of two-dimensional delay-Doppler correlation power, given by:

$$SNR = 10 \log_{10} \left(\frac{P_{\text{peak}} - N_{\text{floor}}}{N_{\text{floor}}} \right) \quad (9.28)$$

where P_{peak} is the DDM peak correlation power and N_{floor} is the floor-noise correlation power, which can be obtained by averaging the correlation power in the delay-Doppler region of the no-signal [10]. Quality control of peak signal-to-noise ratio data is crucial for enhancing surface water body identification. If the signal-to-noise ratio is too low, indicating excessive noise in the reflected signal, it becomes difficult to extract valid information, decreasing the identification rate. Additionally, antenna gain changes with the navigation satellites' movement and the movement of the reception platform (such as low-earth orbit satellites or aerial vehicles), affecting data quality if the received signal comes from a direction with lower antenna gain.

2. **Processing of gridded data**

Similar to the gridding in Sect. 9.2, the main purpose here is to categorize and count the unevenly distributed data in space by gridding, to find representative or trend values in a region, and to reduce the influence of chance data. The size of the grid is first determined according to the requirements of the resolution of the water body identification, and commonly used grids are equidistant grids with equal margins, e.g. 1 and 3 km, etc. [11]. The data within the same grid is used for arithmetic calculation. Data within the same grid are determined by arithmetic averaging. Based on the spaceborne GNSS reflection signal recognition of water bodies, the determination of the grid distance also needs to take into account the moving speed of the navigation satellite, the moving speed of the receiving platform satellite, as well as the sampling frequency of the specular reflection point track and other values, as far as possible, to ensure that the coherent integration and non-coherent integration results in the same grid range, to fully reflect its mean value characteristics. Secondly, the geometric position corresponding to each data sample point and the standard grid are mapped one by one to obtain the spatial distribution of the reflection signal data.

3. **Water body identification method**

The identification of surface water bodies is a typical binary segmentation problem, and the key is the selection of the threshold value. By counting the range of signal-to-noise ratios of reflected signals from non-water bodies and water bodies, and finding the common value of the intersection between the two, the threshold can be determined; it is also possible to traverse all the signal-to-noise ranges with equal

9.3 Surface Water Body Identification

spacing, segmenting each threshold to be traversed and comparing them with the known results to derive the statistical values of the correct rate, the false alarm rate, and so on, so as to obtain the approximation of the threshold-correctness curve, which is used to determine the threshold for subsequent recognition. The former method is used when the statistical characteristics of water and non-water reflections are known; the latter method requires artificial setting of thresholds and their intervals, which will be different from the actual threshold-correct rate relationship, but if the threshold interval is set small enough, good approximate results can be achieved.

Regardless of which method determines the threshold value, the setting can be used to generate a binary segmentation image at that threshold value to determine the result of the water body identification. For example, values below the threshold are noted as 0, indicating a non-water body; conversely, values above the threshold are noted as 1, indicating a water body.

9.3.2 CYGNSS Data Validation

As mentioned in Chap. 1, CYGNSS is a small 8-star constellation launched by the United States for tropical cyclone monitoring that makes publicly available four levels of data, including: raw digital intermediate frequency data (L0), bistatic radar scattering cross-section correlated data (L1), sea surface wind speed and mean square slope orbit-level products (L2), and sea surface wind speed and mean square slope gridded products (L3). The data corresponding to Formula (9.28) is from L1 [12].

Publicly available data for January-December 2020 are selected here, with comparative data being the global land surface water product, which has a resolution of 500m [13]. Since the selected grid scale is 3km, the standard water body data need to be downsampled to complete the spatio-temporal matching. Due to the selected grid scale of 3km, standard water data needs to be downsampled for spatiotemporal matching. To demonstrate the applicability of the water body identification method under different terrain conditions, areas such as the northern part of the African continent around the Nile River, the lake groups around the East African Rift Valley, the Congo River Basin in Africa, and the Amazon River Basin in South America were selected as identification objects. The Nile River in the northern part of Africa flows through climates such as the tropical savanna and desert, with a relatively small flow volume, and the surrounding terrain environment is mainly bare desert and savanna; the lake groups in the East African Rift Valley are south of the equator, with the surrounding terrain environment mainly grassland and rock; the Amazon and Congo rivers are both located on the equator, situated in the world's first and second-largest tropical rainforests, respectively, with dense forests, abundant rainfall, and primarily forested surroundings.

Fig. 9.23 Standardized 3 km grid data for the Congo river basin

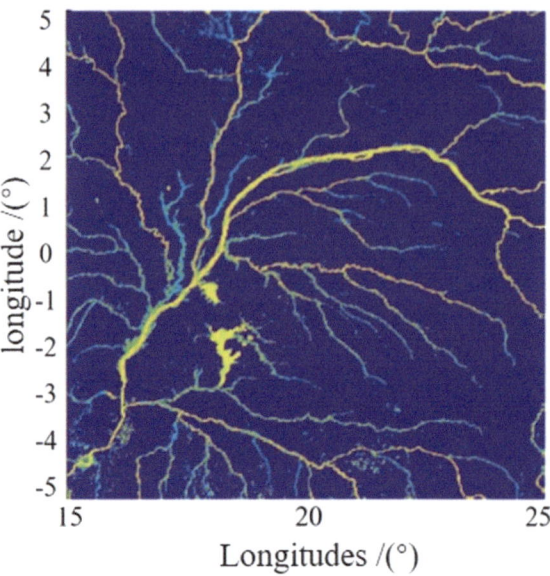

1. Congo River Basin

The Congo River Basin is located on the equatorial line, with latitude and longitude ranging from 15 to 25°E and 5°N to 5°S, surrounded by typical tropical rainforest terrain. After downsampling terrain data, a standard 3 km grid data representation is shown in Fig. 9.23 for the Congo River Basin.

Due to the lack of statistical information on water and non-water bodies, finding the optimal threshold for water body identification is challenging. Thus, the second method of threshold determination is used, traversing the 0 to 20 dB range, with a threshold interval set to 0.1 dB. Figures 9.24a–d show the water body identification results for thresholds of 3 dB, 6 dB, 9 dB, and 12 dB, respectively. When the threshold is lower, larger land areas are identified as water bodies; when the threshold is higher, only the larger main rivers are successfully identified, while smaller tributaries or seasonal rivers are identified as land. At a threshold of 8.7 dB, the water body identification in the Congo River Basin is optimal, achieving an accuracy rate of over 90.64%.

2. Amazon basin

The latitude and longitude ranges of the Amazon River basin are set from 50°W to 70°W and 0° to 10°S, as shown in Fig. 9.25, a schematic diagram of the water bodies in the Amazon River basin. The Amazon River flows from east to west, with a clear distribution of main rivers and tributaries. Besides linear rivers, there are also patch-like lakes. At some tributary ends, there are unclearly bordered speckle-shaped features due to the overall humid climate of the basin, high soil moisture content, or the presence of seasonal lakes and wetlands.

9.3 Surface Water Body Identification

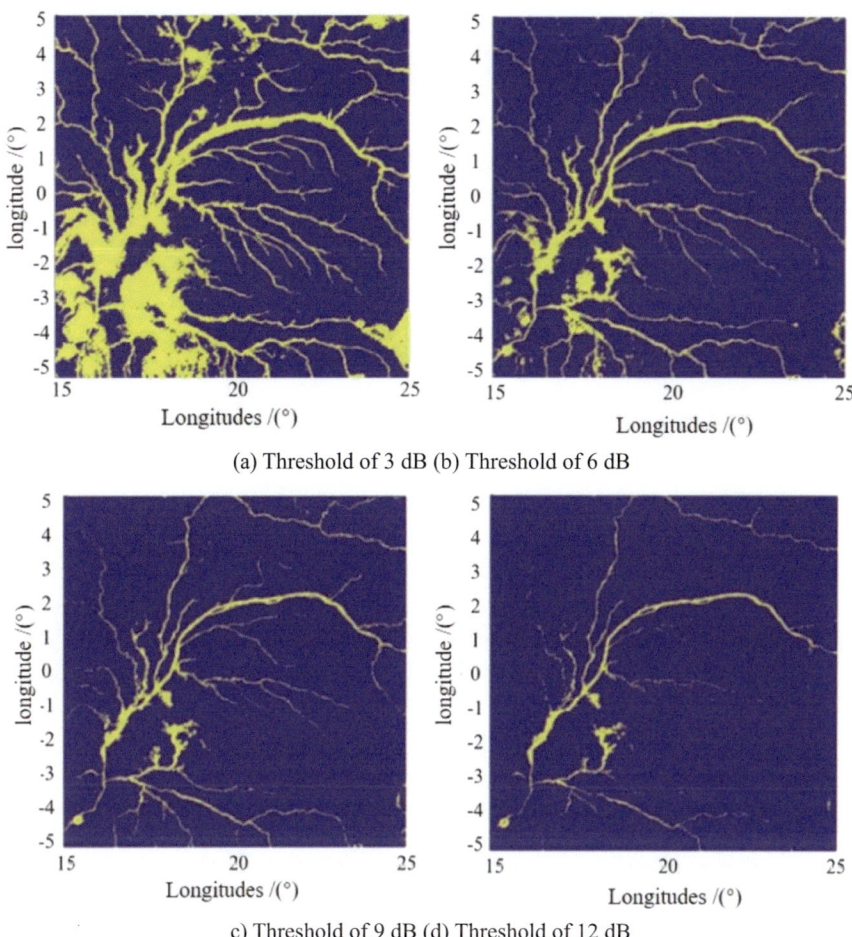

Fig. 9.24 Results of water body identification in the Congo river basin

The results of water body identification for different thresholds in the Amazon River basin are shown in Fig. 9.26. Similar to the Congo River Basin, the Amazon River Basin shows that more land is determined to be water at lower thresholds, while the main river becomes clearer at higher thresholds, but tributaries are determined to be land. The optimal threshold determination follows the same process as the previous section, based on statistical accuracy, false positive rate, and false negative rate data. At a threshold of 3.8 dB, the best results are achieved with a water body identification accuracy rate of 90.93%.

Congo River and Amazon River Basin both are typical tropical rainforest climate, high precipitation, high river flow, and clear distribution. The binary segmentation method based on the signal-to-noise ratio of the GNSS reflection signal has better water body identification results. Both can obtain more than 90% correct rate under

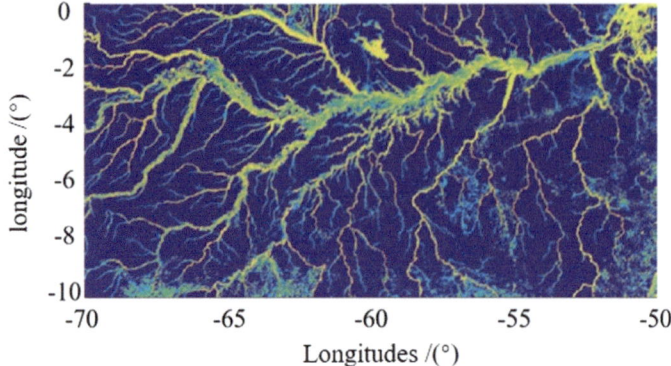

Fig. 9.25 Schematic of water bodies in the Amazon River Basin

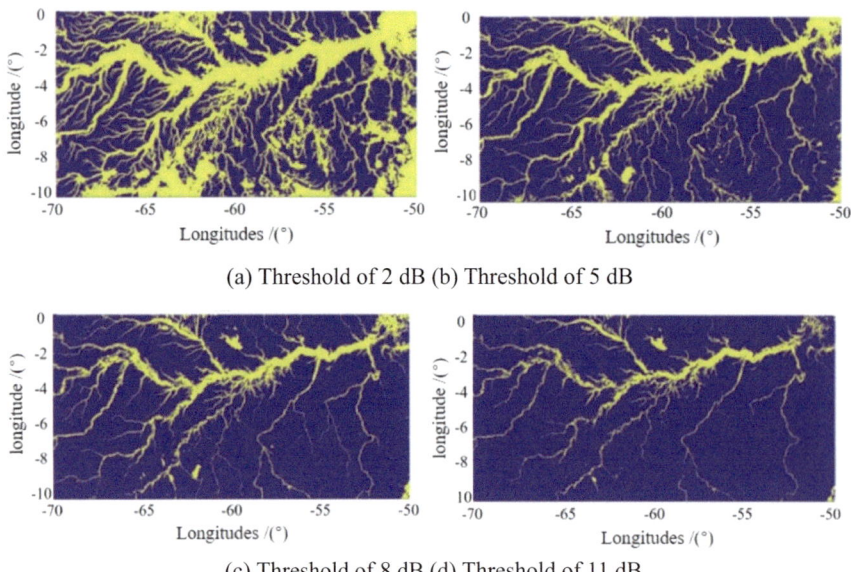

(a) Threshold of 2 dB (b) Threshold of 5 dB

(c) Threshold of 8 dB (d) Threshold of 11 dB

Fig. 9.26 Results of water body identification in the Amazon River Basin

the premise of optimal threshold determination. However, the optimal thresholds of the two are different, the former is about 8.7 dB, while the latter is about 3.8 dB. The reasons for the difference are mainly:

(1) In addition to being controlled by the equatorial low-pressure belt, the Amazon River is also affected southeast trade winds blow moisture from the Atlantic Ocean, leading to greater precipitation;
(2) The Amazon River flows through the plains, and the Congo River flows through the Congo Basin, the different terrain leads to different angles of incidence and

9.3 Surface Water Body Identification

reflection of satellite signals; at the same time Additionally, the Amazon River has more tributaries, while the Congo River has relatively few tributaries;

(3) Although both are typical tropical rainforests, they have different vegetation cover.

3. East African Rift Valley

In the region to the east of southern Africa (27°E to 37°E, 0° to 20°S), surrounding the East African Rift, there are several large lakes, such as the largest crater lake, Lake Victoria; the world's second-deepest lake, Lake Tanganyika; and the narrow Lake Malawi to the south, collectively known as the East African Rift Valley lake group. These lakes are mostly formed by fault subsidence due to crustal movements of the East African Rift, as shown in Fig. 9.27, with the lakes' shapes being elongated and their bottoms deep.

Using the same threshold as the Congo River Basin, the water body identification results are shown in Fig. 9.28. It is clear that at a threshold of 6 dB, the boundaries of several lakes can be clearly seen, and some rivers' flow directions are indicated. However, the SNR in the center of Lake Victoria in the north is lower than at the lake's boundary, with some grid points mistakenly identified as land. At a threshold of 6.5 dB, the water body identification results are optimal, with an accuracy rate of 82.94%.

4. Nile Basin

The Nile River Basin roughly spans from 30 to 40°E and from 10 to 30°N. The Nile flows from north to south, covering a wide latitude range with significant climate

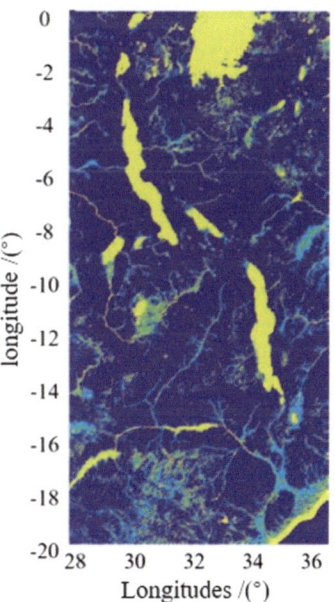

Fig. 9.27 -> East African Rift Valley Lake complex

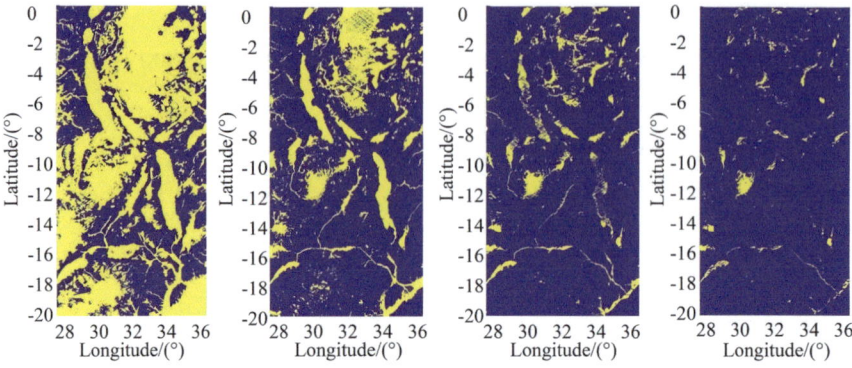

(a) Threshold of 3dB (b) Threshold of 6dB (c) Threshold of 9dB (d) Threshold of 12dB

Fig. 9.28 Results of water body identification for the East African Rift Valley lake complexes

differences, as well as variations in rainfall and terrain. As shown in Fig. 9.29, the southern part of the Nile has many tributaries, forming a delta.

Figure 9.29a–d are the results of water body identification when the thresholds are 4 dB, 7 dB, 10 dB and 13 dB, respectively. It can be seen that, at lower thresholds, the main stream of the Nile River in the north is clearly identified, with an obvious flow direction, but in a large area on the south side of the watershed is misjudged. As the threshold increases," the trend of the main stream in the north becomes less obvious, and the south side of the tributary area reveals a clearer river network. The

Fig. 9.29 Water bodies in the Nile Basin

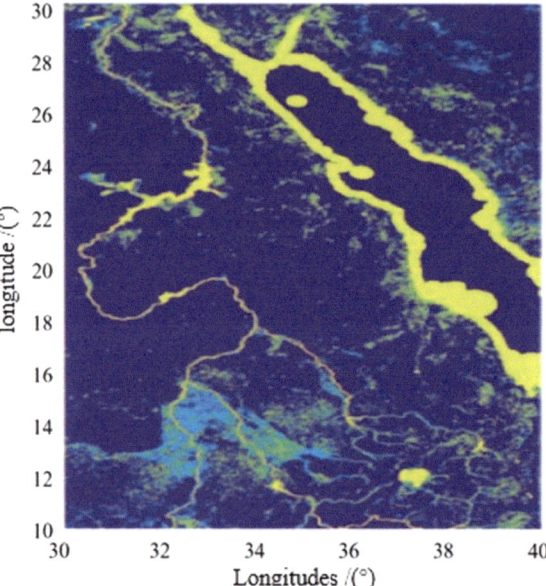

9.3 Surface Water Body Identification

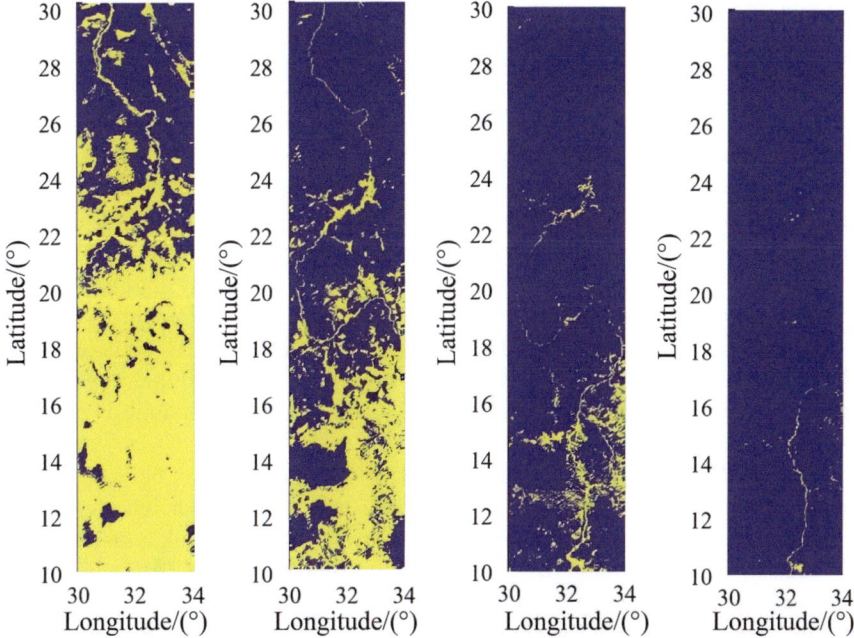

(a) Threshold of 4dB (b) Threshold of 7dB (c) Threshold of 10dB (d) Threshold of 13dB

Fig. 9.30 Results of water body identification in the Nile Basin

best identification results can be reached when the threshold value is about 11.8 dB, and the correct rate is approximately 82.96%. The results are similar to those of the East African Rift Valley and lower than those of the Congo and Amazon River basins (Fig. 9.30).

The Nile River Basin has smaller water flow and higher sediment content. Its tropical savanna climate, which alternates between dry and wet seasons, causes strong seasonal variations in water flow. The flood season can also lead to regular flooding, leaving fertile soil behind as the waters recede. Since the method described is based on averaging data over the year, it is challenging to identify these seasonal changes, and it is only possible to more accurately identify areas downstream of the Nile where the signal-to-noise ratio is higher. This means that the binary segmentation method has certain limitations in this region; more accurate identification results might be achieved with more data or by using other algorithms.

In addition, different topographies also have an impact on the results of the identification; the Congo River is located in a basin, and the Amazon River is in a plain, which makes it easier to clearly identify river network.

9.4 Summary

The applications of Reflected GNSS Signals are vast, both in areas traditionally covered by remote sensing means and in some new research areas. This chapter selectively introduces three novel applications: ground-based airborne target detection, ground-based river boundary detection, and spaceborne surface water body identification, all of which are major research directions being pursued by our research group.

Aiming at ground-based airborne target detection, the feasibility of utilizing backward and forward scattering configurations was preliminarily verified through experiments and data processing results. Despite challenges such as weak signals and direct clutter interference affecting its technical maturity, the advanced conceptual technology, with improvements in signal processing and antenna technology, is expected to progress towards practical application.

GNSS-I/MR, as a unique branch of GNSS-R technology, has been applied to the measurement of parameters such as soil moisture, vegetation height. As typical terrestrial entities, the distinction between river water and river bank is an important direction for the application of GNSS reflection signals, and the determination of the river boundary derived from it is the second focus of the discussion in this chapter. The process of determining the river boundary based on the observation output from the ground-based GNSS reflection signal receiver is investigated from a methodological perspective of view, and its feasibility is verified by combining with the actual scenario and simulation analysis.

The L-band has stronger vegetation penetration, which is favorable for the identification of water bodies in vegetation-covered areas, and is an effective complement to optical and microwave remote sensing. In this chapter, water bodies, including the Congo River, the Amazon River, the Nile River, and the East African Rift Valley Lake complexes," are identified using CYGNSS satellite data, which reveals the great potential of the satellite-based GNSS-R technology in the identification of inland water bodies.

References

1. Cherniakov M. Bistatic radar: principles and practice[M]. John Wiley & Sons, Ltd; 2007.
2. Melvin WL, Scheer J. Principles of modern radar: advanced techniques[M]. IET Digital Library; SciTech Publishing Inc; 2012.
3. Zhang D, Wang H, Wu D. Toward centimeter-scale human activity sensing with wi-fi signals[J]. Computer. 2017;50(1):48–57.
4. Zhang R, Jing X. Device-free human identification using behavior signatures in wifi sensing[J]. Sensors. 2021; 21(17).
5. Kinoshita R, Utami T, Ueno T. Image processing for aerial photographs of flood flow[J]. J Jpn Soc Photogramm Remote Sens. 1990;29(6):4–17.
6. Coats JE, Cheng RT, Haeni FP, et al. Use of radars to monitor stream discharge by noncontact methods[J]. Water Resour Res. 2006.

References

7. Lyzenga DR. Passive remote sensing techniques for mapping water depth and bottom features[J]. Applied Optics. 1978; 17(3).
8. Brakenridge GR, Nghiem SV, Anderson E, et al. Space-based measurement of river runoff[J]. Eos Trans Am Geophys Union. 2013; 86(19).
9. Otsu N. A threshold selection method from gray-level histograms[J]. IEEE Trans Syst Man Cybern. 1979;9(1):62–6.
10. Rodriguez-Alvarez N, Akos DM, Zavorotny VU, et al. Airborne GNSS-R wind retrievals using delay-doppler maps[J]. IEEE Trans Geosci Remote Sens. 2013;51(1):626–41.
11. Al-Khaldi MM, Johson JT, Gleason S, et al. Inland water body mapping using CYGNSS coherence detection[J]. IEEE Trans Geosci Remote Sens. 2012;59(9):7385–94.
12. Ruf C, Chang PS, Clarizia M, et al. CYGNSS handbook[Z]. Natl Aeronaut Space Adiministration; 2016.
13. Pekel JF, Cottam A, Gorelick N, et al. High-resolution mapping of global surface water and its long-term changes[J]. Nature. 2016;540:418–22.

Open Access This chapter is licensed under the terms of the Creative Commons Attribution-NonCommercial-NoDerivatives 4.0 International License (http://creativecommons.org/licenses/by-nc-nd/4.0/), which permits any noncommercial use, sharing, distribution and reproduction in any medium or format, as long as you give appropriate credit to the original author(s) and the source, provide a link to the Creative Commons license and indicate if you modified the licensed material. You do not have permission under this license to share adapted material derived from this chapter or parts of it.

The images or other third party material in this chapter are included in the chapter's Creative Commons license, unless indicated otherwise in a credit line to the material. If material is not included in the chapter's Creative Commons license and your intended use is not permitted by statutory regulation or exceeds the permitted use, you will need to obtain permission directly from the copyright holder.

MIX
Papier aus verantwortungsvollen Quellen
Paper from responsible sources
FSC® C105338

If you have any concerns about our products,
you can contact us on
ProductSafety@springernature.com

In case Publisher is established outside the EU,
the EU authorized representative is:
**Springer Nature Customer Service Center GmbH
Europaplatz 3, 69115 Heidelberg, Germany**

Printed by Libri Plureos GmbH
in Hamburg, Germany